MANAGING THE OCEAN

RESOURCES, RESEARCH, LAW

JACQUES G. RICHARDSON
Editor

LOMOND PUBLICATIONS, INC.
Mt. Airy, Maryland 21771
1985

Most of the chapters comprising the present volume appeared originally in
impact of science on society 1983, No. 3/4, and 1984, No. 4, © by the
United Nations Educational, Scientific and Cultural Organization, 7 place de
Fontenoy, 75700 Paris, France. This material is reproduced with the
authorization of Unesco. For other chapters, copyright as indicated.

Composition by Jeanne Moran and Ruth Ruane
Printed in the United States of America

Published by
Lomond Publications, Inc.
P.O. Box 88
Mt. Airy, Maryland 21771

Acknowledgements

The editor wishes to thank the following colleagues and friends for their patient guidance and unfailing assistance in the preparation of this volume: Mário Ruivo, Dale Krause, Robert H. Maybury (World Bank), Natalie Philippon-Tulloch, Bert Thompson, Kazuhiro Kitazawa, Yves Tréglos, Youri V. Oliounine, Vladimir Sibrava, Eckart von Braun, Mstislav I. Rusinov, Louise Oguse and William J. Whelan.

Publisher's Foreword

Managing The Ocean: Resources, Research, Law is the third volume in a series inaugurated by Lomond Publications with Unesco in 1979. This book deals with an area of growing importance in a world ever smaller, more communicative and interdependent. The editor has invited and selected material related to science, law, economics and engineering as applied to the ocean and coastal environment, and the book should hold substantive appeal to everyone working 'with' the sea.

This book on ocean resources and policy should also be attractive and useful to the lay reader. 'Managing' the sea represents a major world challenge, one involving the country next door or just down the coast, countries with whom we have poor or intermittent economic-technical relations, and even nations with whom we find ourselves at direct odds politically. Surmounting such barriers and doing it effectively and elegantly enough so that we can profit from the great common we call the ocean is what the thirty-eight authors in this book address and describe.

Core chapters are presented as they appeared originally in Unesco's *Impact of Science on Society,* which address to British orthographic style. British style has been extended to supplementary text as well.

I take note here that the United States Government has withdrawn from official participative relationship with the United Nations Educational, Scientific and Cultural Organization. The rich history of Unesco's contribution to world knowledge and understanding is illustrated by this book, so much of which emanates from its comprehensive effort in marine research and management. I look forward hopefully to a resumption of the accommodation necessary for Unesco's mission to be continued with universal participation.

I want to say a word of appreciation concerning this book's editor, who can assemble the expression of some of the best minds of the world to illuminate the most complex issues that face mankind. A major key to his success is that he so informs and immerses himself in the subject that he becomes uniquely master of it. This is illustrated convincingly by his lead chapters in this book and in the recently published *Models of Reality.* To explicate and clarify complexity so well without superficiality is a rare ability. We are therefore especially pleased to note that in his well-earned retirement he is already committed to the editorship of our next collaborative book from Unesco on the subject of creativity and innovation.

Lowell H. Hattery

Managing The Ocean

Table of Contents

PART 1. INTRODUCTION

Chapter 1 The Challenge of Managing One of Nature's
Greatest Resources

PART 2. RESOURCES

Chapter 2 Marine Geology: Mineral Resources of
the Sea

Chapter 3 Geotectonics of the Arctic Seas:
A New Survey

Chapter 4 Return to the Sea – Not as Hunter
But as Farmer

Chapter 5 Marine Pelagic Organisms

Chapter 6 Aquaculture: Realities, Difficulties
and Outlook

PART 3. RESEARCH

Contents

List of Figures

List of Tables

Table   **Page**

PART 1.

INTRODUCTION

Mankind has long explored and used the ocean environment as a source of food and means for transport, trade and occasionally war. Our fundamental understanding of the sea as an integral part of the biosphere and the geosphere is relatively recent, however, and there remains enormous progress to be made in this direction. The proposed international Law of the Sea will regulate, sometimes drastically, how we apply the knowledge acquired—even how we will expand our investigation of the sea around us.

Chapter 1

The challenge of managing one of nature's greatest resources

Jacques G. Richardson

From 1972 until 1985, the author edited Impact of Science on Society, *a quarterly journal published in seven languages by the United Nations Education, Scientific and Cultural Organization (Unesco). Now a consultant in R&D managment, he can be reached at Cidex 400, Authon la Plaine, 91410 Dourdan (France).*

1

By the year 2000, half the world's population will be living on or within 75 kilometres (less than 50 miles) of a seashore—probably about 3 thousand million of us. Yet for every square metre that we terrestrials call solid ground, there will still be 2.45 square metres of sea surface. We are surrounded by the ocean medium, but most of us know little about the marine 71 per cent of the planet's life-supporting envelope that we designate as the biosphere.

Mankind's first use of the sea was probably as both source of food (see box)—and thus of commerce—and a means of transport between two points on dry land. Whether at work or play, man learned early in history to establish a fruitful, relationship with the sea waters washing, swirling or storming past his continental abode. Since then, our species has explored progressively the global expanses of ocean, investigating its surface, its massive resources of water and life and energy as well as its seabed, and becoming increasingly conscious of their intimate interactions with the rest of the ecosphere and human activity.

In this book, we have invited 37 distinguished specialists representing highly diverse domains of ocean knowledge to remind the specialist and enlighten the layman about how research and exploitation of the sea can be expected to develop during the coming generation. The volume is thus concerned with knowledge, applications and the judicial context of what man will be doing with the sea.

2

There exists, over the course of time and near the western coast of South America, a slow upwelling of lower layers of cold sea-water rich in nutrients. This slow surge of water induces a great growth of microscopic plankton which, in turn, sustain an abundance of fish (chiefly anchovy) at the surface. These, in their turn, support a large population of birds of many kinds, and this region of the eastern Pacific Ocean has been one of the major fisheries of the world.

One of the oceanic phenomena that captured world interest in the 1960s and 1970s, accompanied by ecological disaster, happens to occur off this western coast of South America every few years, near Christmas. The phenomenon is known, hence, as El Niño: Spanish for the Christ child. Sited mainly off the Peruvian littoral, El Niño is an oceanological-meteorological anomaly caused by warm waters arriving in the area, thereby suppressing the upwelling of nutrients and playing havoc with the biological chain.

The most recent El Niño was unusual in its excessive duration and severity. The tepid waters were not only lacking in the life-sustaining nutrients, but they induced heavy rains and flooding along this normally arid coast. In addition, El Niño's warm behaviour seems to have been associated with weather changes in 1982-1983 occurring as far distantly as North America and Europe, western Australia and Asia, even Africa. China was affected by drought in the north, and floods in the south; and Mozambique was touched by drought.

Second-order consequences of ecological and social character were also attributed to these world-scale meteorological modifications that engendered the strange current off Peru's shores. Higher incidences than normal of attacks by sharks and unusual flooding were noted in the United States. Australia suffered more duststorms and brushfires than customarily, while several African countries had to deal increasingly with dry spells and brush-fires of their own.

Interestingly, this 'once per century' world-wide El Niño occurred at a time when the oceanographic and meteorological sciences were mature enough to study the phenomenon on a grand scale. Observations from satel-lites, drifting buoys, oceanographic ships and land stations were analyzed through very large, very fast computers. The evolution of this El Niño is well known, and we are beginning to understand the underlying causes on at least a Pacific-wide scale, if not yet at the global level.

For a poor, developing nation such as Peru, the sporadically good catches of anchovy are a boon. But when the fishing fleets must return to port with meagre hauls, underemployment and other labour dislocations inevitably follow—as well as deterioration in the value of the country's already pre-carious monetary and trading positions. It is vital to understand why El Niño behaves the way it does,[1] when it does, erratically, whether providentially or catastrophically—this 'child of both the ocean and the atmosphere'[2]—for the sea and its bounty have significant and obvious societal impact.

3

As already suggested, imagery captured by satellite is helping us to under-stand complex systemic occurrences of the El Niño variety. At Princeton University, George Philander and Anne Seigel have used advanced computer techniques to study surface-temperature effects, derived from remote sensing by satellite, throughout the tropical Pacific latitudes. Their work resulted in a film requiring only a few minutes to view but summarizing by way of the time-lapse technique temperature changes at the Pacific Ocean's surface, extending over many months during the time of the 1982-1983 El Niño. Adrian E. Gill of the Clarendon Laboratory, Oxford, has stressed the value of such a technical approach in order to predict similar events that may recur in the future.[3]

Beginning in 1985, a ten-year research effort called TOGA (for 'tropical oceans and the global atmosphere') is being undertaken by the World Meteorological Organization, Unesco, the Intergovernmental Oceanographic Commission and the international Committee on Climate Changes and the Oceans, a body sponsored jointly by the Scientific Committee on Oceanic Research and the Intergovernmental Oceanographic Commission. TOGA's aims reinforce the hope that climatic conditions may become predictable many months in advance.

Fisheries development and management

Most traditionally acceptable species are now fully exploited and many are over-fished. As a result, the rapid growth in the world catch witnessed in the 1950s and 1960s has given way to a much slower rate of increase. The total catch has [averaged around 71] million t during the past decade, although the catch of food fish has continued to increase by about 1 per cent per annum. Excluding aquaculture, the final yield from currently exploited species is unlikely to be much more than 110 million t. Production could eventually be increased above this limit, however, by catching so-called unconventional species such as the shrimp-like Antarctic krill, lantern fish, oceanic squid and small pelagic fish.

Fisheries in many developing countries are already a vital source of food, employment and income. The changed circumstances of marine fisheries and advances in aquaculture and inland fisheries make it essential, however, to persuade countries accustomed to devoting the bulk of their national resources to agricultural development and the pursuit of industrialization that fisheries are also of considerable value to them.

Fisheries contribute about 6 per cent of the world supply of protein and about 24 per cent of its animal protein, if allowance is made for the use of fishmeal in animal feeds. On a regional basis, the percentage contribution of fish to animal protein in the diet is greatest in Asia. In Southeast Asia, for example, 55 per cent of the animal protein consumed is derived from fish. In Africa, the figure is 19 per cent. . . . Such regional statistics reveal little of the true contribution made by fish to food supplies in developing countries: in some communities, often the poorest, it may be the only source of 'meat'. This dependence on fish is made all the more important because the growth in demand, which is expected to double by the turn of the century, is likely to be greatest in the developing world.

It has been estimated that the fisheries provide employment, some of it part-time, for about 16 million fishermen in developing countries. Many more people are engaged in associated activities, such as processing and marketing. The greater part of this work force is associated with small-scale or artisanal fisheries. Taking into account dependents, tens of millions in the Third World must rely completely, or in a major way, on fisheries for livelihood.

Excerpted from
Fisheries Development in the 1980s,
Food and Agriculture Organization of the
United Nations, Rome

TOGA is one of the largest international co-operative research projects ever mounted, calling for ever closer collaboration between oceanographers and meteorologists. It has been designed to permit scientists everywhere to understand such mechanisms as that of El Niño, to help foresee their re-appearance, and to warn governments and other social institutions of their possible scale and consequences.

Roger Revelle, director-emeritus of the Scripps Institute of Oceanography at the University of California, has underscored the fact that the support of as many governments as possible will be required during the 1985-1994 decade to ensure the success of this project, and that Unesco remains in an unusual position to help obtain such support.

Somewhat contingent on the success and multidisciplinary outcome of TOGA is an even more ambitious scheme called the International Geosphere and Biosphere Programme (IGBP). This is a plan being nurtured through the International Council of Scientific Unions to launch, in about 1990, a long-term study to capitalize on (a) accumulated knowledge of the interactions between biota resident in the troposphere (the atmosphere nearest the earth's surface) and the land and water sitting on the lithosphere, and (b) the new generation of observation satellites, including the new Chinese ones, orbited during the second half of the 1980s. Two units in the present book, Chapters 5 and 13, deal with various aspects of the IGBP.

4

There is more to be accomplished, however, than mere acquisition of new information concerning the earth's life-sustaining systems. We need to learn more about conservation of all resources, and then apply this knowledge.

We must cope ever more intelligently with the long-term consequences of pollution of the ocean environment. Radioactive waste, toxic residue emanating from the chemical industries and petroleum spills have been joined by the effects of non-biodegradable plastics. Fish and whales, seals, sea turtles and even birds ingest bits of plastic material or become ensnarled in abandoned fishing nets and other snares made of the ubiquitous polymers, and then these fauna die prematurely.

Dolphins have been found to contain, once their remains are washed ashore, very high levels of polychlorinated phenyls (PCB) and DDT. Continental runoff from sewage systems tends to contain PCB remnants (no longer used in some industrialized countries) leached from the soil where they were once thought to be stored safely for many years. Both PCB and DDT can cause, furthermore, disturbances of both the reproductive and immune systems of small animals both in the sea and on land.

Ocean pollution is a product of the industrial revolution, a major economic factor affecting virtually every nation in the world having even indirect access to the estuarine and coastal zones of its neighbours. Only the continuing industrial revolution can find the corrective measures to deal with problems of pollution.

Farming the sea for edibles

By the turn of the century, the yield from fish culture will be as important a source of protein as the capture fisheries, according to a leading Indian oceanographer.

Dr S. Zahoor Qasim, in his Bruun Memorial Lecture celebrating the 25th anniversary of the Intergovernmental Commission at Unesco in March 1985, called mariculture the best alternative to enhancing fish yield. Qasim is Secretary to the Government of India in the Department of Ocean Development.

'Sea-farming has been practised traditionally in many Asian countries for several centuries, but in recent years it has become increasingly. . .a rich protein source for mankind. The phenomenal growth. . .can be seen from the fact that fish production rose from 1 million tonnes now,' Qasim said.

'Of the total world production of fish by culture (crustaceans, molluscs and seaweeds), about 68 per cent is contributed by the developing countries and about 17 per cent by the countries bordering the Indian Ocean. . .India, Indonesia, Thailand and Singapore contribute more than 70 per cent of shrimp and prawn production by sea-framing.

'It has been estimated that if the present pace is maintained fish production by mariculture may increase by five times by the end of this century. . . .'

Challenges remain in this field, Dr. Qasim added. Fast-growing, high-yielding and disease-resistant varieties of fish have yet to be developed. Selection of the right type of food in captivity remains the single greatest cost-effective component of mariculture. Most sea-farmers still depend on natural supplies of seeds because there are few commercial hatcheries, and techniques for mass production of quality seeds are still in their infancy.

The Intergovernmental Oceanographic Commission/FAO scheme called Ocean Science in Relation to Living Resources is of particular importance, Qasim commented. This activity has been well received by many nations, and it will increasingly involved developing countries.

David Spurgeon, in
Unesco Features (1985)

5

The Joint Oceanographic Assembly, a professional congress of the world's oceanic specialists convening every six years, held its most recent meeting in the summer of 1982 in Halifax (Canada). This was at almost the same historic moment when eight years of arduous negotiation at the United Nations culminated in the formulation of the Convention on the Law of the Sea. This treaty proposal now awaits ratification in order to become international law. (See below.)

The purpose of the oceanographers' world assembly was to examine how science and technology could be expected to evolve during the coming twenty years: How our knowledge of the sea should appreciate, how we might expect to 'master' the world ocean, how we could hope to do so with maximal benefit to mankind.

The ocean experts presented their forecast in the form of a draft, too, a report called *Future Ocean Research* (FORE, for short)—a projection of great breadth, technical detail and intelligent analysis. FORE inspired, to a considerable extent, the conception and compilation of the present volume, many of whose chapters appeared originally in a special, double issue of Unesco's quarterly journal, *Impact of Science on Society,* at the end of 1983.[4] The most durable of the journal's articles have been retained in this book, and some unpublished material has been added in order to round out the oceanic picture for the non-specialist as well as for the technician or economist.

The draft Convention on the Law of the Sea was initialled in the winter of 1982-1983— when El Niño, incidentally, was shocking the world and puzzling the scientists—and then began its rounds of the signatory governments in order ultimately to obtain ratification. The ratified acceptance of the draft treaty by the first 60 governments to sign the document will make the proposed legislation a binding international agreement. Ratification is a slow process and, in the case of the draft law on rights to and uses of the sea, formal adoption can be expected to take some years. A few of the governments involved have already indicated that they will not ratify the instrument because of basic disagreement with one or another aspect (usually economic in character) of the jurisdictional proposal.

The United States, the Federal Republic of Germany and the United Kingdom have decided not to ratify this Convention. Instead, these governments, together with Belgium, France, Italy, Japan and the Netherlands are signatories to the Understanding Regarding Deep Seabed Matters (Preliminary Treaty, November 1984). The United States has excluded itself from association with the Convention on the Law of the Sea largely because it is the only country to have developed technology suitable to the exploitation of mineral deposits available on the sea floor.

6

In the chapters that follow, the *U.N. Chronicle* presents a succinct profile of the Convention (see Chapter 26), and authors Maria E. Gonçalves, René-Jean Dupuy and M.C.W. Pinto concentrate on various substantive facets of the landmark document (in Chapters 23, 27 and 28). (We would have said 'seamark' if the expression existed.) Author F.W.G. 'Mike' Baker of the International Council of Scientific Unions describes in Chapter 20 the actions of non-governmental organizations concerned with understanding or exploring the oceans. These have represented, for more than a century, scientific impetus for investigation of the sea itself, the seabed, as well as coastal and island areas and their protection.

Our authors have tried to interpret the book title's 'management' in the broadest way possible. Hans Ulrich Roll describes the intricacies of inter-governmental co-operation (Chapter 25), while Ulf Lie (Chapter 29) discusses the research, planning and control required to preserve marine ecosystems. The seas' total ecosystem is reviewed in the sections on Resources and Research in the chapters on marine geology by Peter Rothe and Igor S. Gramberg (see Chapters 2 and 3), on the ocean's living resources–T.V.R. Pillay, Tadehisa Nemoto and David Swinbanks, Albert Sasson, R.G.B. Brown and Luigi Minale, in Chapters 4 through 8. Then, 'Mr Sand,' David Spurgeon and the editor have something to say about the functions of wind and ice in the climatological circuit (Chapters 9-11), to end the rubric on natural resources.

The section called Research was set up especially to underline the role of scientific investigation in man's growing understanding of the ocean mass and how this research can best be organized and managed. Authors S.Z. Qasim and Agustin Ayala-Castañares examine the problem at the national level (Chapters 14 and 21), Klaus Voigt exposes a regional approach dealing with the Baltic Sea (Chapter 15), while the two (unrelated) Bakers take the global route to explain co-operation and D. James Baker, B.A. Nelepo and G.K. Karataev describe remote-sensing technology within the co-operative context (Chapters 16, 17 and 20).

Luo Yuru and G.L. Kesteven recount some of the problems deriving from the vast amount of data being collected and processed on questions oceanic (Chapter 12 and 22), whereas Akihito Hattori and A.S. Monin review the biogeochemical cycle and the accumulation of petroleum resources in the ocean system (Chapters 13, 18). Then, to vary the technical diet a bit, my colleague Selim Morcos tells of some recent underwater findings of what happened when Napoleon, Lord Nelson and General Sir Hugh Abercromby all met their match roundabout the peninsula of Abu Qir (Aboukir) nearly two centuries ago (Chapter 19).

7

From the past we move to a less certain future depicted masterfully by Eugen Seibold, who sees (Chapter 30) the marine environment as a major force in the international life of the coming century. David Doyle takes a realistic look at how we can use the sea increasingly as an energy source (Chapter 31), while Glyn Ford and his associates examine other specialized uses of the sea (Chapter 33).

Frank Barnaby's picture of the ocean as potential battlefield (Chapter 32) reflects the historical role of navies in the development of our understanding of the structure and dynamics of the ocean. Modern oceanology is founded, in fact, on the centuries-old work of naval hydrographers, the creators of oceanography as we have come to know it since the mid-nineteenth century. In this domain of knowledge, the needs of military technology have been the mother of invention as much as in the areas of terrestrial cartography, space exploration, remote sensing, electroacoustics and applied hydrodynamics.[5,6]

Barnaby's accounting for the naval forces of the superpowers is counterpoised by Elizabeth Mann Borgese's essay on music and literature whose inspiration came from the sea (Chapter 34).

These are the final chapters, and the editorial portfolio is completed by a brief postface and a selected bibliography for the reader wanting to know more about a vast, detailed and highly interdisciplinary subject.

8

Before history was recorded as systematically as it has been for the past millennium, exploration of the seas was undertaken by the earliest civilizations: Chinese, Indian, African, then Egyptian, Greek, Roman and Nordic, as well as Arab and Polynesian. Only the Amerindians of the northern and southern hemispheres seem to have eschewed the scouting of the planet's large, watery masses offshore.

Almost six centuries ago, Portugal's Prince Henry, the Navigator, established his oceanic research centre at Sagres, Europe's southwestern extremity 'where in the explorer's world the unknown was simply the not-yet-discovered. . . . Prince Henry. . .knew that the unknown could be discovered only by clearly marking the boundries of the known. . .and this required an incremental approach.'[7] Such an approach led, as we know, to brilliant new 'discoveries' off the African coast and ultimately in the vast sea-spaces of the Indian Ocean, the seas of the Americas, the Pacific Ocean and the polar regions.

This philosophy of systematic knowledge-gathering prevails in today's world of oceanological investigation, of research and development, and it is an approach strongly promoted by the Intergovernmental Oceanographic Commission and Unesco's Division of Marine Sciences (see Mário Ruivo, Chapter 24). The new Law of the Sea seeks to adjust inequalities[8] in the global effort to learn more about the world's ocean, to regulate equitably and sanely its exploitation—in brief, to help everyone plan, manage and protect the use of the precious natural resource for the generations and years to come.   ■

Notes

1. The investigative literature on El Niño is considerable. For a multidisciplinary appreciation of this complex phenomenon, see *Oceanus*, Vol. 27, No. 2, 1984. Almost the entire issue of this excellent quarterly journal is devoted to El Niño.
2. R.A. Kerr, Computer Models Gaining on El Niño, *Science*, Vol. 225, 6 July 1984.
3. Dr. Philander and Ms Seigel can be reached at GFDL/NOAA Laboratory, Forrestal Campus, Princeton University, Princeton NJ 08540 (United States of America).

4. See *Impact of Science on Society,* Vol. 33, Nos. 3-4, 1983. *Cf.* Intergovernmental Oceanographic Commission, *Ocean Sciences for the Year 2000,* Paris, Unesco, 1984. This 95-page booklet makes a comprehensive presentation of the needs for research, training and applications in the oceanological disciplines.
5. See Ashton B. Carter, The Command and Control of Nuclear War, *Scientific American,* Vol. 252, No. 1, January 1985. Se esp. the illustration appearing on p. 24 dealing with the problem of communication with submerged missile submarines.
6. The complex, mathematical four-dimensional field known as hyperspace is being used to plot complicated equations of four variables. In Washington, the Office of Naval Research is using this technique to establish the relationships between such oceanic parameters as current strength, salinity, temperature and biomass content in order to be able to trace accurately acoustic signals passing through sea-water.
7. D.J. Boorstin, *The Discoverers,* New York, Random House, 1983, p. 161-2.
8. In the mid-1980s, only two countries (France and the United States) have submersibles capable of plunging to depths of 3,000 m. And soon this range will be extended to depths of 6,000 m, according to J. Aubouin of the Laboratoire de Géologie Structurale, Université Pierre et Marie Curie, Paris. Indeed, in early spring 1985 the United States Navy announced the successful dive to 6,400 metres (21,000 feet) by the deep-submergence research vessel *Sea Cliff* in the Pacific Ocean's Middle America Trench. This operating level provides access to 98 per cent of the sea bed.

To delve more deeply

ADAM, John A., Probing beneath the Sea (special report on ocean engineering), *IEEE Spectrum,* April 1985.
Aquaculture continentale (theme), *Energie et environnement,* No. 11, Autumn 1983.
CHEMRAWN IV, Chemistry and Resources of the Global Ocean–1985, conference sponsored by the International Union of Pure and Applied Chemistry at Woods Hole, MA, Autumn 1985.Contact: Prof J. Robert Moore, Marine Science Institute, The University of Texas at Austin, Austin TX 78705 (United States), telephone (512) 471-4816.
Civilizations of the Sea (theme), *The Unesco Courier,* December 1983.
DOROZYNSKI, A. Les poissons font aussi les océans (It Takes Fish, Too, to Make the Sea), *Science et Vie,* August 1984.
DUCE, Robert A., International Research Needed in Tropospheric Chemistry, *NewsReport* (National Research Council, Washington), August 1984.
Entre l'inerte et le vivant (theme), a series of public lectures, February-May 1985, offered by the Muséum National d'Histoire Naturelle, Paris, and dealing with oceanic research.
FORD, G., NIBLETTE, C., WALKER, L., Ocean Thermal-Energy Conversion, *IEEE Proc.,* Vol. 130, Pt. A, No. 2, March 1983.
HENDRICKSON, R., *The Ocean Almanac,* Garden City, Doubleday, 1984.
KULLENBERG, G. *The Vital Seas: Questions and Answers about the Health of the Oceans,* Geneva, United Nations Environment Programme, 1984.
Les nodules polymétalliques: Faut-il exploiter les mines océaniques? (Report of the Académie des Sciences), Paris, Gauthier-Villars, 1984.
MARTIN, C., PEARSON, C., HARRISON, R., PROTT, L., O'KEEFE, P., *Protection of the Underwater Heritage* (Technical Handbooks for Museums and Monuments 4), Paris, Unesco, 1981.
ROGERS, M., Fish on Land: Aquaculture in Sri Lanka, *The IDRC Reports* (Ottawa), July 1984. Sea Law: Oddfellows, *The Economist,* 22 December 1984.
SOURY, Gérard, Les mémoires de Caroline, *Océans,* No. 136, November 1984; deals with the tenacity and longevity of sea turtles.

SULLIVAN, K. Overfishing and the New Law of the Sea, *The OECD Observer,* No. 129, July 1984.

SULLIVAN, W., Can Submarines Hide in An Increasingly Transparent Sea? *The New York Times,* 11 December 1984; *International Herald-Tribune,* 13 December 1984.

Tropical Oceans and the Global Atmosphere (theme), Round-table with Drs Adrian Gill, Antonio Moura, Angus McEwan, Roger Revelle, interviewed by David Spurgeon, Unesco Radio Recording No. 22.018, November 1984 (duration 29'50"); transcript available.

VOROPAEV, G., Diversion of Water Resources into the Caspian Seabed, *Options* (IIASA, Laxenburg), No. 2, 1984.

WOODS, J.D., The World Ocean Circulation Experiment, *Nature,* Vol. 314, 11 April 1985.

YOO, K.-I., The World of Microbes in the Sea Plankton, *Proc. VIII Internat. Symp.,* National Acad, Sci., Rep. of Korea. September 1980.

PART 2.

RESOURCES

In the beginning, there were trade and fishing, and for centuries man looked at the ocean bottom only to locate its shallow and thus dangerous spots, harmful to sailing in inshore waters. Entries in ships' logs such as 'Black pebbles are found only off Scilly...' mark the years of search for suitable passages through the English Channel, findings related to the geological composition of the sea floor. And although the deep ocean remained a geological enigma for a long time, today we have a fairly clear idea of its structure and dynamics.

Chapter 2

Marine geology: mineral resources of the sea

Peter Rothe

Peter Paul Albrecht Rothe, currently pro-rector of the University of Mannheim, gained his university degrees at the University of Frankfurt-am-Main, earned his Habilitation *at the University of Heidelberg and worked aboard the* Glomar Challenger. *The author of sixty-five different scientific works, Dr Rothe is professor of geology in his university and a consultant in marine geology to Unesco's Intergovernmental Oceanographic Commission. His address is: Lehreinheit Geologie, Universität Mannheim, Postfach 2428, D-6800 Mannheim 1 (Federal Republic of Germany).*

Introduction

Alexander von Humboldt (1769–1859) mentioned, as late as the mid-nineteenth century, that the sea's depth was unknown. If one takes into account the time necessary to lower a sounding-line to a depth of a few thousand metres, using a modern deep-sea winch, several hours are required. It becomes obvious that soundings throughout the world's oceans can unravel the mystery of the sea floor only very slowly—as the topography becomes progressively evident.

The process became more efficiently possible during the 1920s, with the invention of echo-sounding. During the first expedition of the research ship *Meteor* in the South Atlantic, it also became evident that the ocean's basins are by no means simple bathtubs; they have a rugged topography, with mountain chains, submarine cones, and table-shaped guyots. Vast areas of the sea bed were detected which were completely flat. (These were to be called, later, the abyssal plains.) Submarine canyons, some of them being a continuation of continental rivers, seem to be connected with these plains. They form pathways along which continental material is transported to the deep sea. Except in the inshore areas, almost nothing was known about the material of the sea floor. Occasional dredging brought pieces of basalt to the surface; some of the soft sediments then collected revealed for the first time, during the second half of the nineteenth century, the enormous variability of skeletons of marine organisms (of which most of these sediments are composed).

From 1872 to 1876, the famous *Challenger* expedition under the direction of Sir C. Wyville Thomson carried out, during a three-year cruise, the first systematic, overall study of the world's oceans. More than 12,000 geological samples were collected and the now famous manganese nodules were detected on the sea floor. This marked the beginning of studies on underwater minerals, but only today have we begun to think about their possible economic importance. One fact seems evident from *Challenger*'s data: that manganese nodules are mostly restricted to the deepest parts of the ocean. On the other hand, sea salt, extracted from the ocean since antiquity, or found in bodies of rock salt formed in the geological past, is restricted to the very shallow marginal seas, where the evaporation of small bodies of water is feasible. Both manganese and the constituents of salts form part of a very large reservoir of elements which occur in solution with sea-water. The ocean is a big reservoir of metals, but there is a slight problem: the concentrations of most of the elements are so low that we may never be able to recover these resources. Some organisms are known to concentrate specific elements: e.g. uranium is enriched in the calcareous skeletons of coccolithophorids (a group of algae), certain sponges concentrate nickel, and specific radiolaria build their shells of strontium sulphate, but these mechanisms are inadequate for the formation of valuable deposits. We must, therefore, look at additional natural mechanisms which concentrate elements to such a degree that the latter become interesting from an economic point of view. Whether or not a deposit becomes economically mineable depends largely on the world market price of the metals contained therein. Science, however, points out the rules of nature, which govern the formation of valuable deposits, which are immediately dependent on geology.

Geological features of the sea floor

The *Challenger* samples were essentially samples from the sea-floor's surface. With the development of coring techniques after the Second World War, man began to penetrate this surface, and up to fifty metres of thickness can now be taken. During the past fifteen years, valid drilling techniques have become available via the *Glomar Challenger*, and more than 1,000 metres of sea-floor sediments or basalts have been penetrated. The results of the Deep-Sea Drilling Project have caused a revolution in the geological sciences, offering new bases for the search for resources, both on land and under water.

Drilling has enabled geologists to unravel the history of the ocean basins. These are rather young when compared to the age of the earth. The oldest sediments found so far by deep-sea drilling are from the Upper Jurassic, about 150 million years old. (In comparison, the planet Earth is about 5,000 million years old.) The material below the sediments is basalt of similar age and, in some of the basalt, veinlets of copper and zinc minerals were encountered.

Both basalts and the overlying sedimentary sequences are youngest in the central parts of the oceans; they become older the closer they are situated to the continents. Symmetrical patterns of basalt ages were found on both sides of the so-called mid-ocean ridges, with the ages increasing towards the continents. Similarly, the ocean floor becomes deeper, and the sediment cover thicker when approaching the continents. This is partly because of the increased input from land and partly because of the age of the ocean floor: the 'rain' of particles from surface production of organisms, which contributes to the sedimentary cover, was active longer in the older marginal parts. The sediments there are about ten times as thick as on the deep-sea floor. 'Sea-floor spreading' means that the oceans evolved from small and narrow fractures within former continental areas. This is an ongoing process, and we shall, within a short time, be able to measure directly the movement of the continents relative to each other by satellite techniques. This is the dream of Alfred Wegener, who developed the 'continental-drift' concept at the beginning of the century.

TABLE I. Main types of mineral resources on the sea floor

Type of deposit	Material or elements	Geological setting	Political Law of the Sea setting
Construction materials	Pebbles, quartz and carbonate sand, shells	Coast and shelf	Exclusive Economic Zone (EEZ)
Placer deposits	Iron Gold Platinum Tin Diamonds Rare earth elements Zirconium Titanium and others	Coastal and nearshore	EEZ
Hydro- carbons	Petroleum Gas	Mainly passive continental margins and back-arc basins	Mostly EEZ
Hydrothermal ore deposits	Iron Manganese Copper Zinc Lead Silver and others	Fracture zones, spreading centres back-arc basins(?)	Mostly area
Manganese nodules	Manganese Iron Cobalt Copper Nickel Titanium Molybdenum and others	Deep sea (4,000 metres)	Mostly area
Phosphorites	Phosphorus Uranium Rare earth elements and others	Coastal (locally continuation of land resources) and nearshore: submarine plateaux	Mostly EEZ

The new global tectonics are slightly different, however. It is not the continents that move like ships through oceans of semiplastic basaltic material; instead, the continents, of essentially granitic composition, are frozen within rigid basaltic plates, and the two move together, probably driven by plastic undercurrents of convection cells in the earth's mantle, with a speed of the order of several centimetres a year.

Plate boundaries can be separated into (a) divergent, where the plates drift away from each other, and new oceanic crust is formed, and (b) convergent, where the plates meet and are destroyed, and their material is consumed or piled up to build mountain ranges (as in the Andes, the Rocky Mountains or the Himalayas). Both types of plate boundaries are interesting from the resource point of view. Plate boundaries can, but should not, coincide with boundaries between continent and ocean; the latter are called continental margins. Where a continent and an ocean rest on the same plate, we speak of 'passive' margins, such as around the Atlantic. Where the boundary between continent and ocean is a plate boundary, we speak of 'active' margins, such as around the Pacific Ocean. Both types of continental margins have the characteristic, in general, that their situation between continents and the deep oceans is suitable for the accumulation of relatively thick sedimentary layers. Passive margins, however, are more favourable in this regard. In some areas, these simple types of margins are more complicated owing to the presence of islands and smaller ocean basins between these islands and a continental land mass: the so-called back-arc basins, which is the case in the south-west Pacific. At the different types of margins, different types of resources can be expected.

Geological settings and resources

The so-called porphyry copper ores of the west coasts of North and South America have been known for a long time, but it is only recently that the global concept of plate tectonics has provided the explanation of their possible origin. All these deposits are situated at active margins, where a certain fractionation of rocks has provided pathways for ore-forming solutions. It was then only a small step towards predicting similar ore bodies in, for example, Papua New Guinea, which is situated in a similar plate boundary zone. Japan, the Philippines, and other areas fit into this global pattern as well. General geology, thus developed from its basic scientific purpose, was and is able to guide prospection for mineral deposits. Pure science thus often leads immediately to applied science.

What applies to copper also applies *inter alia* to a specific type of gold deposit. This gold is found at a higher level within the geological column, above the copper. The famous gold-rush regions of Alaska and California are situated on or close to active plate boundaries. By prediction from the general picture, the gold in Papua New Guinea was recently found on top of the copper deposits there.

Petroleum, however, seems to be more common on passive margins. Its origin has two prerequisites. One is a certain type of organic matter deposited together with a thick layer of sediments. The other is heat: a certain temperature is required to mature the organic substances to hydrocarbons. Finally, reservoir rocks and certain structural features must be present wherein the hydrocarbons can be contained. Such reservoirs rocks must be sealed by impermeable clay so that the hydrocarbons cannot escape. It is expected that such conditions exist at passive margins where old reefs are buried, as their porous limestones would provide excellent reservoirs. The sediment cover of the deep sea is not thick enough to provide the necessary conditions for petroleum formation. Hydrocarbons are thus most likely to be found at continental margins, and probably within some back-arc basins.

On the other hand, the deep sea is the most suitable place to find manganese nodules. Their formation is confined to extremely low sedimentation rates. At water depths below 4,000 metres, the carbonate particles present in the higher levels of the water column are dissolved and the typical sediments of such areas are red clays. The origin of the manganese is manifold: volcanism,

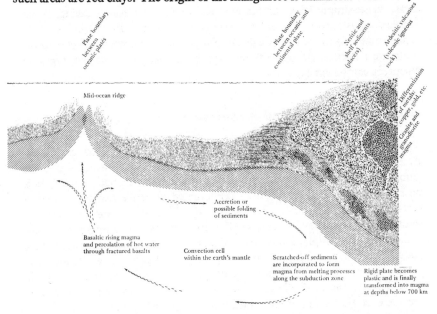

FIG. I. Highly simplified section explaining the geological situation of mineral deposit formation at active margins and mid-ocean ridges

FIG. 2. Blown-up mid-ocean ridge area from Fig. 1. Fractures by faults. Mid-ocean ridge zone or other active spreading centres. Possible development of ponds between basaltic rocks favourable to sediment accumulation. Reducing conditions or hot brines may develop locally and ore-bearing hydrothermal solutions precipitate to form deposits as in the Red Sea or the Galapagos area.

At water depths below 4,000 metres, the carbonate particles present in the higher levels of the water column are dissolved and the typical sediments of such areas are red clays. The origin of the manganese is manifold: volcanism, leaching of basalts and preconcentration by organisms from the ocean water are all now regarded as its possible mechanisms.

During the 1960s, there was much enthusiasm about the tremendous resources which these nodules could provide to mankind, with their contents of copper, nickel, cobalt and other elements. Yet the manganese itself is not worth mining under present economic conditions. One has to keep in mind that very few of the deposits investigated so far contain sufficiently high concentrations to justify a pilot mining effort. Manganese nodules are restricted to the surface and uppermost layer of the deep-sea sediments. Unlike most other ores which occur as 'bodies', these must be regarded as two-dimensional. Due to the conditions of their formation, the manganese oxides and hydroxides are dissolved in the deeper layers. So mining of manganese nodules means 'harvesting' the sea-floor surface, and manganese is only a by-product.

Heat, muddy waters, placers

Hydrothermal ore deposits are the most recent discoveries among underwater minerals. Some of them are spectacular, and most are far from being understood. The first deposit was found in the central Red Sea in an area of hot brines at a depth of about 2,000 metres. The multicoloured, muddy sediments contain iron sulphides and hydroxides, manganese, copper, zinc, lead, and even silver and gold. The geological setting is in the centre of a sea-floor-spreading area where Africa drifts away from the Arabian plate.

Comparable muds were discovered on the East Pacific Rise, at the mouth of the Gulf of California and near the Galapagos archipelago where enormous quantities of hot water escape from the ocean floor, carrying similar metal sulphides, which are then deposited around the so-called 'black-smokers'; they form fields of small-scale submarine mounds which may grow together to build massive deposits in the geological future.

Volcanism was originally thought to be responsible for this type of ore formation, but specialists are beginning to speculate about metals being leached from the surrounding rocks by circulating hot water. Support for the idea came from the Deep-Sea Drilling Project, where considerable quantities of water were found to percolate through one of the boreholes drilled in a thick pile of basalts; this suggests that the circulation of seawater through porous rocks might provide a powerful transport mechanism for ore-forming solutions. Pyrite, zinc and copper sulphides were found at such a locality. Possibly the veinlets of copper and zinc sulphides, mentioned above, have a similar origin, since cracks within rocks provide favourable pathways which can be filled by the precipitation of minerals from ore-bearing solutions.

It is not yet clear how far the tectonic pattern controls such conditions, but fracture zones seem to be favourable places for the formation of such hydrothermal deposits. An additional prerequisite seems that reducing conditions be present to avoid oxidation of the sulphides formed. Once the basic geological conditions of such deposits are known, the search for ores in the ocean will be less hazardous—and less costly—than at present.

The least speculative types of underwater deposits are placers of different composition. These deposits of alluvial and other material containing economic quantities of valuable minerals are concentrated by sorting processes resulting from resistance to weathering and because of their specific weights. Gold, platinum, tin (cassiterite), iron (magnetite), zirconium, titanium (rutile), rare earth elements (such as zircon and monazite) and diamonds are examples. Some of these were mined long ago (for example, the tin of South-East Asia), but many are still unexplored. Since placers occur in coastal and nearshore waters at depths usually not exceeding several tens of metres, they will be explored and exploited within economic zones falling within national jurisdiction.

Sand and gravel, phosphorites, other non-living resources

Sand and gravel deposits are the least spectacular, but probably the most mined, underwater resources from the quantitative point of view. Their geology is at first glance simple, but their mining needs to take into account local currents as well as environmental impact. (The mining of sand for construction purposes might have to compete locally with the use of sand for

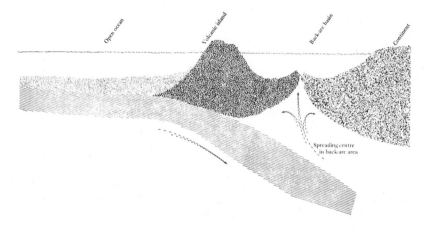

FIG. 3. Situation regarding back-arc basins. Active margin with volcanic chain between open ocean and continent, separating back-arc basin with separate spreading centre. Accumulation of a thick layer of continent-derived sediments and gradually reducing conditions together with high heatflow make such regions suited for hydrocarbon formation and hydrothermal ores.

touristic purposes, such as beaches.) Abundant quantities of carbonate sand are formed by organisms in the shallow-water environment at or near coasts and around islands of the tropical or subtropical oceans. Where carbonates on continental parts and islands are lacking, the accumulation of seashells can be used in the production of cement or, directly, as fertilizer.

Special conditions apparently govern the formation of phosphorites on the sea floor. Phosphorus is concentrated, within the marine environment, particularly in vertebrate bones and fish teeth. Because of the food chain, beginning with algae, phosphorite deposits are formed in areas of high

FIG. 4. Passive continental margin, favourable for hydrocarbons and placers. Accumulation of as much as several thousand metres of sediments, including organic substances, is favourable to hydrocarbon formation; there is input from both continent and open ocean.

primary productivity, such as upwelling zones. Although the conditions are not yet understood in detail, the concentration of valuable deposits seems to require a shallow-water environment. Favourable conditions prevail in the shelf areas and probably also along shallow submarine ridges and plateaux.

To the various examples of underwater minerals cited above, others may be added: diatom muds (diatoms, the siliceous remains of microscopic plants, are used to make filters), or the complicated silicate structures of zeolites, a group of silicates of in-place, or authigenic, origin used for a variety of purposes.

Thus the principal non-living resources belong to genetically different types. Of the first type are placers, consisting of reworked material, most of which is of continental origin. In most cases, mechanical transport to the marine environment is necessary. The second type consists of hydro-carbons, manganese nodules, hydrothermal deposits and phosphorites, all of authigenic origin. They are formed within the marine environment itself by geological or biological processes, or by a combination of both.

Exploration methods and perspectives

Every type of resource I have mentioned is related to special geological conditions. It is therefore necessary, prior to exploration, to study the laws of nature leading to these conditions. Such study, requiring expensive research vessels, navigation by satellite, geophysical equipment and modern laboratories makes cost-sharing co-operation between countries indispensable for years to come. In the instrumentation currently available, devices for detailed bathymetric mapping (e.g. *Gloria* and *Sea Beam*) are of the first order. So far, only a small percentage of the sea floor has been charted with these instruments, but the results are fascinating. Recently, even laser techniques have been applied successfully to depth-sounding from aircraft.

Among the methods used for geophysical exploration of the sea floor, techniques are now available not only to detect the depth of the boundary layer between the basaltic ocean crust and its overlying sediments, but also to find stratification patterns within these sediments. Changes in their lithology form seismic 'reflectors', which can be verified by drilling, and the results extrapolated over large areas. It is easier and less costly to carry out geophysical exploration than to drill.

Special vessels are required for the study of high-latitude regions. The Arctic Ocean and the seas surrounding Antarctica are among the least-known portions of the earth. Icebreakers equipped with geophysical instruments, side-scanning sonar and modern sampling devices will help to unravel the mysteries of the geology and related resources of these waters. For local, very detailed observation and sampling, promising results can be expected from research submersibles such as *Alvin* and *Cyana*, which have already been used in highly successful ventures: studies of the Mid-Atlantic rift valley and the Galapagos hydrothermal vents, for example.

Besides this field equipment, laboratory instrumentation has been improved considerably. This includes X-ray diffraction techniques to analyse the mineralogical composition of sediments and rocks as well as the electron scanning microscope; both are now used routinely to study the shape and composition of various materials, with resolutions at a scale beyond that of optical microscopes. Such devices require very small amounts of sample material. The chemical analysis of large-sample series has also become routine, using atomic-absorption spectrophotometry and X-ray fluorescence. Although most concentrations of elements are measured in 'trace' quantities, such basic results lead finally to the formulation of laws governing the formation of all geological resources.

Exploitation is a different matter

An entirely different proposition is the exploitation of such resources. There are known deposits on land which continue towards and under the sea: coal, salt, sulphur and ores have long been mined from deposits closely adjacent to land masses, using techniques similar to those common to terrestrial mining.

Drilling for offshore hydrocarbons, using stable platforms, is also now routine within the shallow waters of the continental shelf. New ventures are advancing towards ever deeper waters. Dredging for placers and construction materials (sand and gravel) is now problem free, except for a few, very local environmental consequences. Little has been done so far, however, in the exploitation of marine mineral resources situated within deeper waters.

Pilot mining of metalliferous muds in the Red Sea is under way, as are trial projects applying different techniques (e.g. pumping, or the line-bucket system linking two ships) in order to obtain manganese nodules from the deep-sea floor. The mining of manganese nodules will have to deal with a specific problem: the accumulation of hard clay is not easily removed from the nodules. This clay must be washed off during open-ocean mining and separation, so, here again, environmental problems created by the vast clouds of red clay from the nodules, which are left suspended in surface waters, need to be considered. It must also be stressed that, given present world market conditions, none of these ventures is economically viable at the moment. Yet man needs to think about the future and other potential resources which might be added to our known reserves.

Charles Lyell, a famous nineteenth-century geologist, established one of the fundamental laws in our discipline which is now also applicable to mineral resources: 'The present is the key to the past.' From studies of the recent geological environments affecting ore formation, we shall be in a better position to understand the formation of comparable deposits now found on land. This is a precondition to their further exploration and mining.

Marine geology and geophysics are thus necessary and useful, even if the young, metalliferous deposits of the oceans are never mined on a large scale.   ■

Recent investigations of the continental shelf of the USSR have considerably broadened our understanding of the geological structure of the Soviet Union's continental borderlands. A particularly significant amount of new information has been obtained from study of the water area of the USSR's extensive Arctic shelf.

Chapter 3

Geotectonics of the Arctic seas: a new survey

Igor S. Gramberg

Igor Sergeevich Gramberg holds a doctorate in geological and mineralogical sciences; he is a corresponding member of the Academy of Sciences of the USSR, a specialist in the regional geology and petroleum and gas resources of the Arctic and Pacific Oceans. He currently serves as Director-General of the Sevmorgeologiya Industrial Geological Combine, and can be reached at N.P.O. Sevmorgeologiya, Naberezhnaya Moiky 120, Leningrad 190121 (U.S.S.R.).

Tectonically, the seas of the Soviet Arctic belong to what are known as passive continental borderlands. A geological map of the Arctic seas, compiled largely from seismic and geo-acoustic data on (a) the structure of the bottom layers stratified over a base of bathyal rock material and (b) the geology of islands and coastal land areas, and also from the findings obtained from the shallow drilling of different sections of the sea floor from ice-floes and ships, provides evidence that the strata lie in gently sloping, almost horizontal planes near the surface of the sea floor. Judging from the configuration of extensive rock outcrops and the comparatively small age interval between the layers in the bottom complexes, the Arctic shelf resembles a series of platforms.

Seismic studies show, however, that the deep-sea tectonics of the sedimentary mantle are affected by intensive movements accompanied by simultaneous sedimentation and characterized by large, linear depressions filled with thick layers of sediment of complex lithographic profile. The amount of sediment deposited over a period of a million years is here as much as 35 to 50 metres, which is close to miogeosynclinal conditions of development.[1]

These large depressions developing on the Arctic shelves along with those of adjacent littoral lowlands form genetically linked families of structures that accumulate sediment, i.e. sedimentary basins. In the continental borderland these basins are so large as to constitute first-order negative tectonic elements, whose autonomy is accentuated by the rises that flank them.

The sedimentary basins of the Arctic borderlands of Eurasia and North America and the basins of the Arctic Ocean make up the Arctic geodepression, a superordinal structure of global scale. At its present stage of development, the Arctic geodepression finds distinct physiographic expression in the drainage area of the Arctic Ocean and is characterized by symmetrical zonation. From the periphery to the centre, three main zones may be distinguished: a limiting orogenic (or mountain-building) belt; a centroclinal belt of epicontinental sedimentary basins that include the shelf depressions; and the oceanic or bathyal core of the geodepression. Palaeo-geographical and palaeotectonic reconstructions show that the Arctic geodepression formed on the earth's surface at the juncture between the Early and Late Permian, and owes its present-day structure to four stages of structural transformation: Late Permian–Triassic, Jurassic–Early Cretaceous, Late–Cretaceous–Palaeogene and Neogene–incomplete Quaternary.

Some general features

The major changes in the Arctic geodepression occurred during the Late Cretaceous period. Until that time, in its core there were no ocean deeps but an updoming of the crust, from which sediments spilled into the surrounding basins. In the Late Cretaceous, this domelike structure underwent rift-forming rotation which led to the formation of ocean deeps with a spreading basic rock basement (the Norwegian-Greenland and Eurasia sub-basins of the Arctic Ocean). On the other hand, the Amerasia sub-basin of the North Pacific resulted from the active flexure of the Hyperborean Platform and the folded rocks that surrounded it, which began as far back as the Early Mesozoic. Judging from seismic sections on the margins of the Canadian Deep, the formation of the Amerasia sub-basin, as an abyssal zone of the ocean, did not, however, occur before the Palaeogene and was due to the

---

Glossary of terms

(Geological time is expressed in mean years)

Anthropogene	Most recent geological time, with man as factor; Quaternary
Baikalian	About 560 million years ago
Bathyal	Part of continental slope, 200–4,000 metres from littoral
Caledonian	570 to 370 million years ago
Cenozoic	From 66 million years ago to present
Centroclinal	Dipping towards a common centre
Cretaceous–Eocene	About 60 million years ago
Early Cretaceous	About 135 million years ago
Early Mesozoic	About 220 million years ago
Early Permian	About 275 million years ago
Epicontinental	Within the limits of continental mass
Facies	Rock beds differing in characteristics from comparable beds
Flexure	Turn, bend or fold in rock stratum
Gneiss	Banded, coarse-grained rocks formed during high level of regional metamorphism
Hyperborean	Extremely northern
Jurassic–Early Cretaceous	About 141 million years ago
Karelian	About 2,000 million years ago
Kimmerian	About 4 million years ago
Kungurian	About 251 million years ago
Late Caledonian	About 400 million years ago
Late Cretaceous	About 100 million years ago
Late Cretaceous– Palaeogene	About 65 million years ago

Late Permian	About 250 million years ago
Late Permian–Triassic	About 235 million years ago
Lias (Lower Jurassic)	195 to 172 million years ago
Lower Cretaceous	About 120 million years ago
Lower Palaeozoic	About 550 million years ago
Massif	Principal mountain mass
Mesozoic	230 to 66 million years ago
Middle Palaeozoic	About 400 million years ago
Neogene–incomplete Quaternary	About 2 million years ago
Oligocene–Miocene	About 22.5 million years ago
Orogeny	Period of mountain building
Palaeogene	65 to 22 million years ago
Palaeozoic	570 to 230 million years ago
Paralic	Developed in marginal marine environments (such as lagoons)
Permian	280 to 230 million years ago
Permian–Carboniferous	About 281 million years ago
Permo–Triassic	About 232 million years ago
Pliocene–Anthropogene	About 1.8 million years ago
Polycyclic	Multiringed
Precambrian	From the earth's formation to about 600 million years ago
Riphean deposits	Of 1,600 to 650 million years ago
Synoceanic	Main period of ocean formation
Tectogenesis	Evolution of earth's structures in a given place
Triassic	230 to 195 million years ago
Triassic–Jurassic	230 to 141 million years ago
Upper Cretaceous	About 80 million years ago
Variscan	Molluscan
Virgation	Fan- or sheaf-like arrangement (of fault lines)

failure of the sediments to compensate for the flexure. From this standpoint, the Amerasia sub-basin of the Arctic Ocean may be considered as a very deep marginal sea. Here there may be seen blocks of continental, sub-oceanic and oceanic crust with an anomalous magnetic field of generally mosaic pattern.

The Late Cretaceous is the dividing line between the pre-oceanic and synoceanic (or concurrent) stages in the development of the Arctic geo-depression. The sedimentary infill of the Arctic geodepression forms a single complex throughout and may be divided into two sub-complexes—pre-oceanic and synoceanic. The Arctic shelf as a geological structure in the ocean-continent series began to develop in the Late Cretaceous on the basis of the epicontinental basins[2] of the first stage in the development of the Arctic geodepression. This is why the shelf inherited structural features from the previous stage in the history of the development of the geodepression.

At the same time, the general dip of the surface of the centroclinal belt from the orogenic frame towards the ocean determined the diffusion and thickness of the synoceanic complex on the shelf. It was an increase in the angle of dip of the fault wall of the geodepression in the Cenozoic period that led to the formation of step faults, as a result of which the continental centrocline took on the shape of a multi-tiered amphitheatre, with a sharp discontinuity between the shallow- and deep-water zones. At that time, the steps that had subsided most inside the epicontinental sedimentary basins took on the form of marginal continental plates that became the main area for the building up of shelf synoceanic formations.

The levels of sediments

The deep-water structure of the sedimentary basins is characterized by a division into three levels. The upper level corresponds to the structure-forming complex of the developing basins. The middle level is represented by relicts of the platform cover which was formed before these regions became part of the Arctic depression. It can be called the intermediate level and has quite often become thinned. Finally, the lower level is represented by fold formations; it becomes the folded basement of the continental borderland. The surface of the basement comprises both folds in sedimentary rocks and metamorphic complexes, representing the granite-gneiss layer of the earth's crust.

The three-level model adopted for the sedimentary mantle of the earth's crust in the continental borderland is consistent with the limits of resolution of the regional geological and geographical methods for the study of water areas, which made it possible to obtain for each of the levels representative geological characteristics with important practical applications.

Within the Arctic shelf of the USSR, four sedimentary basins may be distinguished—the Barents, the West Siberian, the Laptev and the East Siberian–Chukchi Basins.

The Barents Basin includes the water areas of the Pechora and Barents Seas and of the northern half of the Kara Sea. It is open towards the Lofoten Trough while its coastal side takes in the Bolshaya Zemlya tundra. The Barents Basin is separated from the Russian platform by the Timan Range and the mountains of the Kola Peninsula, and from the West Siberian Basin by the Polar Urals and the Novozemlya-Taimyr rise. It is separated from the Nansen Deep by the uplifted islands of Spitzbergen, Franz-Josef Land and Severnaya Zemlya. The thickness of the earth's crust in the Barents Basin ranges from 25 to 35 kilometres. The crust is typically continental in structure.

The Barents Basin has been found to be characterized by a distinct three-level structure. Study of the folded basement complex has revealed the existence of areas of Karelian, Baikalian and Caledonian folding. In the narrow zones adjacent to the orogenic borders of the basin, dislocations of the Kimmerian stage of the earth's development can also be distinguished. Here Triassic–Jurassic movements entrapped the previous geosynclinal complexes, the Palaeozoic platform deposits and the lower layers of the upper level and deformed them into linear folds. This process of deformation is still occurring today in conjunction with the growth of mountain ridges. The surface of the folded-rock basement sinks from its mountain flanks towards the central areas of the basin, reaching its maximal depth in the South Barents Deep.

The intermediate level contains sedimentary formations of epi-Karelian, epi-Baikalian and epi-Caledonian platform structures. Palaeogeographical reconstructions suggest that the terrigenous (or land-formed) red and evaporate sediments that fill the deeps resulting from the Late Caledonian stage in tectonic activation were probably widely distributed near the basin's western and eastern borders.

Details of the deposits

The upper level of the Barents Basin began to develop between the Early and Late Permian, the date being fixed by the beginning of the development of the orogenic belt. In sedimentation terms, the date is marked by the substitution of terrigenous sediments for carbonate formations in the intermediary layer. This substitution occurred over the greater part of the basin area without causing any unconformity in the stratification, but the contrasting lithologic differences between the layers in contact makes the boundary of the upper level a good baseline for seismic investigations. Migrations of the formation boundary from the sides of the basin towards the centre may be expected within the range of the Kungurian and Ufan stages.

The upper level, represented by terrigenous marine and continental deposits, consists of two sub-complexes: a pre-oceanic complex of depressions and deeps and a synoceanic complex of gently sloping depressions, troughs and grabens.[3]

The West Siberian Basin comprises the West Siberian depression and the southern half of the Kara Sea. It is divided into two large oval-shaped areas of sedimentation, northern and southern.

In the northern oval (the Yamal and Gydan peninsulas and the Kara Sea between Novaya Zemlya and Taimyr), the Precambrian (probably Karelian) basement complex predominates. The intermediate level, represented by terrigenous carbonate Riphean deposits and Lower and Middle Palaeozoic deposits, becomes thinner towards the centre of the basin.

The upper level of the northern oval began to develop in Permian and Triassic time with the accumulation of fine-grained marine sediment with an admixture of igneous rocks. The layer of sediment that conformably covers this layer is made up of a series of terrigenous marine and continental deposits, from the Lias to the Anthropogene, common to the whole basin.

The southern oval, lying on the Baikalian, Caledonian and Variscan folded structures of the Uralo-Mongolian belt is not as thick as the unfolded deposits; its crust is between 35 and 40 kilometres thick. In the Jurassic a system of negative structures of different sizes, represented by depressions, troughs, grabens and taphrogeosynclines (or rift valleys) of Permian–Carboniferous, Permo-Triassic and Triassic provenance was brought together here by common folding.

The sedimentary infilling of these structures forms a transitional complex, representing the substitution for the geosynclinal Variscan (molluscan) cycle of a new cycle of intensive platform flexure. Formations, corresponding in time to the initial Permo-Triassic stage in the development of the northern oval, are represented here by sedimentary–igneous (basic-rock) structures. These, of course, show a different relationship with the underlying and overlying deposits. They show evidence of close association with the Carboniferous–Permian structural formation complexes of epivariscan depression, showing unconformity with the Jurassic deposits and also the reverse relationship (i.e. conformity with the Jurassic and unconformity with the Permo-Carboniferous), characteristic of a discordant system of taphrogenic structures (i.e. those causing rift valleys).

The Laptev Basin's characteristics

The Laptev Basin is open towards the ocean, but from the continental side it is bounded by rises (Severnaya Zemlya, Taimyr, the Pronchishchev ridge, the Kharaulakh and Kular ridges and the series of rises from Svyatoi Nos to the Novosibirsk Islands). The Amundsen Basin is also a continental limb of these geological structures. Gakkel's median ridge, which divides the ocean deeps, ends at the continental slope, but the seismically active zone of its central rift valley traverses the Laptev Basin and extends as far as Buorkhaya Bay, where it is marked by a chain of grabens filled with Upper Cretaceous and Cenozoic deposits.

The Laptev Basin is characterized by the continental structure of its crust, which attains a thickness of 35 kilometres. In the grabens that form an extension to the ocean's rift system, the crust is thinner and does not exceed 30 kilometres in thickness. The mantle beneath the grabens is of loose structure and is characterized by lower seismic wave velocities of 7.5 km/sec.

The underlying part of the folded-rock basement of the Laptev sedimentary basin consists of an extensive massif, composed of Karelian metamorphic complexes and probably of complexes dating back to an earlier period. On the west, north and east, this massif is flanked by zones of epicontinental and miogeosynclinal Kimmerian fold complexes. On the south, the narrow virgation (or fan-like arrangements) of Late Kimmerian linear folds in the Olenek zone separates the Laptev massif from the Siberian platform.

The intermediate level is represented by deposits from the Riphean to the Lower Cretaceous inclusive. In structure and composition it is similar to the cover of the north Siberian platform. Judging from palaeogeographical reconstructions, the intermediate level is thickest in the peripheral sections of the massif and may be assumed to taper down to nothing in the centre of the massif and near the continental slope. Here there is a stable distributive province, transporting the materials resulting from the erosion of metamorphic rocks into the surrounding Early Mesozoic depressions.

The upper level of the Laptev Basin began to develop in the period between the Early and Late Cretaceous in conjunction with the development of the rises surrounding it. The relationship between the complexes comprising it and the underlying rocks is not a uniform one; together with angular and structural unconformity in places where erosion had occurred earlier, on the periphery of the massif gradual transitions may occur as the result of co-sedimentary changes in the situation of the sources of the deposits.

The nature of the internal structure of the upper level and the distribution of the different thicknesses of Mesozoic and Cenozoic deposits are determined by the presence of a central chain of grabens and two marginal trenches (the Taimyr-Anabarsk and the Novosibirsk Islands trenches) which developed against the background of general moderate subsidence. The trenches and grabens are compensated for by sedimentation and are found on the sea floor by means of gravimetric and seismic data and also by examination of the silt facies of recent bottom deposits. Three stages can be distinguished in the development of the upper level: Late Cretaceous–Eocene, Oligocene–Miocene, Pliocene–Anthropogene. The first stage is characterized by the development of trenches and grabens, filled primarily with silt and clay deposits with a thin basal layer of sand and shingle. In the course of the two following stages, a common mantle was formed and, at the same time, trenches and grabens continued to develop actively. The deposits forming the mantle were characterized by a facial zonation conforming to the contours of the basin: a narrow littoral belt of sand and shingle and a vast inner zone of sandy clay sediments.

Geological formations in eastern Siberia

The East Siberia–Chuckhi Basin comprises the East Siberian and Chukchi Seas and the Yana–Kolyma plain. The continental side of the basin is squeezed between the flanking mountains of the north-eastern USSR. The basin is open towards the ocean deeps of the Amerasian part of the Arctic Ocean.

In this vast region, the earth's crust is marked by minor variations in thickness but on the average measures 35 kilometres. Despite this regularity, the folded-rock basement of the basin has a complex heterogeneous structure. In its northern areas are the remnants of the ancient Hyperborean Platform, composed of Karelian and pre-Karelian metamorphic rocks. To the south it contains zones of Baikalian and Caledonian fold complexes: these plunge into Kimmerian fold complexes to form the polycyclic systems of Mesozoic rocks of the north-eastern USSR.

The intermediate level is represented by terrigenous-carbonate Riphean deposits and also by Lower Middle Palaeozoic and partly Upper Palaeozoic deposits. Palaeogeographical reconstructions show that the spreading of the mantle was curtailed by pre-Mesozoic erosion and the transport of material into the Verkhoyano–Chukchi geosyncline. It can be assumed that relicts of the mantle that were part of the make-up of the intermediate level are to be found in the shallow depressions to the north of the line between the Novosibirsk Islands and Wrangel Island.

In the basin, the upper level is characterized by differences in age composition caused by the migration of the divide between folded and unfolded complexes in time. In areas unaffected by intensive Kimmerian movements in the unfolded beds, Triassic–Jurassic and, possibly, Permian deposits may be present. It may be anticipated that thin terrigenous paralic carboniferous (i.e. developed in marginal marine environments) and marine layers of these deposits fill the small depressions in the region of the Anzhu Islands and between the De Long Archipelago and Wrangel Island. To the south of the line formed by the Anzhu Islands and Wrangel Island and extending to the foothills, the upper level is represented primarily by Cenozoic deposits.

Thus, the contrast brought out by modern geological and geophysical findings between the platform-like surface structure and the subgeosynclinal deep structure of the sedimentary mantle can be regarded as the first distinctive feature of the structure of the Arctic seas of the USSR. The second distinctive feature lies in the fact that the large depressions of the shelf and adjacent land area form families making up sedimentary basins, first-order geological structures of the continental borderland. At the same time, linear depressions and rises show conformity with ocean deeps and together with them make up a single Arctic geodepression, flanked by an orogenic belt of water-divide crests and ridges.

The third important feature of the geology of the continental border of the Arctic Ocean is the complementarity of the shelf, ocean and dry land within the limits of the geodepression, i.e. their genetic interrelationship, together with the fact that they fully complement one another to form a single whole.

A few conclusions

The unity of the sedimentary mantle and the sequence in which it was formed indicate that the development of the ocean and the continent is interlinked, and that a long period of prehistory was involved in the formation of the Arctic Ocean and its shelfs.

The whole set of complementary structures that arose during the development of the Arctic geodepression are readily explicable from a geotectonic standpoint if that geodepression is viewed as the head section of the nascent Atlantic mobile belt of the earth at the initial stage of its development. Moreover, the oceanic zones can be viewed as an orthogeosynclinal region and the bordering orogenic belt and continental centroclinal as areas of geosynclinal tectogenesis, with continental borderland basins presenting analogies with certain forms of subsidence.   ■

Notes

1. A geosyncline is a principal structural, sedimentational unit of the earth's crust. A miogeosyncline is a thin development of sediments, with no volcanic rock, forming adjacent to a Precambrian shield.
2. Epicontinental basins are those found within the limits of a continental mass.
3. A graben is a depressed block found between two parallel faults.

Conditions governing man's access to food from the sea have changed, and the empirical methods of fishing have changed. The laws devised to nationalize access have had adverse effects on many countries. These factors are compelling us to pay increasing attention to farming and ranching in the marine environment, a growing field in which research and training are badly needed.

Chapter 4

Return to the sea — not as hunter but as farmer

T.V.R. Pillay

Dr Pillay was co-editor in 1979 of the major reference volume, Advances in Aquaculture, *published by Fishing News (Books) Ltd. He is programme leader in the Aquaculture Development and Co-ordination Programme, attached to the Food and Agriculture Organization (FAO) of the United Nations, Via delle Terme di Caracalla, 00100 Rome (Italy). The author acknowledges that the views expressed are his own and not necessarily those of FAO.*

Introduction

Besides being the primordial source of life, the ocean has also been the source of sustenance for living beings from very early times. Although in the course of evolution man has distanced himself from water, through his ability to breathe atmospheric air by means of lungs, thus becoming terrestrial, he has always tried to return to the sea as sailor, explorer or fisherman. Knowledge gained through years of observation and exploration revealed to him the immense potential of the sea and its resources for a variety of human needs, ranging from food and recreation to defence and industry. Human subsistence, up until the Neolithic period, had been dependent on hunting and gathering and so, naturally, man turned to the sea to hunt for, and gather, its living resources. In the course of time he developed suitable implements to help him in his hunting and gathering. These were obviously the beginnings of the modern fishing industries of the world, and there is no doubt that hunting and gathering on the seas steadily expanded. For example, in a fifteen-year period (1952–67) the total marine fish catch increased by 146 per cent, while the human population increased by only 34 per cent.[1] The increased contribution of this catch to human nutrition, directly or indirectly, is estimated to be around 74 per cent. In recent years, however, the contribution of marine production to human nutrition has slowed down greatly. The total landings, particularly of the favoured food species, remain more or less stationary or have shown a decline. Most objective predictions for the future are not too optimistic owing to the present state of the exploited fish stocks and the problems relating to their management.

To place the story of fishing in its proper perspective, we may look at the evolution of other forms of human subsistence. Over the years human societies have adopted forms of cultivation, pastoralism and ranching that were expected to stabilize production and bring it under greater human control. Of the resources hunted and gathered by man for food, only the marine or aquatic resources remain largely unchanged in respect of the basic form of exploitation. The reasons for this are not too difficult to find. Agriculture and animal husbandry obviously developed from a need to adopt more productive means to feed increasing populations. In the case of fishery resources, man sought to meet this need during the years by discovering new resources and by adopting more efficient methods of hunting (fishing) and utilization. Unlike agriculture, common access rights generally prevailed. However, conditions have now changed and the methods so far widely adopted to obtain increased production are seen to be counterproductive. Furthermore, the new laws of the sea greatly restrict common access, with adverse effects on many fishing nations. These factors have naturally resulted in greater attention being directed towards farming and ranching as means of meeting human needs of aquatic products, particularly for food. Thus, man's return to the sea as a farmer, either in support of the 'capture economy' or to start a complementary farming economy, has now started.

Capture versus culture

Although not as primitive as hunting, farming of fish or other aquatic animals is not a recent innovation. It has existed on a small scale in inland areas for centuries, in some countries from ancient times, most likely from the time of the evolution to pastoralism and land cultivation. But, probably as a result of the hopes and fears that the watery depths have always aroused in man, he tended to concentrate his efforts in catching what nature provided. What is really new today is the widespread recognition of the need to establish aquatic farming as a major means of resource development and to acquire the knowledge and skills necessary to bring the production of suitable aquatic animals and plants under human control in marine and inland waters.

Is there any incompatibility between the two diverse means of production, namely hunting and farming? Some significance has been attributed to the nature of the activities involved. In the past the hunter/fisherman could roam the seas and catch fish wherever these occurred—provided he had the necessary skills and equipment. There was little need, or even purposeful attempts on his part, to tend the resource. Some social scientists believe that the hunter perceives a certain occupational prestige in the adventures of fishing and that he looks with some contempt on those who adopt land- or coast-based methods of production, devoid of the excitements of open-water hunting. Obviously, this concept ignores the attitudes of the millions of part-time fishermen-farmers of inland and coastal areas. Neither does it take into account the numerous oyster fishermen who gradually transformed themselves into oyster farmers, or the present-day cage farmers of yellowtail, groupers and sea-bass.

Most changes in occupations or modes of production have occurred out of necessity and a good many small-scale fishermen around the world, especially in the developing world, have few alternatives if they want to continue as producers of aquafoods. It is only natural that they are initially somewhat sceptical and hesitant, but experience so far would seem to indicate that there is no major resistance among fisherman to the adoption of farming by themselves or by other community groups, if it does not interfere with or affect their fishing operations.

Some conflict has arisen in areas where farming is based on seed collected from the wild. Fisherman tend to ascribe poor fishing to the collection of larvae or juveniles for farming. Although there is hardly any adequate evidence of such adverse effects, the adoption of controlled breeding and seed production is gradually eliminating this possibility. Another kind of conflict can arise when increased production through farming leads to depressed market prices. Since this situation cannot be ruled out, for the present the advantage remains largely with the fishermen. Where culture takes the form of ranching or stocking of water bodies, it can contribute substantially to capture fishery and thus be supportive of this.

Although the entire, or greater part, of production of certain species of fish and shellfish is now the result of culture of one type or the other, world-wide production through aquatic farming is not expected to reach the magnitude of that by hunting in the near future. So the concern of certain capture fishery interests who see aquaculture as a threat seems unwarranted.

In an allied field, silviculture and reforestation have developed in support of forestry without antagonism and conflict. There is no reason why the same relationship should not be maintained between farming and capture in the fisheries sector. Undoubtedly a newly developing science like aqua-culture would need relatively higher efforts to develop technologies, train manpower and establish viable farming operations. But this can be expected to be well compensated by the higher rates of increased and sustained production. Such efforts, however, need not be at the expense of those required for the management of capture fisheries.

State of aquaculture

Because of the dispersed nature of operations and the lack of an organized system for the collection of statistics, precise figures of aquaculture pro-duction are not available. But estimates have been made at FAO on the basis of data provided by fishery agencies and information collected from available reports. The estimated production in 1980 can be seen in Table 1.

From the total estimated world production of over 8.7 million tonnes, 37.1 per cent are finfish, 36.7 per cent molluscs, 25.4 per cent algae and 0.8 per cent crustaceans. About 75 per cent of this comes from the farming of the seas, especially in coastal areas. Compared with 1975 when the total production was about 6.1 million tonnes, there was an increase of some 42 per cent in overall aquaculture production in five years. The significance of these figures in respect of future expansion becomes clearer from the fact that Asia and Europe account for over 84 and 13 per cent respectively of the total production, and the contribution so far of the vast continents of Africa and Latin America is only 0.9 per cent, where farming or 'return to the sea' has only just begun.

The estimated aquaculture production currently forms only some 12 per cent of the capture fishery production, but available information shows very clearly that it is a growth industry. Its future will largely depend on contributions that science can make to technology development, the skills that can be imparted, the investments that can be made and the organizations that can be built to undertake production and distribution. In other words, it is within the control of man and largely dependent on his efforts.

While production of food is the primary objective of aquaculture, in many countries its development is also directed towards the solution of other social and economic problems. The creation of employment in rural areas in order to arrest the drift of populations to urban areas is now a major social objective. The effectiveness of aquaculture in this regard has been much greater when incorporated in integrated rural development, covering all the major rural needs. Yet even as a purely sectoral development, well-organized aquatic farming has served to provide employment for large numbers of unemployed or underemployed people.

TABLE 1. Estimated aquaculture production by region and commodity groups, 1980 (in tonnes)

Region	Finfish	Molluscs	Crus-taceans	Seaweeds	Total
Africa	4 061	471	—	—	4 532
Asia and Oceania	2 516 203	2 568 925	60 260	2 206 434	7 351 822
Europe	630 196	505 516	30	50	1 135 792
Latin America and Caribbean	25 480	44 404	5 360	—	75 244
North America	57 386	76 993	5 596	—	139 975
TOTAL	3 233 326	3 196 309	71 246	2 206 484	8 707 365

The problem of surplus small-scale fishermen has prompted an attempt to convert them into farmers on a part-time or full-time basis. Because of dwindling fish stocks in the seas or the excessive number of 'hunters and gatherers' for the limited stocks, the catches and therefore the incomes of such fishermen have remained low and, in many cases, below the poverty line. Where appropriate tested technologies have been used and farming properly organized, fishermen have adjusted, both in attitude and in occupation, to farming. There has also been the added advantage that farming allows greater involvement of women and other family members in productive activities.

Aquacultural science

There can be little doubt that the development of aquatic farming will depend largely on scientific support. Although aquaculture, including mariculture, has existed for not less than 2,500 years, its emergence as a science is very recent. Most of the existing practices have been based on traditional skills. It was only some fifty or sixty years ago that biologists began to pay some attention to the large-scale rearing of aquatic animals and plants. Even then, the emphasis was on the life history and biology of cultivated organisms. This bias continues to dominate research related to aquaculture in a large majority of institutions concerned with the subject. Though biology of the cultivated organism, including its reproduction, nutrition and health form the basis on which a farming technology has to be developed, it has become very evident that aquaculture is a multidisciplinary science.

For the development and testing of an aquaculture technology, expertise in a variety of disciplines such as engineering, physiology, nutrition, pathology, ecology, microbiology, feed technology and economics would be necessary. Research advances in any of the constituent disciplines would certainly contribute in the long run to better farming practices but, unless a system-orientated multidisciplinary approach is adopted, production technologies may take an unduly long time to evolve. The trial-and-error methods that characterized early aquacultural development would continue, causing probably more failures than successes. Experiences in shrimp and prawn farming in some industrially advanced countries exemplifies the long-term handicaps caused by failures of enterprises based on unsound and untested technologies.

An important lesson that has been learnt in recent years is that the concept of appropriate technology is very real and vital in the development of aquaculture. There is no doubt that the adoption of aquatic farming and significant increases in production will occur worldwide as a result of the transfer of technologies. However, experience so far clearly shows that the technologies need to be modified and made appropriate to the conditions where they have to be adopted. A good example is that of shrimp-farming technology developed in Japan, which inspired attempts to adopt it in several other countries, both developing and developed. In almost every case where a transplantation of the technology, as such, was tried it resulted in failure. Those who persevered and modified the system have now succeeded in evolving technologies appropriate to their situations.

More models of suitable technology

Another example of appropriate technology is the integration of fish farming with agriculture and animal production which is the strong-point of Chinese aquaculture. Although the Chinese experience relates to fresh-water fish farming, it should be relevant to marine situations as well. The Chinese practise the polyculture of nine or more species in their ponds, integrated with the raising of pigs, poultry, cattle, silkworm, or with the farming of aquatic weeds, fodder grass, grains, vegetables and sugar-cane. The combination of species of fish is designed to make maximal use of the biological production in ponds, which receive treated or untreated manure and other wastes from animals and plants grown in the integrated farms. The silt from the ponds is used as fertilizer for the crops, and in times of need the pond water can be used to irrigate the crops or to water the cattle; ducks feed in and also fertilize the ponds. This sort of integration in China results in high production at apparently low operating costs. But it is apparent that this system cannot be transferred as such to every other country, not even every developing country. Integration of pisciculture with animal or crop production may be appropriate, but obviously the same patterns of integration may not be acceptable or economical in other countries.

As is evident from the above examples, technologies suited to typical conditions based on locally occurring and acceptable species have to be developed through multidisciplinary research. While laboratory and field experiments form the first steps in technology development, pilot production has to be undertaken to test the economic viability of the system before it is recommended for adoption. The requirements mentioned above should serve to indicate the extent of facilities and the variety of expertise necessary for effective aquaculture research.

An increasing number of scientific institutions and agencies are now devoting attention to such research, but comprehensive scientific knowledge on cultivated species and their rearing is limited to only a small number, and that too to species and systems suited to temperate climates in industrially advanced countries. Experience has clearly shown that high production at low costs can be obtained in tropical and subtropical regions which have higher water temperatures throughout the year, and it is likely that these are the areas where aquaculture can be relatively more profitable. Most of the developing countries are in these regions and are the ones that are in greatest need of aquaculture as a source of food and employment. A number of discussions and consultations have been held on aquaculture research in developing countries. The Consultative Group on International Agricultural Research (CGIAR), sponsored by the World Bank, the Food and Agriculture Organization of the United Nations (FAO) and the United Nations Development Programme (UNDP), have considered possible international research on aquaculture since 1973. They have not yet been able to come to a decision on supporting it.

Besides being outside of the common agricultural sectors, aquaculture has also the drawback that it has to depend on a large variety of species and systems of production suited to different regions. This makes it difficult to establish concentrated research in centralized institutions on single crops, as at the International Rice Research Institute and the International Maize and Wheat Improvement Centre. A network of regional centres has therefore been established in Asia, Africa and Latin America under FAO/UNDP auspices, and these centres are organizing multidisciplinary research on selected aquaculture systems of common interest and wide application. There is much to be learnt and adapted from allied disciplines to build up aquacultural science on a sound basis. In this respect aquaculture has to turn to agriculture and animal husbandry, rather than to capture fisheries. Agriculture or irrigation engineering, agricultural economics, animal and plant genetics, veterinary science, animal nutrition and animal feed technology are some of the fields that could contribute experience and practices of value to the aquaculture scientists.

Education and training

Apart from tested technologies, the most crucial element in successful aquaculture development is qualified manpower. There are very few institutions that have the facilities and multidisciplinary faculties to offer training in aquaculture at the university or vocational-school level. As countries expand their aquaculture activities, there is a continuously increasing demand for qualified personnel, particularly those with practical experience.

An assessment made in 1975/76 of the requirements of trained personnel in the developing regions has already been exceeded several times as a result of new programmes started in both public and private sectors. The FAO/ UNDP regional centres referred to earlier are offering specialized interdisciplinary courses for senior personnel, but the needs, particularly of aquaculturists with 'hands-on' experience is now so large, and expanding steadily, that only a part of the requirements can be met through these centres. Institutions on the model of agricultural or forestry colleges and schools may be needed to turn out the required number of trained personnel.

Although a number of books and journals are being published today on aquaculture, there are probably none that could be used as comprehensive textbooks for instruction in the subject as a whole or in the constituent disciplines. The production of textbooks and other teaching material deserves high priority attention.

Conclusions

There is little doubt that the farming of the seas and inland waters will expand in years to come. Since there is growing competition for land, it is the seas and coastal waters that are likely to offer greater opportunities for development. Many projections of likely future production have been made.

A fivefold increase by the turn of the century has been considered a reasonable prediction, though this will be contingent on the required scientific, financial and organizational efforts. Because of the nature of the activities involved, aquaculture projects have longer gestation periods than, say, capture fishery projects. For example, a large-scale fish farm may take two or more dry seasons (years) to build and, depending on the nature of the soil, the farm may require another year or two of management to become productive. Then, even if a well-tested technology is used, the farm may take a further year or two before fully satisfactory production is achieved. However, once it becomes productive, the unit can be maintained in this state almost indefinitely. (A well-maintained farm can have almost an indefinite economic life.)

Considering the massive inputs and scientific support needed to make greater use of the farming potentials of the seas, it will be necessary and even advantageous to seek changes in the attitudes of marine hunters (or fishermen) and marine scientists. Even though the antagonism that is often present will disappear in the course of time, the period of conflict can be minimized if these specialists, as well as others concerned with the management and use of marine resources, realize that the return to the sea as a farmer is a natural evolutionary process. It has to be remembered that it is only the exploitation of marine and other aquatic resources that has remained largely a hunting and gathering activity. Till such time that man's needs for aquatic resources can be met by farming alone, the two forms of production can and should coexist. For many years to come we can expect both fishing and aquatic farming to remain complementary and therefore mutually supportive, contributing substantially to human food production. ■

Note

1. S. Holt, Marine Fisheries, In: E. Borgese and N. Ginsburg (eds.), *Ocean Yearbook*, Vol. 1, Chicago, University of Chicago Press, 1978.

To delve more deeply

FAO/UNDP. *Aquaculture Development in China. Report on FAO/UNDP Aquaculture Study Tour to the People's Republic of China.* Rome/Nairobi, FAO/UNDP, 1979.
McVey, J. Current Development in the Penaeid Shrimp Culture Industry. *Aquacult. Mag.*, Vol. 6, No. 5, 1980, pp. 20–5.
Pillay, T. The State of Aquaculture Development 1976. In: T. Pillay and W. Dill (eds.), *Advances in Aquaculture*, pp. 1–10, 1979.

———. Research and Extension Services for Aquaculture Development. In: T. Pillay
and W. Dill (eds.), *Advances in Aquaculture*, pp. 84–9, 1979.
———. *Planning of Aquaculture Development, An Introductory Guide*. Farnham,
Fishing News (Books) Ltd, 1979.
PILLAY, T.; WIJKSTRÖM, U. Aquaculture and Small-Scale Fisheries Development.
(Paper presented at Symposium on the Development and Management of
Small-Scale Fisheries (19th Session), Kyoto, 1980.)
POLLNAC, R. Socio-cultural Aspects of Implementing Aquaculture Systems in
Marine Fishing Communities. Providence, R.I., University of Rhode Island,
1978. (Anthropology Working Paper, No. 29.)

The importance of the study of marine pelagic organisms in relation to an International Geosphere and Biosphere Programme (IGBP) is outlined. The lack of knowledge about feedback mechanisms between living pelagic organisms and the geosphere is stressed. The IGBP should attempt to analyse changes in the marine pelagic system of the biosphere brought about by the effects of changes in the geosphere and, further, should extend the results of such analysis to prediction of interactions between the bio- and geospheres.

Chapter 5

Marine pelagic organisms and the International Geosphere and Biosphere Programme

Tadahisa Nemoto and D.D. Swinbanks

Professor Nemoto (a Japanese) and Dr Swinbanks (a Briton) are researchers at the University of Tokyo's Ocean Research Institute, 1-15-1 Minamidai, Nakano-ku, Toyko 164 (Japan). Their paper was prepared originally for the book Global Change *(Thomas Malone, ed.), published in 1985 by the ICSU Press and Cambridge University Press, and with whose generous permission it appears in the present volume.*

Introduction

The ocean contains 97 per cent of all the water on earth, and the majority of the rest of the water is locked up in ice in the polar regions in which most organisms cannot live for any length of time. Thus, the ocean is the most important biological domain for aquatic organisms and has the greatest biomass carrying capacity amongst aquatic environments.

The surface of the ocean is about 2.5 times as large as the land surface of the earth, and the average depth is about 4,000 metres. It has already been established that living organisms are widely distributed throughout all depths of the ocean. The inhabitable zone for terrestrial organisms may extend from up to several hundred metres above the land surface down to some 10 metres below the surface in mud and rocks. The ocean has a much wider inhabitable zone as indicated by the sampling of swimming vertebrate fish at as much as 8,000 metres' depth in the sea. Not only is the terrestrial inhabitable zone much narrower, but also the seasonal, ontogenetical and diurnal vertical movements of living organisms on land are far less pronounced on the whole than in the sea-water column.

Marine pelagic organisms are widely dispersed throughout the whole water spectrum, but the high pressures at depth in the ocean present a major obstacle to scientific investigation of these organisms in the deep sea. However, pelagic organisms should also be one of the targets of study in the IGBP project, because we cannot make a complete quantitative analysis of the geosphere beneath the ocean without taking into account the contribution of pelagic organisms. For example, vast areas of the ocean floor are blanketed with thick sequences of 'ooze' composed almost entirely of the tests of pelagic organisms.

In this chapter, we would like to stress the importance of both 'bio-interaction' between living organisms and interactions of these organisms with the physical environment of the geosphere. (We use 'geosphere' here as an inclusive term for the earth and its hydrosphere and atmosphere.)

At first we will describe characteristics of the distribution of phytoplankton and zooplankton in relation to such physical factors as water temperature and ocean circulation, looking not only at the present-day situation but also taking account that of the past, in particular the last ice age which has had a profound influence on the present-day situation. Next we will discuss larger predators, in particular whales, and point out how man's whaling influenced the populations of other organisms and habits in polar regions. Finally we will touch upon other examples of human interference with the marine environment and other natural catastrophes. Throughout our paper and at the end, we will point the way to future possible lines of research by IGBP.

Phytoplankton

The main primary producer in the sea is phytoplankton. Total net, primary production in the marine system is about half of that in the continental domain (namely, 25 thousand million tonnes per year for the former, compared with 50 thousand million tonnes for the latter[1]). The biomass of plants in marine areas is, however, less than one four-hundredth of that on land because tiny phytoplankton with rapid rates of turnover are the main components of primary production in the sea. Giant kelp are restricted in distribution, although their biomass in certain areas is very high.

Temperature of the surface waters within the shallow-sea system has a direct effect on the distributional depth, primary productivity and even the composition of phytoplankton. Global climatic changes in the ocean from the time of the last ice age to the present day have been well described by works in the Climap project.[2] The climatic variations directly affect the sea's surface temperature in the shallow-water system above 200 metres' depth where phytoplankton are living.

The geohistorical climatic changes since the last ice age have been a major factor determining the present patterns of distribution of living organisms. Estimates of old sea-surface temperature patterns can explain some of the patterns, such as the antitropical distribution of the same species in the northern and southern hemispheres, but the global distribution of phytoplankton also changes according to the distribution of water masses and ocean circulation. Some pelagic *Ceratium* dinoflagellates are at present divided into temperate and polar zones in both the northern and southern hemisphere, with transition zones between each of these,[3] and these zones need to be compared with the distribution of secondary producers (i.e. zooplankton).

The thickness of the euphotic zone (penetrated by daylight), where almost all primary production occurs, differs according to the solar irradiance and the types of organism present in the sea water. The decrease in solar radiation on a global scale (because of the eruption of dust from a volcano, for example) can cause a large effect on primary production even if the extent of the change in thickness of the euphotic zone is small. The biomass and species composition of phytoplankton, the primary producers, have a close link to CO_2 dynamics in the sea water, and the link may extend to the atmospheric cycle of CO_2; however, this has still to be confirmed.

Zooplankton

The distribution and biomass of the main secondary producers of the sea, zooplankton, are closely correlated with phytoplankton distribution in the shallow-sea system, and the extent of the vertical diurnal movement of zooplankton is also affected by the concentration of their food, phytoplankton. It has been suggested that the changes in surface-water temperatures had a pronounced effect on the habitats of phytoplankton.

The present horizontal distribution pattern of zooplankton and micronekton [less active swimming organisms than nekton, as opposed to the

non-swimming plankton] is ascribed mainly to water-temperature distribution, or the horizontal zonation of water masses.[4] If a water-temperature
change of at least 2 degrees C has occurred in the upper 200 metres of the
surface-water system since the Pleistocene [the second youngest or glacial
stage of the Tertiary period], a variation in the quantity, composition and
distribution of zooplankton must have occurred.[5] In the case of some species
of zooplankton, their distribution at present coincides with the distribution
of water masses; but some common cosmopolitan species also occur across
water fronts and lines of convergence in different water masses and even in
different water currents.

The vertical migration of zooplankton and micronekton is apparently an
adaptation to variations in water temperature or light intensity, and is
performed for feeding, resting or reproduction purposes.

The major effects of recent morphological changes on the ocean have
been compiled by Van der Spoel and Heyman,[6] and they conclude that the
separation of plankton into northern and southern hemispheres occurred
13,000-15,000 years ago. Close contact of each northern and southern
plankton group occurred during the Würm glaciation 30,000 years ago; and,
during other ice ages of the earth, contact and exchange of northern and
southern hemisphere zoo- and phytoplankton occurred.

Precise examination of separate groups of the same species existing at the
present has revealed some interesting features. For example, some
euphausiids, such as *Euphausia similis*, which have been separated into
northern and southern antitropical zones, already show some slight
morphological differences. Zooplankton and micronekton feeding mainly on
phytoplankton often form dense swarms in the sea. Cold-water species, such
as Antarctic krill *(Euphausia superba)* and subarctic small krill *(Thysanoessa
inermis)* are rather stenothermal ['seeking a constant ambient temperature']
animals, and form swarms.

Thus, from the above it is clear that present-day changes in water circulation, water masses and the extent of vertical mixing of waters should be
examined and compared with variations in the abundance, composition and
distribution of organisms belonging to trophic levels higher than and
including secondary producers on a global scale. The feedback of these
animals to nutrient cycles in the ocean and even in the terrestrial domain are
also very important subjects to be studied, yet to date little research has been
carried out in this field.

Interactions of organisms through food webs

The analyses of Hubbs,[7] Brinton,[5] and Reid *et al.*[8] strongly suggest that if
coastal waters are cooled by eight degrees C in winter or three degrees in
summer, the distribution of each antitropical species overlaps, leading to a
mixing of the two populations. This overlap of separate groups in the northern and southern hemispheres, in the case of certain euphausiids such as

Thysanoessa gregaria, may occur when the temperature drops by only two degrees C at 200 metres' depth.[5]

Geographical isolation has produced many subspecies of marine organisms. Water masses are separated by a frontal zone, but this boundary is apparently a less effective barrier than land barriers for marine organisms. Nekton and even the larger micronekton can cross these fronts; however, the strong convergence at such fronts often results in the congregation of organisms at the front due to stimulation of the growth of frontal species and because mobile animals stay there to feed.

Organisms in the marine ecosystem closely interact with each other through food webs, and there is also a strong feedback mechanism operating after death in the marine environment. In the shallow sea system, the secondary producers include nanno-, micro-, meso-, and macroplankton. The regeneration of nutrients, including ammonia, urea and phosphate, has a strong stimulating influence on phytoplankton blooms. Sometimes this nutrient-rich water has a patchy distribution that replicates the swarms of secondary producers. For example, this is often true of herbivorous species which are abundantly distributed and form dense swarms in the sea.

The mass occurrence of krill in the Antarctic is apparently connected with blooms of several phytoplankton species, such as *Fragilariopsis (Nitzschia) kerguelensis* and other smaller diatoms. Interglacial sediments along the polar front and at 65 degrees south latitude are 'diatom-rich', which suggests the extensive distribution of herbivorous *Euphausia* species and salps, because the fecal pellets of these organisms transport these diatom shells to the bottom.

Larger predators

Changes in the horizontal distribution of organisms belonging to higher trophic levels that feed on zooplankton should also be considered. Marine pelagic organisms move themselves or drifts with current. They also respond very quickly to changes in physical phenomena as well as interact with other organisms, mainly through predator-prey relationships.

In the case of eurythermal organisms [those withstanding a broad range of temperatures], which have a broader range of movement, changes in water temperature and in irradiance, do not regulate their movement much in the short term (i.e. several decades). Amongst the interactions between and within such species, migration and distribution are determined mainly by two factors, feeding and reproduction. The feeding behaviour of baleen whales is determined mainly by the food-carrying capacity of the area and by the swarming behaviour characteristics of each food species and its developmental stage.

The size of certain species of baleen whales in the southern hemisphere is larger than that in the northern hemisphere, but the toothed whales, such as sperm whales, show little difference in size between northern and southern hemispheres. Sperm whales do not separate clearly into northern and southern zones, and reproduction areas cover tropical and subtropical waters. The lack of a size difference apparently is a result of the fact that their feeding region, in the mesopelagic zone and on the shelf floor, does not differ much between northern and southern hemispheres.

The formation of some local stocks amongst organisms that actively migrate in the sea also occurs that is inconsistent with any interpretation based on geographical or physical conditions, such as the presence of land barriers or water-termperature fronts. For example, pygmy blue whales *(Balaenoptera musculus brevicauda)* are concentrated around Kerguelen Island and are far smaller than ordinary Antarctic blue whales and even those of the northern hemisphere. This result suggests that local stocks respond rather quickly to isolation, when compared with the geological time scale.

An expanding of the feeding ranges of whales to the south and north should have occurred after the last ice age. The retreat of the ice front and the summer pack-ice allowed extension of baleen whales to the south, and then they established feeding zones according to the 'biological strength' of each species.[9] The largest blue whales established feeding zones closest to the ice edge, while fin and sei whales occupied the adjacent zones. The decline of the blue whale's stock due to whaling allowed the penetration of other whales to higher latitudes, where they can feed on Antarctic krill. This phenomenon simply resulted from thirty years of whaling by man, although the formation of feeding zones of baleen whales probably took hundreds of years to be established under natural conditions. Thus, feedback among organisms in the marine system is rather rapid, but of course it goes completely unnoticed if no appropriate monitoring system is set up.

Changes in the abundance of top predators, regardless of whether they are high or low in the food chain, also affect organisms at higher trophic levels than the prey. For example, the decrease in the main krill feeders, baleen whales, caused an increase in lesser predators such as crabeater seals, penguins, and pelagic fish. These changes occurred rather quickly after 1960, a time when sea conditions did not undergo any critical change in the Antarctic ocean.

The prey of baleen whales, namely, krill and other plankton, may be subject to less predation pressure, and the surplus is fed on by predators at the same or lower trophic levels than the whales. The rapid increase in population of these other predators also had an impact on the terrestrial domain. Some natural antibiotics in phytoplankton (such as acrylic acid) passed through krill and were brought, by increased numbers of penguins, to the land around rookeries. As a result, bacterial and fungal activity appears to be changed in these areas.

Other sea bird populations have also been affected by fluctuations in prey population due to human intervention and disturbance. Many sea birds have increased in population by feeding on the waste of fishing boats, especially in upwelling waters. The rookeries of these sea birds are sites of increased accumulation of guano. The traces of prey these birds feed on can be detected by the presence in the guano of fish scales, bones and even the diatoms that were fed on by the zooplankton on which the fish were feeding.

Other examples of human influence

The influence of man on the sea already extends over a vast area. Many studies have been undertaken on the pollution caused by the dumping of wastes in rivers that run into the sea and on the direct dumping of wastes in the ocean itself. The expansion of the human population has caused much destruction of beaches, brackish lakes and marshes that form the interface between the land and the ocean. In addition, small bays and inlets are being destroyed by reclamation activities. Perhaps, the biggest impact of man on the ocean has been through the input of man-made chemicals such as DDT and PCB.

Another effect of man is to bring about the transportation of toxic red-tide phytoplankton into new areas. Some species of *Protogonyaulax* have been detected in the central coastal areas of Japan, and *Chattonella* species originally described in the waters around India are now present in Japanese coastal waters in some summer seasons and have damaged fish in aquaculture systems. This internationalization of endemic toxic phytoplankton should be examined to assess the changing of species composition in the biosphere.

Some catastrophes

Catastrophic change in the population of marine organisms, such as swarming of fish, is often observed on the small and large scale and may be ascribed to changes in sea temperature, advection of cold currents or blooms of toxic plankton such as red tides. The El Niño phenomenon not only destroyed the upwelling system in the waters off Peru, but it also produced the southern ocean oscillation which affected water temperatures in the western Pacific, thereby bringing about fluctuations in fish populations.

The time scale of these changes in population sometimes extends over fifty years. The swarming of fish such as herring, sardine and anchovy may suddenly increase or decrease in many water areas. For example, herring decreased in 1930 in the western Dover Channel while sardine increased. The reverse happened in the 1970s, and herring increased again. California sardine and anchovy also showed similar turnover in biomass in waters off California. The analysis of fish scales in sediment cores from the ocean floor, and in guano deposits on land, could provide information about such cycles on a geological time scale. This type of study should also carried out in waters around Japan and South Africa.[10]

A case-study site for IGBP and related programmes needs to be selected for intensive research. If global changes in sea level caused by melting ice occur due to the warming up of the earth as a result of the increase in atmospheric CO_2 (the so-called greenhouse effect), drastic changes in faunal and floral distribution can be expected to occur most rapidly in polar waters, as organisms in this region are sensitive to even slight changes in water temperature. Thus, the Antarctic region should seem to provide an ideal location for a case-study site.

Lines of research involving marine pelagic organisms

As marine organisms are diverse and widespread, scientific programmes that have aimed at studying the biosphere *(sensu stricto,* the work by Roederer[11]) have inevitably been rather restricted, and still much remains to be done. Recent information has revealed interesting and important fields of study on marine organisms. All of these fields of study cannot, of course, be covered directly by IGBP; however, they should be incorporated as far as possible in this project, as marine organisms are a very important part of the whole biosphere.

We describe here eight lines of research involving pelagic organisms as examples of what might be studied by the IGBP.

Distribution, stratification and swarming of plankton
in relation to physico-chemical factors in the sea

Within the three-dimensional space of the sea, living organisms including plankton exhibit non-random distribution. The aggregations and swarms of zooplankton and micronekton range from a few centimetres in radius to several kilometres, and the time scales of these patches differ, ranging from minutes to years.

This patchy distribution of living organisms is also observed in non-motile phytoplankton if their distribution is examined precisely. Phytoplankton blooms often occur synchronously with or immediately following the input of nutrients. The sources of nutrient input are various. Upwelling and vertical mixing of water are generally believed to provide the main supply of nutrients. However, patchy distribution of certain nutrients such as urea and ammonia has also been detected after swarming zooplankton or micronekton swim away. The nutrients are derived from excretion products of the zooplankton and play a significant role as a trigger of phytoplankton blooms.

The stratification of plankton distribution, including subsurface phytoplankton maximum, is often considered to be related to the presence of internal waves and a pronounced thermocline. Many recent studies have clarified the species composition and mechanism of maintenance of the sub-surface (deep) maxima of phytoplankton and its dynamics. The analysis of biomass and species composition should be carried out to provide basic information for the study of stratification. In addition, further information on diurnal, seasonal and ontogenetic vertical migration is required.

*Mega-scale distribution of plankton and micronekton
in the ocean*

Studies of the mega-scale distribution of plankton and micronekton began after 1945. Most research on this topic has been carried out in the North Atlantic and North Pacific and has included interdisciplinary work covering climatic changes, ocean circulation and development of zones of particular species in certain areas on a geological time scale.

Ocean circulation divides the world into a set of quite different biological provinces.[8] The distribution of nutrients suggests that divergence and upwelling bring nutrients into the euphotic zone, but nutrient concentration also varies greatly from one circulation system to another.[8] This mega-scale distribution of plankton shows decade-to-decade and century-to-century variability.[12]

Vertical transport of organic matter

Recent research has revealed the rapid vertical transport of organic matter by biological processes in the ocean. The main sampling techniques employed to analyse such vertical transport are sediment traps and the sampling of particles in each stratum of the water column by filtering large volumes of water. Fecal pellets of zooplankton are considered to be largely responsible for this rapid vertical transport of organic matter. 'The rain of detritus' in the sea water is an old concept, but now it has been revived to explain the abundant organic flux from the shallow to deep-sea systems. The detection of surface-produced chlorophyll pigments at great depths and the presence of many undissolved fragments of planktonic organisms in detritus support the idea of rapid vertical transport. Decomposition of sinking organic matter has been studied to some extent already, and geochemical analysis of components of the vertical flux and counting of sinking biogenic particles have been carried out.[13] However, it is still uncertain what kinds of organisms are the main contributors to this rapid vertical transport, and it is still open to question whether there is any 'ladder of migration',[14] accelerating vertical transport.

The settling rain of detritus is consumed by Holothurians and other benthic organisms living in the sediments of the sea floor. Further quantitative analysis of such interactions between sediments and bottom dwellers is also needed.

Climatic impacts on the population of pelagic swarming fish

In recent analysis of the fluctuation of sardine resources, Tomosada pointed out[15] that meteorological conditions such as the amount of cloud, duration of sunshine, and precipitation in the spawning grounds and nursery grounds of sardines have a strong impact on sardine larval development. These meteor-ological factors affect phytoplankton growth, which, in turn, has a significant effect on copepods, the most important food for sardine larvae. As a result, the survival of larvae can be changed drastically by meteorological conditions.

The meander of the Kuroshio also has a strong impact on the transporta-tion of larvae to suitable nursery grounds. Similar observations of sardines, anchovy and mackerel have also been made in California waters and off the coast of Peru in the Pacific. Analysis of long-term and short-term data and experiments are needed to clarify further the impact of climate on pelagic organisms.

The analysis of growth rings in long-living organisms

On land, annual rings of trees (i.e., growth rings) have often been used to assess variations in air temperature, average sunshine and also some unusual effects due to harmful parasites and insects. In the sea, a few organisms such as shellfish and corals have also been used to assess temperature changes through the study of stable isotope ratios and through studies of daily, monthly or annual growth rings in fish, shellfish, nautilus and corals. This approach offers a promising way of assessing parameters of the environment in which organisms lived.

Long-living organisms such as whales, seals, deep benthic molluscs, turtles, and corals are the best specimens for such studies because they live for some tens of years up to a hundred years, and have annual rings which are affected by environmental conditions. Annual rings in the ear plug of baleen whales and teeth of toothed whales are well laminated, and the space between the lamina may be used to assess nutritional conditions as well as other environ-mental parameters, including temperature.

Recent studies have also revealed that these daily, monthly and annual growth increments occur in many marine organisms including fish otoliths, corals, nautilus and others. Analysis of these rings is unusually linked to daily, monthly or even annual changes of rhythm within animals. However, some such studies may possibly be used to look at long-term geohistorical changes and for short-term prediction. Perhaps we can use this technique, coupled with knowledge of recent changes on the short-term time scale of say ten years, to predict trends of global change in the near future.

Biological systems under extreme conditions, such as polar waters

The ecosystems in polar regions where ice abounds have particular character-istics, and marine organisms are highly adapted to inhabiting such regions. These animals play an important role in geochemical cycles in polar waters, but they have not been studied in detail from the geochemical point of view. For example, in shelf waters of the Bering Sea during winter when there is no sun, there are few primary producers to be found; but when the sun starts to shine, algal mats develop on the bottom surface of sea ice and sink to the shelf bottom. This organic matter is very quickly utilized by bottom feeders, including bivalves and amphipods. In summer larger predators, including gray whales, seals and birds, in turn feed on the benthos. This perturbation of the bottom by such animals and the dynamics of nutrients in the sea water have been examined in programmes such as Probes, but the carriers of organic matter and the dynamics of living organisms remain poorly understood and require systematic investigation.

Through the recent advancement of observation technology, it has become possible to assess polar geosphere conditions on a global scale. The polynya (ice lake) in the Weddell Sea in the Antarctic, revealed by satellite observations of passive microwaves, almost certainly has a great influence on recruitment of Antarctic krill and other organisms. This lake varies greatly in size from year to year. The trend of the variability in size may be very important if it is positively coupled with the production of primary producers and animals of higher trophic level in the Antarctic. Study of this matter will be an appropriate successor to the BIOMASS (Biological Investigations of Marine Antarctic Systems and Stocks) project of the International Council of Scientific Unions.

Interface regions between the ocean and land

The interface between the land and the sea is a very important environment for study that is constantly undergoing change because of natural and man-made processes.

The first thing that needs to be carried out is a statistical assessment of what has been lost at the interface between the ocean and land, in such environments as sandy beaches, coral reefs, mangrove swamps, salt marshes, rocky shores and inland seas and bays. Such study sites have often been examined for environmental protection studies. Now, we must look at these sites as typical interface regions where mankind meets the sea. Studies should include examination of biological and geochemical processes operating at the interface of complete coastal units, as well as studies of geomorphological change.

Red-tide phenomena induced by man

The formation of deserts in terrestrial regions can result from over-population of grazing domestic animals, over-cultivation, or ineffective irrigation. This type of human interference with the environment is perhaps comparable to the over-setting of aquaculture systems and over-cultivation of fish and shellfish in coastal areas.

One response of the marine environment to over-cultivation is the formation of thick deposits of fecal matter and unused food that quickly make the shallow sea bottom anoxic. High eutrophication induces red-tide phenomena, and as a result mass fish kills have sometimes occurred. Blooms of toxic dinoflagallates and the internationalization of red-tide species should also be observed carefully in this context.   ∎

Notes

1. G.M. Woodwell, Aquatic Systems as Part of the Biosphere, In: R.K. Barnes, K.H. Mann (eds.), *Fundamentals of Aquatic Ecosystems,* Oxford, Balckwell, 1978.
2. Climap Project Members, The Surface of the Ice-Age Earth, *Science*, Vol. 191, 1976, p. 1131.
3. H.W. Graham, N. Bronikowsky, The Genus *Ceratium* in the Pacific and North Atlantic Oceans, Washington, *Sci. Res. Cruise VII–Carnegie Biology V* (Carnegie Inst. pub.), 1944, p. 565.
4. Sverdrup, Johnson and Fleming, *The Oceans, Their Physics, Chemistry and General Biology,* New York, Prentice-Hall, 1942.
5. E. Brinton, The Distribution of Pacific Euphausiids, *Bull. Scripps Inst. Oceanogr.,* Vol. 8, No. 2, 1962, p. 51.
6. S. Van der Spoel, R.P. Heyman, *A Comparative Atlas of Zooplankton: Biological Patterns in the Oceans,* Berlin-Heidelberg-New York-Tokyo, Springer Verlag, 1983.
7. C.H. Hubbs, Antitropical Distribution of Fishes and Other Organisms, *Proc. 7th Pacific Sci. Congr.,* Symposium on Problems of Bipolarity and of Pan-temperate Faunas (Vol. 3), p. 324.
8. J.L. Reid, E. Brinton, A. Fleminger, E.L. Venrick, J.A. McGowan, Ocean Circulation and Marine Life, In: H. Charnock, G. Deacon (eds.), *Advances in Oceanography,* New York, Plenum Press, 1976.
9. T. Nemoto, Food of Baleen Whales with Reference to Whale Movements, *Sci. Rep. Whale Res. Inst.* (Tokyo), Vol. 14, 1959, p. 149.
10. M. Uda, Problems and Perspectives in Fisheries Oceanography, *Marine Sci. Quart.,* Vol. 1, 1978, p. 106.
11. J.G. Roederer, The Proposed International Geosphere-Biosphere Programme (IGBP): Special Requirements for Disciplinary Coverage and Programme Design, paper prepared for ICSU Symposium on Global Change, Ottawa, September 1984.
12. A.S. Monin, V.M. Kamenkovich, V.G. Kort, *Variability of the Oceans,* New York, John Wiley, 1974.
13. S. Honjo, Material Flux and Modes of Sedimentation in the Mesopelagic and Bathypelagic Zones, *J. Mar. Res.,* Vol. 38, No. 1, 1980, p. 53.
14. M.E. Vinogradov, *Vertical Distribution of the Oceanic Zooplankton,* Moscow, Nauka, 1968; Israel Program for Scientific Translations, 1970.
15. A. Tomosada, Estimation of Climate Impact to the Decrease of Sardine Resources in the Prewar Period, *Bull. Tokai Regional Fisheries Res. Lab.,* No. 111, 1983, p. 1.

Salt-water aquaculture or mariculture makes it possible, by reason of the completely satisfactory food-conversion efficiency which it ensures, to produce a high yield of high-quality proteinic foods. It has been practised for centuries but is evolving as the result of modern researches and methods: ability to control the reproductive cycle and hybridization of the species concerned, control of parasites, more economical feeding and appropriate technology. It is time that developers of ocean shores and lagoons considered promoting mariculture; since it can be developed even without large investments and can be adapted to the existing economic and social structures it offers great promise if it is combined harmoniously with intensified fishing.

Chapter 6

Aquaculture: realities, difficulties and outlook

Albert Sasson

Albert Sasson, who is a specialist in applied microbiology, is engaged in long-range planning and management activities in the Unesco Bureau of Studies and Programming. His address is: Unesco, BEP, 7 place de Fontenoy, 75700 Paris (France).

Introduction

It is usual to bring together under the name of aquaculture the various forms of more or less intensive rearing of species of fish, crustaceans and molluscs in fresh, brackish and salt water. Mariculture, which is salt-water aquaculture, has a very long history. In East Asia oyster culture dates back to before the Christian era. In the West the Romans were probably the first to cultivate oysters. A treatise on fish-farming was written in 475 B.C. by the Chinese Fan-Li and over 1,000 years ago the Chinese had developed a complex system of fresh-water fish-farming in which several species of fish were introduced into an aquatic environment in which each species occupied a specific ecological niche and had its own distinct food. Thus six varieties of Chinese carp were able to co-exist in the same pond: one was herbivorous and fed on the plants floating on the surface of the water, the second and third varieties lived on phytoplankton and zooplankton respectively while the three others lived on the bottom of the pond and fed on molluscs or detritus. This trophic complementarity made it possible to obtain high yields. Furthermore, organic wastes could be eliminated, since the three bottom-living varieties of carp were partly saprophagous or detritivorous. The Chinese are past masters in designing simple and ingenious installations for recycling waste matter of every kind while producing a useful biomass. On the basis of their remarkable experience in this respect they have developed brackish-water and salt-water aquaculture.

Fish-farming was common in Eastern Europe and South-East Asia in the thirteenth and fourteenth centuries but only since 1965 has it become a real science. In some countries, such as China or India, almost 40 per cent of foods of marine origin came from fish-farming. In South-East Asia fish-rearing in salt-water or brackish water has been a traditional activity for centuries. The mangrove vegetation is felled and the marshy ground sub-divided into ponds separated by mud banks; the spawn of several species of fish and crustaceans is then introduced into the ponds; after being raised in a special pond the larval and juvenile forms are transferred into rearing ponds where they are fed on algae, plankton and invertebrates; their growth period lasts for a year at most and the average yield is of the order of 540 kg per hectare, which is altogether comparable to protein yields in agricultural ecosystems.

Advantages of aquacultures: food chains and energy flux

To obtain the animal protein he needs man consumes the products of food chains that are generally long and are characterized by considerable loss of energy at each trophic level. From a production of phytoplankton estimated at 7×10^{11} tonnes per year, annual production will be of the order of 7 to 14×10^{10} tonnes in the case of herbivorous species (zooplankton, invertebrates, anchovies, mullet and other species feeding on phytoplankton) and 7 to 14×10^9 tonnes in the case of primary carnivorous species, 7 to 14×10^8 tonnes in the case of secondary carnivorous species and 7 to 14×10^7 tonnes in the case of tertiary carnivorous species. Energy conversion efficiency from the trophic level of the primary producers to the upper level of the herbivorous species ranges from 10 to 20 per cent in various types of aquatic ecosystem, which means that 100 kg of phytoplankton (trophic level 0) will produce only 10 to 20 kg of herbivorous organisms (trophic level 1) and then assuming conversion efficiency of the order of 10 per cent, 1 to 2 kg of primary carnivorous species (trophic level 2), and 0.1 to 0.2 kg of secondary carnivorous species (trophic level 3). The energy needed to produce fish at trophic level 3 is therefore considerably greater than that required for production at trophic level 2.

A production of 740,000 million tonnes of phytoplankton per annum provides only 60 million tonnes of fish and other marine creatures, i.e. less than 0.01 per cent of the primary production. The production of a natural ecosystem comprising several types of fish and other animals feeding at different trophic levels is the resultant of the samples taken at those levels: for example, mussels, oysters, mullet and certain species belonging to the *Tilapia* genus are harvested at trophic level 1; anchovies which feed both on phytoplankton and on zooplankton are harvested between trophic levels 1 and 2; herrings are taken at trophic level 2, because they feed on zooplankton; small cod are harvested at trophic level 3 and large cod at trophic level 4.

The production of animal protein for human nutrition requires an amount of vegetable protein six times as great as the quantity of meat produced by herbivorous domesticated animals. By shortening the food chains aquaculture could produce animal protein more economically. For example, mussels and certain species belonging to the *Tilapia* genus feed at trophic level 0 and their production is therefore more efficient. It is also possible to select a species or subspecies of high food-conversion efficiency and in which the ratio of body weight increment to the amount of food consumed is high. This efficiency and the trophic level must be borne in mind when a species is being selected for monoculture or when several species are being bred to occupy all the niches in an ecosystem on the basis of their physiological characteristics and their conversion efficiency.

Efficiency of production in aquaculture

The proportion of edible tissue is greater in fish than in beef, cattle, pigs or poultry (Table 1). For example, more than 80 per cent of the dressed carcass

TABLE 1. Dressing percentage and carcass characteristics
of channel catfish, beef, pork and chicken

Flesh	Dressing per-centage[1]	Characteristics of dressed carcass			
		Refuse[2] (%)	Lean[3] (%)	Edible fat (%)	Food energy (kcal per 100 g)
Channel catfish					
(*Ictalurus punctatus*)	60	13.7	80.9	5.4	112
Beef, choice grade	58	15	51	34	323
Pork, medium fat	65	21	37	42	402
Chicken, broiler	72	32	64.7	3.3	84

1. Marketable percentage of animal after slaughter.
2. In fish, bones only; in beef and pork, bones, trim fat and tendons; in poultry, bones only.
3. In fish, beef and pork, muscle tissue only; in poultry, muscle and skin.

Source: R. Lovell, 'Fish Culture in the United States', *Science*, Vol. 206, No. 4425, 1979.

of the channel catfish *Ictalurus punctatus* is edible and only 13.7 per cent
represents bones. The food value of fish is equal to 83 if that of eggs is taken
to be 100. It is therefore slightly higher than that of meat (80). Its essential
amino-acid composition bears witness to its high nutritive value.[1]

Fish convert what they eat into flesh more efficiently than farm animals
because of their ability to assimilate a protein-rich diet. It should be noted
however (Table 2) that poultry converts food protein into tissue protein
with comparable efficiency.[2]

TABLE 2. Feed conversion in four different animals

Animal	Feed conversion ratio based on	
	Dry-weight feed: live weight	Dry-weight feed: shredded weight (flesh)
Cow	7.50:1	12.6:1
Pig	3.25:1	4.2:1
Chicken	2.25:1	3.0:1
Trout	1.50:1	1.8:1

Source: H. Ackefors and C. Rosén, 'Farming Aquatic Animals. The Emergence of a
World-wide Industry with Profound Ecological Consequences', *Ambio*, Vol. 8, No. 4, 1979.

Carnivorous fish need food with a protein content of 35–40 per cent, for
example in the form of fish-meal, scallop meat and dehydrated blood. With

such a diet, food-conversion efficiency in the carp ranges from 1.3:1 to 1.5:1; the corresponding value for soybeans is 3–5:1, for the fish are then fed products of vegetable origin, i.e. they feed at trophic level 0. The herrings used to make fish-meal are basically primary consumers (trophic level 1); assuming that conversion efficiency from one trophic level to the other is 10 per cent the assimilation of soybeans as a proteinic food is thus three or four times as efficient as the assimilation of fish-meal.

The energy needed to meet metabolic requirements is lower in fish than in warm-blooded animals, since fish do not have to keep a constant body temperature. Moreover, since fish excrete ammonia instead of urea less energy is expended on protein metabolism. They can thus synthesize more protein per food calorie consumed than poultry or cattle.[3] It is there rather than in more efficient food conversion that the essential advantage of fish-farming over stock-raising lies.

The role of micro-organisms

Micro-organisms play an important role in the direct or indirect nutrition of aquacultured animal species. They help to eliminate the nitrogenous wastes in the case of recirculating systems. They increase the oxygen demand as a result of overfeeding of the cultured species and the accumulation of nutrients and organic residues. Certain micro-organisms are also pathogenic agents and pose problems that are all the more serious in that the animal populations are extremely dense.

The larvae and juvenile forms of cultivated animals often feed on unicellular algae or on small invertebrates such as rotifers, crustaceans in the nauplius (or first) stage and young molluscan larvae. These invertebrates must also be reared on unicellular algae.

In 1934 Hudinaga was the first to succeed in inducing the kuruma shrimp, *Penaeus japonicus*, to spawn and then in rearing the larvae. This result had been obtained by feeding the penaeid larvae with diatoms and flagellates. Shrimps are often reared at 30 °C, which limits the number of algal species that can be used. After a week, when the penaeid larvae undergo metamorphosis, they are fed with nauplius larvae of *Artemia salina* and with diatoms. The technique developed by Hudinaga is still used in Japanese and North American hatcheries.

Fish are in most cases fed on small crustaceans and other organisms in the microzooplankton as soon as the fry have absorbed their yolk-sacs and are able to feed themselves. It is the nauplius larvae of *Artemia salina* that are most often used for this purpose; when the very young fingerlings are unable to eat these larvae they are given rotifers. Rotifers were artificially reared for the first time in 1968 in Japan. In several British laboratories these rotifers are used as food for the larval forms of plaice (*Pleuronectes platessa*), sole

(*Solea solea*) and turbot (*Scophthalmus maximus*) until such time as the fingerlings can eat *Artemia nauplius* larvae.

Fish that feed on phytoplankton are eminently suitable for aquaculture, since the phytoplankton constitutes a good source of proteins. Moreover the micro-organisms in phytoplankton and zooplankton as well as other natural aquatic nutrients supply the necessary food at lowest cost.

Conditions for optimal production in aquaculture

Aquaculture may consist in rearing fingerlings or larvae until they reach a certain size and then returning them to the sea to resettle formerly densely populated zones or to colonize new zones. It may also consist in using ponds or submerged cages to rear eggs (taken from captive fish), larvae and then juveniles and adults up to marketable size. Fingerlings may be placed to grow in lagoons or in restocked natural ponds. It is also possible to facilitate the entry of juvenile forms of fish into brackish or salt-water ponds, lagoons or enclosures in the sea and then to let them grow under natural conditions while promoting their development by eliminating competitive species. This form of extensive aquaculture is different from the intensive rearing that takes place in ponds or submerged cages when the fish are supplied with all their food.

In addition to the nutritional conditions essential for optimal production, the ways in which the cultured species reproduce and the pathological conditions to which they are subject are crucial factors in the management of fish-farms. In oyster and mussel culture the molluscs reproduce under natural conditions, whereas in shrimp-rearing, females with eggs are caught at sea and the young are then produced in captivity. At the Brittany Oceanological Centre near Brest, French biologists succeeded in 1975 for the first time in the world in obtaining the complete reproductive cycle of the kuruma shrimp *Penaeus japonicus* in captivity.

Some species of fish will not spawn at all in captivity. In some cases, such as the sturgeon and the salmon, artificial fertilization is possible. To rear most sea-water or brackish-water species and the catadromous eel, spawn must be obtained from natural waters. (A catadromous species comes from fresh water but spawns at sea.) In the case of bass research workers at the Marine and Lagoon Biology Station of the Languedoc Scientific and Technical University at Sète in France have been using other methods to induce this species to produce young since 1971. In 1980 several million fingerlings were produced and over half of them delivered to public or private experimental rearing centres. In the same year the Deva-Sud pilot station, belonging to the National Centre for Utilization of the Oceans (CNEXO), at Maguelonne, near Montpellier, and the Sepia hatchery at Martigues, near Marseilles, had each produced about 4 million eggs and 122,000 juvenile bass.

Selectivity, hybridization, profitability

One of the best known experiments in selective breeding is Donaldson's rainbow trout. After almost forty years of selective breeding this species of trout shows astonishing growth: the fish weigh about five kilogrammes at the age of one year and are sixty-seven centimetres long after three years. Generally speaking, selective breeding in the salmonids has been highly successful. Research carried out by the Unilever company at Aberdeen in Scotland was concerned with ways of slowing up the sexual maturation of salmon so as to obtain more flesh, i.e. to ensure that the metabolism is directed towards the making of muscle rather than gonads. In Norway at the Animal Genetics Institute tetraploid fish obtained by special treatment of the eggs are crossed, once adult, with normal diploid fish to produce sterile individuals with greater muscular development.

Hybridization between various species of salmon or between strains of the same species is also of great value. Indeed Atlantic salmon show great genetic diversity. Crosses have, moreover, given spectacular results: a cross between the Arctic char and the Atlantic salmon led to fingerlings being obtained whose rate of growth for eleven months is equal to the sum of the growth-rates of the two parents. Donaldson succeeded in crossing anadromous with sedentary strains to produce a hybrid which completes its growth at sea and after an interval of two years returns to spawn in the river where it was born.

The species reared are subject to parasite, viral, bacterial and fungal infections. Good fish husbandry makes the control of these pathogenic agents easier and aquaculture also offers a valuable opportunity for studying the effects of various forms of environmental stress on the health of aquatic populations. The addition of chemoprophylactic substances, the injection of vaccines and the mixing of antibiotics with the fishes' food represent other ways of controlling pathogenic organisms.

However, it is the cost of the feed for the animals and the efficiency with which it is converted that determine the profitability of aquaculture. The high cost of certain forms of fish-farming has rightly been emphasized; a case in point is salmon fed with fish-meal. The operating costs of the installations, the initial investment and depreciation allowances must be added to the cost of the fish food, so that profitability is only attained when the selling price is so high as to be characteristic of a luxury product. The effect of this has been to direct aquacultural operations in the United States of America towards the rearing of oysters, trout and salmon.

In 1976, in the opinion of the White Fish Authority in the United Kingdom a price of £700 ($1,050) per tonne of fish represented the threshold of profitability. The authority then gave up rearing plaice in favour of sole and turbot; turbot grew satisfactorily in submerged enclosures, whereas sole only gave interesting economic results if they were kept in warmer water to stimulate their growth. In Japan in the case of the rearing of the yellowtail, *Seriola quinqueradiata*, the cost in 1972 of feeding the fish represented half the total production cost: for one kilogram of fish produced it had been necessary to spend a little over one dollar on food, which made the selling price extremely high.

Constraints and opportunities in aquaculture

In 1975, 70 million tonnes of fish were caught, of which 60 million tonnes were from brackish or marine waters and 10 million tonnes from fresh-water fisheries. Aquaculture and fish-farming represented 40 per cent of the total fresh-water catch (4 million tonnes) and 3 per cent of sea-water and brackish-water catches (2 million tonnes). The quantities of cultured fish, crustaceans and molluscs harvested were therefore equivalent to almost 9 per cent of the world's catches and according to forecasts from the United Nations Food and Agriculture Organization (FAO) they should double between 1980 and 1990.

Also in 1975, nine Asian countries produced over 3 million tonnes of fish and 2 million tonnes of molluscs and seaweeds. This production was the equivalent of 80 per cent of the world total. The remaining 20 per cent was produced in Europe and North America.

The reasons adduced for the increasing interest in aquaculture and fish-farming include, on the one hand, the fear that catches on the high seas and in coastal waters will fall very markedly as a result of overfishing and pollution and, on the other hand, the fact that the waters in the continental regions, the estuaries and the inshore zones are the only ones where a country can exercise real control over the intensiveness of fishing and the conditions under which its marine resources are being exploited, and above all where, as in agriculture, it can ensure planned and controlled management of marine and fresh-water produce.

According to J. Ryther, of the Woods Hole Oceanographic Institute in Massachusetts, there are some 400 million hectares of coastal marshes in the world.[4] If only one-tenth of that area were devoted to aquaculture it would be possible by using improved methods of rearing and management to extract 100 million tonnes of biomass per annum. Ryther considers that this productivity would be all the more remarkable in that it would come largely from the functioning of relatively simple food chains that do not need supplemental food or large investments in labour or capital. This very optimistic opinion is shared by some who consider that intensive mariculture and aquaculture would make it possible to extract ten times as much protein and food as was contained in the world catches of 1975. They believe that by the year 2005 man will be able to cultivate phytoplankton in sea-farms near the coasts and that on that basis aquaculture, like stock-raising on land, will produce important biomasses. Other specialists, less optimistic, consider that aquaculture will be able to make a contribution towards solving the socio-economic problems of the coastal areas but that it will remain restricted to the rearing of animal species considered as luxury products, that the tonnages it produces will be very much lower than those obtained by sea-fishing but that it could nevertheless reduce the trade deficit recorded by certain countries in respect of marine products.

The economic dimension

In 1979 the United States imported 57 per cent of its food of marine origin, thus accounting for 11 per cent of the deficit in the trade balance. The growing demand for these products and the increase in fishery costs have led to rises in prices and have favoured the development of aquaculture and fish-farming in the United States, which will have to provide a larger proportion of animal protein while improving the ratio between the consumption of fish and that of meat. The average American consumed only 6 kg of fish in 1975 as against 53 kg of red meat. In Japan the production of artificially reared fish rose from 100,000 tonnes in 1971 to 500,000 tonnes in 1976.[5]

With a production of 540,000 tonnes in 1981, France occupied the eighteenth place in the world for fisheries. However, if the value of the fish landed had been taken into consideration she would have been in first place in the European Economic Community and sixteenth place in the world. To these catches must be added 180,000 tonnes of oysters, mussels and crustaceans and of fish produced on salmon farms and by aquaculture. Fisheries and aquaculture provided employment on the coasts for 1.5 million persons and kept in activity 666 high-sea trawlers and 574 coastal fishing vessels. The average consumption of seafood in France is 12.6 kg per inhabitant per annum, whereas the figure in Denmark is as high as 30 kg.

According to E. W. Shell (quoted by R. Lovell),[6] mariculture will evolve more slowly than fresh-water aquaculture, for several reasons: the juvenile forms of marine species are not easy to obtain; mariculture competes with other forms of land and water use in the coastal regions (housing, tourism and agriculture); the produce of mariculture is often dearer than produce caught under natural conditions; and climatic conditions, pollution and erosion often make it difficult to develop the coasts and the estuaries. Incidentally Lovell estimated that most investments in mariculture would probably be made in Latin America, because water temperatures there are relatively constant and the resources of the coastal areas in that region seem to him to be less subject to the influence of conflicting interests.[1]

Nevertheless it is the cost of equipment and operation and the technical and regulatory problems (control of the export and import of fish, fry, crustaceans and spat, quarantine measures and production norms) that will determine the future of aquaculture, since in the case of most species the biological obstacles to culture can be surmounted. If the cost prices of aquaculture products are not too high the future of this ancient but revived way of managing fresh-water and marine animal resources can be considered very bright. It is just as interesting for developing countries as for the industrialized countries. Research on aquaculture represents an investment for the future, since the decline of sea catches, the increase in the cost of energy and the growing demand for seafood are all factors encouraging the development of aquacultural activities.

Conflicts of interest

While it is probable that the number of large-scale aquaculture projects that require considerable volumes of water (like the pumping of sea-water from depths of several hundred metres to enrich the surface layers where the photosynthetic activity of the phytoplankton takes place) will increase, small operations round rivers, ponds and estuaries should attract the interest of planners and decision-makers because of the possibilities they offer for satisfying the immediate needs of the local population. According to Elisabeth Mann Borgese,[7] aquaculture in the East is the basis of existence itself, since its aim is to improve the standard of living, to supply food and to provide work; it forms an integral part of the social and economic infrastructure, whereas in the West aquaculture, based as it is on science and technology, is closely linked with industrial activities and its main object is to bring in money.

There are conflicts of interest between small-scale fishing operations and aquaculture projects of a more or less intensive nature, which worsen still further the difficulties faced by traditional fishermen as a result of the growing industralization of fisheries. Moreover, the production of considerable tonnages of artificially reared fish could make the prices paid for fish caught under natural conditions go down, as has been the case with salmon and eels.

These traditional fishermen make an important contribution to national economies despite their poverty or the economic difficulties caused by the relatively low productivity of small-scale fishing, the seasonal nature of their income, their chronic indebtedness and middlemen's control of the market for fishery produce. For example in South-East Asia the contribution of these fishermen (who number over 6 million, or over 40 million if their families are taken into account) to the economies of Hong Kong, Indonesia, Democratic Kampuchea, Malaysia, the Philippines, Singapore, Thailand and the Socialist Republic of Viet Nam, amounted in 1975 to over $2,300 million. In India in 1980 traditional fishermen represented a community of 6.5 million persons, using 192,000 fishing vessels and catching 70 per cent of the total catch. However, their position was threatened by the uncontrolled activity of 16,500 trawlers and other mechanized vessels operating in shallow coastal waters and belonging to 8,000 persons, who employed directly or indirectly 165,000 wage-earners and were responsible for the remaining 30 per cent of the total catch.

Some final considerations

The situation in regard to small-scale fishing and recent developments in industrial fisheries and intensive aquaculture in the Asian and Pacific countries make it essential to consider measures to safeguard the essential needs and interests of traditional fishermen, to protect existing marine resources and to maintain a balance between the various ways in which those resources are exploited. Fishing grounds must be protected and account must be taken of the differences between full-time and part-time fishermen, between those who run a small-scale operation and those who undertake subsistence fishing, in such a way as to associate the various categories in any development policy suggested. Policy should not systematically favour intensive and industrial undertakings but should rather promote the growth of small-scale enterprises, their grouping into co-operatives and the adoption of technical innovations, particularly in aquaculture.

Modernization of the system of fish distribution inside the countries would also improve the economic situation of traditional fishermen, who would more easily find outlets for their catches and would make a still greater contribution to the protein nutrition of the local populations. For this same purpose the traditional methods of preserving fish (drying, smoking and salting) should be improved.

Finally, certain specialists advocate reforms comparable to agrarian reforms, which would enable small-scale fishermen, fish-farmers and aqua-culturists to ensure better management of their natural heritage and to play a more stable economic role side by side with the industrial enterprises. ■

Notes

1. R. Lovell, 'Fish Culture in the United States', *Science*, Vol. 206, No. 4425, 1979, pp. 1368–72.
2. Ibid.; H. Ackefors and C. Rosén, 'Farming Aquatic Animals. The Emergence of a World-Wide Industry with Profound Ecological Consequences', *Ambio*, Vol. 8, No. 4, 1979, pp. 132–43.
3. Lovell, op. cit.
4. J. Ryther and J. Goldman, 'Microbes as Food in Mariculture', *Annual Review of Microbiology*, 1975, pp. 429–43.

5. Lovell, op. cit.
6. Ibid.
7. E. Mann Borgese, 'Farming the Seas: Toward a "Blue Revolution" ', *Economic Impact*, Vol. 1, 1977, pp. 68–75.

To delve more deeply

ACKEFORS, H. Production of Fish and Other Animals in the Sea. *Ambio*, Vol. 6, No. 4, 1977, pp. 192–200.

HARVEY, B.; HOAR, W. *La reproduction provoquée chez les poissons: théorie et pratique*. Ottawa, Centre de Recherches pour le Développement International, 1980. (IDRC-TS21.)

LOFTAS, T. Day of the Farmed Scotch Salmon. *New Scientist*, Vol. 70, No. 994, 1976, pp. 20–1.

———. Making More of Tropical Fish. *New Scientist*, Vol. 72, No. 1010, 1976, pp. 166–8.

OSAMU, O; MUTSUO, O.; YOSHIYUKI, T.; JUNKO, Y. *Fisheries in Asia—People, Problems and Recommendations*. Nyon, Fondation Internationale pour un Autre Développement, 1979, pp. 31–46. (IFDA Dossier, 14.)

PAYNE, I. *Tilapia—A Fish Culture. New Scientist*, Vol. 67, No. 960, 1975, pp. 256–8.

PRITCHARD, G. *Fisheries and Aquaculture in the People's Republic of China*. Ottawa, Centre de Recherches pour le Développement International, 1980. (IDRC-115e.)

QUAYLE, D. *L'ostréiculture sous les tropiques*. Ottawa, Centre de Recherches pour le Développement International, 1981. (IDRC-TS17f.)

WEATHERLEY, A.; COGGER, B. Fish Culture: Problems and Prospects. *Science*, Vol. 197, No. 4302, 1977, pp. 427–30.

Evolutionarily speaking, we vertebrates are a paradoxically inconsistent group of animals. Our amphibian ancestors first crawled out of the primeval waters more than 300 million years ago, and ever since then we have been trying to get back in again. We have tried it as reptiles of various kinds, from the extinct ichthyosaurs to modern turtles and sea snakes. We have tried it as mammals: whales, seals, otters, manatees, and polar bears. And we have tried it as sea-birds. Here follows an introduction to . . .

Chapter 7

Birds and the sea

Richard G.B. Brown

Richard Brown is a research scientist working with the Seabird Research Unit of the Canadian Wildlife Service at the Bedford Institute of Oceanography, Dartmouth, Nova Scotia. *This article appeared originally in* Oceanus, *Vol. 26, No. 1, 1983, published quarterly at the Woods Hole Oceanographic Institution, with whose kind authorization we reproduce it for interested non-specialists who would not normally have access to this excellent journal of marine science and policy.*

Families from the reptiles, mammals and sea-birds have become, to some degree, readapted to life in the sea. All of them differ from the ancestral fishes in the traces they show of their previous adaptations to life on land—lungs, for example, circulatory systems, and skeletal changes.

In terms of number of species, the most successful of these groups today are the sea-birds. There are at least 284 living species (the actual figure depends on how you define 'sea-bird' and 'species'), whereas there are about 115 living species of marine mammals, 5 sea turtles, one marine iguana, and about 50 sea snakes. 'Sea-bird' is actually a catch-all term that covers birds from several families, each of which has adapted to marine life independently. The process began very early on. Birds as a class diverged from the reptiles 100 to 150 million years ago, and fossil remains from the early Eocene epoch (about 60 million years ago) show that the four principal groups of what we call sea-birds were already in the process of evolution.

These evolutionary lines are:

Pelecaniformes: boobies, gannets, cormorants, tropic birds, and frigate birds, as well as pelicans.

Lari-Limicolae: the ancestral shore-bird stock that evolved into the auks (murres, murrelets, dovekies, and puffins—also known as alcids), jaegers, skuas, gulls, terns, and skimmers.

Tubinares (named for their tube-shaped nostrils): albatrosses, fulmars, shearwaters, prions, and petrels.

Sphenisciformes: the penguins, an offshoot of Tubinares.

Roughly speaking, the gull/shore-bird stock seems to have evolved in the Northern Hemisphere and the albatrosses and penguins in the south, although their distribution today is wider than that. Outside these four main groups, sea-birds also include members of other families: loons; grebes; ducks, geese, and swans; herons; and even a hawk, the osprey.

Variations

The extent of marine adaptations in this heterogeneous assembly varies considerably. Many of the birds, literally and figuratively, have done little more than dip their toes into the water again and cannot even swim. Examples include the herons, the osprey, and almost all shore-birds. Others, such as ducks, cormorants, loons, grebes, and most of the gulls and terns, keep to the shallow inshore zone and seldom or never go out of sight of land; the majority of these species divide their time between fresh water and salt. But the most highly adapted sea-birds, such as the auks, albatrosses, petrels, penguins and gannets, have no representatives on land or fresh water. They are true oceanographers' sea-birds, and they spend much of their lives far out at sea. Once a young albatross has fledged, for example, it may be another five years or more before it sets foot on land again.

Sea-bird species have evolved a variety of techniques for living in their newly reacquired marine environment. Some of them plunge from the air into the water to catch their prey, head-first like the gannets and terns or feet-first like the osprey. Some, like the eiders and other ducks, dive deeply and feed at the bottom on benthic organisms. Others, like the auks, penguins, cormorants, and some of the shearwaters, actively pursue their prey under-water, using their wings and feet for propulsion. Frigates, skuas, and jaegers feed partly by pirating food from other sea-birds. Skimmers and prions feed in flight, skimming their bills along the surface of the sea. Many species dip in flight to catch food at the surface, or sit on the water and feed on living or dead prey there. Small phalaropes, the only swimming shore-birds, pick at the zooplankton trapped in tide rips and along convergences. Giant albatrosses, at the other end of the scale, feed at the surface on squid. Many of these surface feeders—especially the gulls and fulmars—have learned to scavenge on the debris left behind by fishing and whaling vessels. In many cases, the same feeding technique has evolved independently in different stocks of sea-birds. The dovekie of the Arctic and the diving-petrel of the Sub-Antarctic both hunt by diving for zooplankton, and the two have become almost identical in foraging behaviour, size, bill-shape and even plumage; everything except the minor anatomical details that prove they are basically quite unrelated—a classic case of convergent evolution.

The anatomy of sea-birds has evolved along with these foraging techniques. Birds that plunge down from the air have thin, streamlined bodies and pointed beaks and, in the case of the gannets, skeletal features that absorb the shock when they strike the water. The forms of the legs, feet and bills of the various shore-birds allow them to specialize on preys of different sizes, at different depths in the sand or underwater. The phalaropes, for example, have a flat fringe of skin on the outside of each toe which acts as a simple web for swimming. Unlike other hawks, the osprey's foot has two toes pointing forward and two behind, and these, along with the roughened 'sole' of the foot, give it a good grip on a fish. The most specialized divers, the auks and penguins, have compact, streamlined bodies with the legs and feet set well back for steering and propulsion; their short wings act as paddles, and, in the penguins and the extinct great auk, the birds have gone further and altogether lost the power of flight. By contrast the albatrosses and frigates have long wings adapted for gliding, allowing them to cover long distances with a minimum expenditure of energy. More fundamentally, since sea-birds ingest a large quantity of salt when they eat or drink, they have developed a gland that extracts the salt from the bloodstream and excretes it through the nostrils. For this reason most sea-birds have perpetually runny noses.

The halcyon factor

In other words, sea-birds are marine animals, and the more specialized

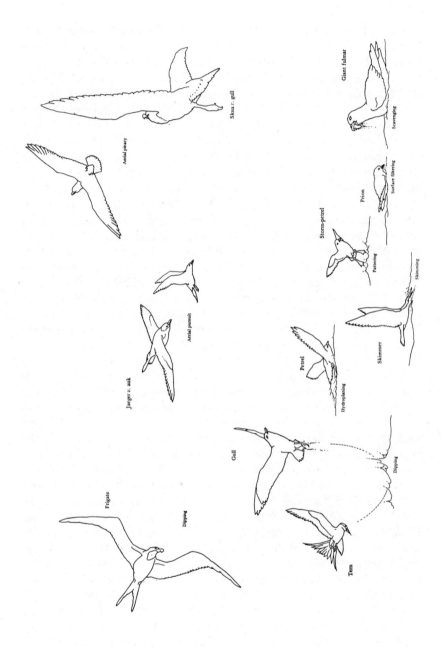

FIG. 1. Sea-bird feeding methods. (After Ashmole.)

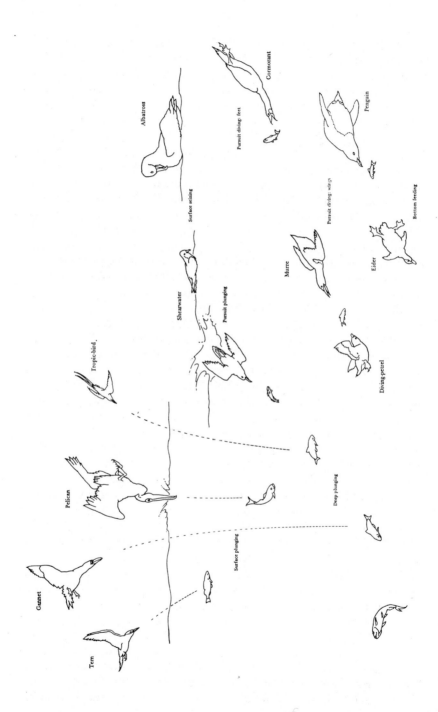

Albatross

Surface seizing

Cormorant

Pursuit diving: feet

Penguin

Pursuit diving: wings

Murre

Bottom feeding

Eider

Shearwater

Pursuit plunging

Diving-petrel

Tropic-bird

Pelican

Deep plunging

Gannet

Surface plunging

Tern

groups like the penguins are as well adapted as most of the higher marine vertebrates for their lives at sea. There is, however, one important difference. Sea-birds can feed at sea, but they cannot breed there. Like seals, sea turtles, and some sea snakes, they thus are always to some extent tied to the land. The ancient Greeks had a pleasant fantasy about a sea-bird called the halcyon, which never came to land at all; it laid its eggs on the sea in a nest of foam during the 'halcyon days', the calmest season of the year. In sober fact, the nearest any specialized sea-birds have come to this, the last logical adaptation to a marine life, are certain murrelets in the North Pacific. These auks incubate their eggs for six weeks and then take their chicks away to sea only two days after they have hatched. Young murres leave their colonies three weeks after they hatch, and often travel several hundred kilometres before they can fly. But this is an exceptionally short time for a sea-bird chick to stay in its nest, and the feeding strategies of most species require the parents to bring the food to their young, and not the other way around. The young of the larger penguins and albatrosses, for example, may remain at their nesting sites on land for a year or more.

The reasons for this link are plain to see. Sea-birds are warm-blooded animals whose eggs require incubation at temperatures of around 40 °C, and the newly hatched chicks require brooding and feeding as well. Clearly the birds cannot cast their eggs into the sea and leave them to develop there, like the eggs of most fishes. Marine mammals are viviparous. Many of the reptiles that have become adapted to life at sea, like the extinct ichthyosaurs and most of the modern sea snakes, have solved this problem by becoming ovoviviparous, retaining the eggs in the mother's body until they hatch. But sea-birds' eggs are quite large, about 10 per cent of the mother's body weight in the murre, for example, and one may doubt whether a flying bird could carry such a weight for the four or more weeks needed for incubation. Conversely, the smaller the egg the more helpless the chick when it hatches and so, presumably, the less its chances would be of surviving birth at sea. On the face of it, sea-birds have no alternative but to retain the pattern of their terrestrial ancestors and lay and incubate their eggs on land. The fact that no sea-bird has become a halcyon has placed important restrictions on the evolution of the group as marine animals.

The most obvious restriction is on their movements. The great advantage that flight gives sea-birds over other marine animals is that it allows them to travel widely in search of food and suitable climatic conditions. Terns and red knots from the Canadian Arctic spend the winter off South Africa and in Patagonia, respectively, and the immature terns move on to Antarctica. Conversely, greater shearwaters from the South Atlantic and Wilson's storm-petrels from Antarctica 'winter' in our summer in the North Atlantic, from Georges Bank northward. The shearwaters are able to exchange the cool-temperate oceanic zone off Tristan da Cunha for the corresponding one off eastern North America, 10,000 kilometres away. Once they are off our shores, they can follow the course of capelin spawning along the New-foundland coast, move on to catch the later capelin season off southern Greenland, and then come back to the euphausiid (very small shrimp) swarms off Nova Scotia on their way back south again.

But only non-breeders can exercise such options. The peak demand for food is during the breeding season, when the adult sea-birds must find food for both themselves and their chicks. At this season, they are restricted to whatever they can find within economical cruising range of their colonies. As an additional complication, the seas may be rich in food but, as in the waters around Antarctica, there may be no land nearby where the birds can nest. The halcyon factor, in other words, is a restriction on the birds' distributions and, as a corollary, on their population sizes as well.

The need to breed on land has had an even wider effect on the adaptive radiation of sea-birds as marine animals. A diving bird that must travel long distances from its colony to find food must retain the power of flight. It therefore cannot reduce its wings to the penguin-like paddle shape that is the most efficient form of propulsion underwater. This in turn sets limits on the depths it can reach and the efficiency with which it can chase its prey. The speed of swimming vertebrates is also related to their size; there is a limit to how much a bird can weigh and still be able to fly, and so a lighter bird is limited underwater in both its speed of pursuit and the size of prey available to it.

Penguins, of course, are not restricted in this way. They are the sea-birds most highly adapted to the marine environment. Emperor penguins, the biggest species, can reach depths of 265 metres and speeds of 9.6 kilometres per hour and can remain submerged for 18 minutes—a performance fully comparable with that of seals. On the other hand, the flightlessness of penguins undoubtedly restricts their foraging range. For example, there have been recent declines in the populations of many sea-birds breeding in South Africa, related to overfishing. The species affected worst is the jackass penguin, the one with the narrowest foraging range. It appears that these penguins now no longer have an abundant, predictable source of food within range of their colonies, whereas the flying sea-bird species are able to respond more flexibly to this new situation.

Vulnerability

Lastly, and most immediately, sea-birds on land are exceptionally vulnerable animals. Their anatomical specializations for life at sea have in many cases made it difficult for them to walk about, to take off, and to avoid or defend themselves against land predators. The radiation of penguins in the Southern Hemisphere has undoubtedly been assisted by the absence there of such

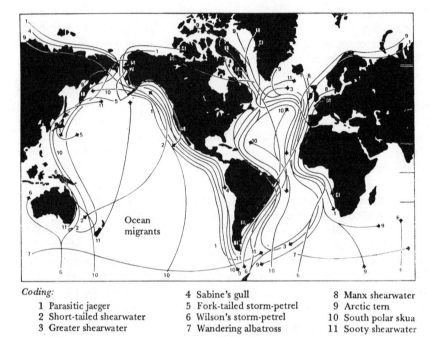

Coding:

1 Parasitic jaeger	4 Sabine's gull	8 Manx shearwater
2 Short-tailed shearwater	5 Fork-tailed storm-petrel	9 Arctic tern
3 Greater shearwater	6 Wilson's storm-petrel	10 South polar skua
	7 Wandering albatross	11 Sooty shearwater

FIG. 2. Some sea-bird migration routes. (After the National Geographic Society.)

northern predators as polar bears, foxes, rats and racoons, all of which
would have been devastating to flightless sea-birds. Even flying sea-birds
tend to nest in trees, on cliffs, or on offshore islands free of ground predators,
or a combination of the three, to minimize risks. We know all too well what
happens when this strategy breaks down. Cats, rats, foxes, mongooses, pigs,
and goats, deliberately or accidentally introduced by man, have for one reason
or another all ruined sea-bird colonies. But the greatest devastation usually
comes about where man himself is the predator.

The fate of the great auk, the flightless 'penguin' of the North Atlantic,
is a case in point. This sea-bird's last sanctuary in the New World, on Funk
Island, 48 kilometres off the Newfoundland coast, was free of every ground
predator until the colony was first discovered by European man in 1534.
From that point on, men came to slaughter the great auk for three separate
reasons during the last 300 years of its existence.

First came the fishermen from the Grand Banks, hunting for fresh meat
like any other ground predator. They pillaged the Funks so regularly that
it is surprising to learn that they were still able to kill a 'boatload' of birds
there as late as 1785. But that was before the New England merchants took
to putting crews ashore to kill the birds and boil them down for their feathers

and oil. In the face of this industrial fishery, the colony collapsed. The only great auks then left anywhere were on a small rock off the southern coast of Iceland, and museum ornithologists finished them off; the rarer a species was, the more necessary it became to collect it. The bird's inaccessible rock sank in an earthquake—another of the perils of land for a breeding sea-bird—and the alternative site was far less secure. The last great auks were killed on 3 June 1844, and that was the end of a very promising evolutionary experiment in the specialized adaptation of a marine bird to its marine environment.

New predators, new dangers

The moral of this tragic history is that in the last 400 years, and especially in the last forty, there has been a radical change in the factors that control the survival of sea-bird populations. Evolutionary strategies depend on long-range probabilities: that the breeding colony will not be invaded by ground predators; that the food supply will remain predictably close and abundant; that mortality from such 'random' events as winter storms and cold breeding seasons is on average low. No species is immortal, and these probabilities are bound to break down in the very long run, of course. But recent events suggest that the tempo of such events has increased drastically, and that sea-birds are faced with novel sources of mortality which are quite outside their 60 million years of evolutionary experience.

For example, it is unlikely that ground predators have ever reached predator-free sea-bird colonies at quite the rate that man and his commensals did during the age of European exploration. The sea-bird populations of Peru have always fluctuated irregularly, following population 'crashes' of their main food, the anchoveta, after intrusions of warm water along that coast (known as the 'El Niño' phenomenon). But excessive fishing has added another dimension, and it is feared that this, combined with the El Niño of 1976, has left the populations of both anchovetas and sea-birds at a permanently low level. The modern emphasis on harvesting the anchovetas, capelin and krill that are the sea-birds' own food, instead of the large predatory fish that are the birds' competitors, is also likely to work to the birds' disadvantage—as it already has for the jackass penguins of South Africa.

There are many direct sources of mortality as well, and they occur more regularly. The effects of oil pollution are well known; our mythical halcyon and its floating nest would be hard-put to survive in the polluted Mediterranean of today. Pesticide residues, working their way up the food-chain and accumulating in the bodies of sea-birds, have affected the fertility of the eggs of ospreys, pelicans, gannets, gulls, and even the cahow petrel of Bermuda, already on the brink of extinction because of introduced ground predators. Monofilament gill-nets, almost invisible and virtually indestructible, have drowned large numbers of auks off Greenland and Newfoundland and in the North Pacific.

Less specialized marine birds, such as shore-birds and waterfowl, are perhaps more endangered on migration and in the winter than they are during the breeding season. The market-hunting that brought meat to the markets of Boston and New York during the nineteenth century put an end to the Labrador duck, which became extinct around 1875, and has virtually wiped out the eskimo curlew as well. But the main risk is the loss of their feeding habitat. It is a paradox that although marine birds can and do travel long distances outside the breeding season, along the coasts and out at sea, the number of places in which they can find the right kinds of food in the right quantities is actually very limited.

Migrating red knots move down to Patagonia and back along a track marked by a few well-defined pit-stops—beaches and mudflats where they can rest and build up their fat reserves for the next leg of their trip. At the end of the breeding season most of the semipalmated sandpipers and northern phalaropes in the eastern Canadian Arctic leave the tundra and migrate, probably non-stop, down to the Bay of Fundy. The sandpipers go to half a dozen very restricted mudflats at the head of the bay, while the phalaropes feed in the tide-rips off the Maine and New Brunswick coasts.

Greater snow geese breed only in the region of northern Baffin Island, and they have only two feeding sites farther south: in winter in Chesapeake Bay, and in spring and autumn at Cap Tourmente, a salt marsh just outside Quebec City. These birds are clearly as vulnerable in their way as was the great auk on its isolated breeding rocks. Cap Tourmente has already been menaced at least once by an oil spill. There are plans to use the Fundy tides to generate electrical power, and if carried out, this project would eliminate at least one of the semipalmated sandpipers' stopovers. If these or any of the other marine birds lose their preferred feeding habitats, it is an open question whether they can find alternatives.

What of the future?

What, then, can we say about the future of sea-birds? If we think only in evolutionary terms and leave man completely out of the picture, how much further can birds go in their adaptations as marine animals? Their next step depends on whether they can solve the riddle of the halcyon, and evolve a means of breeding out at sea.

Could a sea-bird nest on one of the enormous Antarctic icebergs and drift along with it through the dense swarms of krill? Non-breeding sea-birds already use these as bases for their foraging, and ivory gulls have been known to breed on ice islands in the Arctic. Could a murrelet lay its eggs on a mass of sargassum weed and hunt for lumpfish and sargassum fish among the fronds below?

What would happen if a penguin became ovoviviparous? Dougal Dixon has 'preconstructed' a world 50 million years in the future in which the Southern Ocean is populated by ovoviviparous 'pelagornids'—penguin descendants that have taken over the ecological roles of the whales, from dolphin-like fish-eaters to giant birds that feed on krill, as baleen whales do today. It is a fantasy, but on the evolutionary scale of time, who can tell?

Of course the biggest fantasy of all is to leave man out of the picture. We have changed the situation far too much to do this, and the immediate future of marine birds undoubtedly depends on how well they can adapt to our changes. It is not too difficult to make predictions from what we already know about their reproductive strategies. There are two basic patterns, named for the mathematical constants that define them. 'K-selected' species lay only one egg in a season and have a long period of adolescence before they start to breed and a very low annual mortality as adults. This is the strategy that has been evolved by the sea-birds most highly adapted to marine life: auks, penguins, and albatrosses and their relatives. It has proved effective in the past because the sea was normally a safe and predictable place to live, and the birds' life on land was confined to a largely predator-free environment. By contrast, 'r-selected' species lay several eggs in a season and usually have short adolescent periods and relatively high annual rates of adult mortality. The species least adapted to marine life, such as ducks, geese, shore-birds and, up to a point, gulls and cormorants, tend to show this pattern. Clearly an 'r-selected' species, with its higher annual rate of reproduction and population turnover, will be better placed than a 'K-selected' species to absorb and recover from man-induced mortalities, such as net-drownings or oil spills. For example, the short-tailed albatross of the Bonin Islands off Japan was persecuted almost to extinction by feather hunters until the early 1930s; its population is only just showing signs of revival. Herring gulls, similarly persecuted in New England, received protection in the early 1900s, and by 1930 the population had expanded so rapidly that the birds were becoming a menace to terns and other birds.

However, reproductive strategies are not the only key to a sea-bird's chances of survival. The adaptability of its feeding and nesting habits is also important. The herring gull expansion was, literally, fuelled by the availability of edible garbage during the winter, which increased the chances of survival of the first-winter birds, the age-class with the highest natural mortality. Herring gulls also have shown an astonishing versatility in their choice of nest sites: cliffs, sand dunes, moors, salt marshes and now in the United Kingdom, the roofs of city buildings. The kittiwake, a more specialized and oceanic gull, is exploiting fish offal and other human wastes

FIG. 3. Post-breeding migrations of some birds in the north-west North Atlantic.
SG: greater snow geese to Chesapeake Bay, via Cap Tourmente, Quebec.
F: Bay of Fundy; a migratory stopover for most of the semipalmated sandpipers
    and northern phalaropes breeding in the eastern North American Arctic.
M: migrations of thick-billed murres—(M1) from the eastern Canadian high Arctic
    to western Greenland; most of these birds later move on to Newfoundland waters;
    (M2, M3) from western Greenland and Hudson Strait to Newfoundland; (M4)
    from Spitsbergen and Novaya Zemlya to western Greenland.
P, K: Atlantic puffins (P) from Iceland, and black-legged kittiwakes (K) from
    northwestern USSR to Newfoundland.
G: greater shearwaters from the South Atlantic Ocean, initially to Georges Bank
    and the Grand Banks.

at sea, and in England is even nesting on waterfront warehouses instead of its
more usual island cliffs. And who would have believed that as specialized
a bird as the osprey could learn to nest on man-made structures, such as
power pylons, and is on its way to becoming a bird of the suburbs?

Odds and omens

The odds are therefore quite good for many species of marine birds, especially
species with r-selected reproductive strategies that can not only absorb the
increased mortalities caused by man but can go even further to take advantage

of the new opportunities we have created for feeding and nesting. It is encouraging that populations of ospreys and pelicans are recovering from the damage done to them by chemical pollutants. Much of what happens next depends on man, of course. We must be careful to monitor the chemicals we spill into the sea and quick to put a stop to them if their effects are dangerous. We also must preserve the pieces of shoreline habitat that are crucial feeding areas for many of the birds.

But the omens are not nearly as good for the more specialized sea-birds. I hope our species has gone beyond the direct, deliberate slaughtering of a species into extinction. But the effects of our commensal animals continue, and we are still living with the effects of direct exploitation and other man-made mortalities in the not-so-distant past. Hunting, egg-collecting, and net-drownings have caused recent, drastic declines in murre colonies in the Canadian Arctic, western Greenland, northern Norway, and Novaya Zemlya. Murres and other auks were once hunted in Great Britain, and suffered losses from oil pollution, too. It is encouraging that their numbers there are slowly starting to increase again, though it is too soon to say how far this trend will go, or if or when it will extend to the northern populations.

We still seem quite prepared to countenance the extermination of sea-bird populations, and even species, by indirect means. Abbott's booby breeds only in the treetops of the virgin forest on Christmas Island, in the Indian Ocean, but the whole island is rapidly being excavated for its phosphate deposits. Competition with the fishing industry would be slower, but just as drastic in the long run. The puffins in the biggest colony in Norway have had only a single successful breeding season since 1969. There is a scarcity of their principal food, immature herring, attributable to over-fishing of the Norwegian herring stock in the previous two decades.

The developing industrial fisheries therefore will have to be monitored very carefully. The latest of these is for krill, the euphausiid shrimps that are central to the food webs of sea-birds and all the higher marine predators in Antarctic waters. We must limit our catches so that the stock of krill is not damaged and enough is left for our competing marine predators. This will not be easy because man, as a fisherman, is under enormous pressure to find enough protein for fellow members of his own rapidly increasing, and increasingly hungry, species.

In the last analysis, we ourselves, through our technological achievements, have joined the ranks of higher vertebrates that have tried to go back to the sea again. Our trouble is that we have done this so recently that we are still trying—unsuccessfully, so far—to find our place in a balanced marine ecosystem. ■

To delve more deeply

ASHMOLE, N. Sea-bird Ecology and the Marine Environment. In: D. Farner and
     J. King (eds.), *Avian Biology*, Vol. 1. New York, Academic Press, 1971.
BOURNE, W. Sea-birds and Pollution. In: R. Johnston (ed.), *Marine Pollution*.
     London, Academic Press, 1976.
BROWN, R. Sea-birds as Marine Animals. In: J. Burger, B. Olla and H. Winn
     (eds.), *Behavior of Marine Animals*. New York, Plenum Press, 1980.
DIXON, D. *After Man: A Zoology of the Future*. London, Nelson, 1981.
FISHER, J.; LOCKLEY, R. *Sea-birds*. Boston, Mass., Houghton-Mifflin, 1954.
GASTON, A.; NETTLESHIP, D. *The Thick-Billed Murres of Prince Leopold Island*.
     Ottawa, Canadian Wildlife Service, 1981.
MILLS, S. Graveyard of the Puffin. *New Scientist*, Vol. 91, No. 1260, 1981.
NELSON, B. *Seabirds*. New York, Hamlyn, 1980.

See also:

LEMAHO, Y. Le manchot empereur: une stratégie basée sur l'économie d'énergie.
     Le Courrier du CNRS, March 1983.

Between 40 and 45 per cent of all drugs in current use are of natural origin. Until recently, all these came from micro-organisms and plants. Today the number of active products being discovered from these sources is declining and scientists are turning for new ones to marine organisms.

Chapter 8

Medicine from the sea

Luigi Minale

Luigi Minale, an Italian, is professor of Organic Chemistry at the University of Naples. From 1973 to 1981 he was director of the Istituto per la Chimica di Molecole d'Interesse Biologico at the CNR (National Research Council). He has written more than 120 scientific papers on the chemistry of natural products. Since 1969 he has worked almost exclusively on marine natural products. His address is: Dipartimento di Chimica delle Sostanze Naturali, Università degli Studi, Via L. Rodinò, 22, 80138 Napoli (Italy).

The seas and oceans that cover almost two-thirds of the earth's surface and contain approximately 500,000 different species are a vast and virtually untapped source of natural products, many of which are biologically active and potentially useful. Marine organisms live in such different environments from those of terrestrial organisms that it is reasonable to suppose that their secondary metabolites (products obtained from them) will differ considerably. Surprisingly, however, scientists engaged in research on natural products have focused their attention almost entirely on land sources, to the exclusion, at least until about fifteen years ago, of marine natural products.

The age-old belief that the sea is an element hostile to man was undoubtedly one reason why marine chemistry and pharmacology developed later than the chemistry of natural products of terrestrial origin (plants and micro-organisms) and phyto-pharmacy. Another reason for the slowness of progress in research in these fields, closely linked to the first, is the lack of a discipline in marine studies equivalent to ethnobotany, the object of which is to study plants that were used by primitive peoples for therapeutic purposes, and still are today; it has provided an invaluable basis for research into pharmaceutically active molecules. Indeed, until the early twentieth century medicinal drugs were obtained exclusively from natural sources. The development of synthetic organic chemistry subsequently gave a considerable boost to the production of new drugs. Nevertheless, natural products still play an important part in chemotherapy. It is estimated that 40 to 45 per cent of all drugs in current use are of natural origin or directly derived from natural molecules. In the case of antibiotics this percentage is considerably higher. Until recently, the only sources of active natural molecules were secondary metabolites, which were isolated from micro-organisms and higher plants. Today the number of active products being discovered in these sources is declining, and this has prompted scientists to consider marine natural sources.

The rapid growth of the chemistry of marine organisms over the last fifteen years has led to the discovery of a surprisingly large number of new structures, many of which have no precedent among structures of terrestrial origin, and possess previously unknown pharmacological and toxicological properties.[1-4] This field of research has developed along several different lines, and various approaches to the subject have been adopted.

One of the first approaches was to consider biotoxicity as an indicator of general pharmacological properties, and scientists (especially in the United States and Japan) focused their research on toxic marine organisms, especially those which were known to be toxic to man. The marine environment contains many poisonous species,[5] and the research carried out in this field, notably in the identification of various toxic substances, many of which have unexpected structures and remarkable physiological properties,[6] has done much to help solve problems of public health. Hashimoto, one of the pioneers of the chemistry of marine toxins, has reported that every year there are more than 1,000 cases of food poisoning in Japan, between 40 and 50 per cent of which are caused by marine organisms. A knowledge of the structures of the toxins concerned is of the utmost importance for any preventive medicine programme. Moreover, cases of food poisoning caused by marine products, which occur unexpectedly and affect large numbers of people simultaneously, have an adverse effect on the fishing industry.

A more systematic approach to the search for new active marine products has been to subject a large number of extracts from marine organisms to pharmacological tests, and then to examine those that prove to be active. This approach—a costly one—has been adopted by the pharmaceutical industry, often in collaboration with the universities, and has led to the initiation of major research programmes.

Lastly, I consider that an important part is played by fortuitous, sporadic discoveries which are not the result of a closely monitored research project— that is, by what is termed as serendipity, which has often produced highly rewarding and useful results. The isolation in 1969 by Weinheimer and Spraggins, of the University of Oklahoma, of large amounts of prostaglandin from a biological order called a gorgonian, *Plexaura homomalla,*[7] was certainly a case of serendipitous research. Their discovery caused a stir throughout the biomedical community, prompting the organization of increasingly large-scale and ambitious projects in this sphere of research during the early 1970s.

The discovery in an animal of prostaglandins—substances which play a crucial part in the regulation of many vital biochemical processes and are used as therapeutic agents in various fields of medicine—is not in itself remarkable. Prostaglandins in infinitesimal quantities occur in all animal tissues. What was remarkable was the large amount of prostaglandin present in *P. homomalla* (15 per cent of its dry weight) and the fact that the gorgonian is an important source of prostaglandins. The prostaglandin (15R)-PGA$_2$, obtained from the gorgonian, is not physiologically active, but is easily transformable by chemical means into an active form (15S)-PGA$_2$ (Fig. 1).

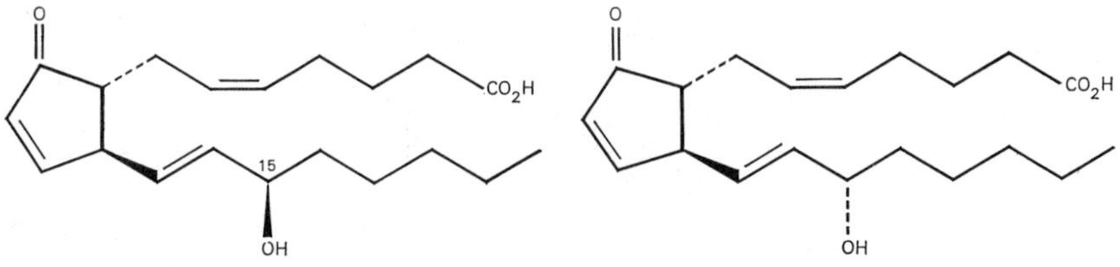

(15R)–PGA$_2$ (15S)–PGA$_2$

FIG. 1. The prostaglandin (15R)-PGA$_2$ was isolated in large quantities (1.5 per cent of the dry weight of the animal) from the gorgonian, *Plexaura homomalla*. It is not pharmacologically active but can be easily transformed chemically into an active form (15S)-PGA$_2$.

Another useful result of serendiptous research was the isolation by Bergmann, a pioneer of marine chemistry, of spongotimidin and spongouridin from the sponge *Cryptothetya crypta*.[8] These molecules differ from natural nucleosides, the constituents from which nature makes the nucleic acids that determine the genetic processes, purely by the nature of the sugar (arabinose instead of ribose and/or deoxyribose) linked to the pyrimidine base. Consideration of these facts has encouraged organic chemists to undertake a series of syntheses, based on the model of Bergmann's compounds, the Ara-C (D-cytosine arabinoside) and Ara-D (D-adenine arabinoside), which possess powerful antiviral and antitumoral properties (Fig. 2). Ara-C is used successfully, combined with other drugs, in the treatment of leukaemia, while Ara-A is used in the treatment of the herpes encephalitis virus. The mechanism by which these drugs act is attributable to their close structural similarity to the nucleosides that constitute nucleic acids, which inhibit the synthesis of DNA and, consequently, cell growth.

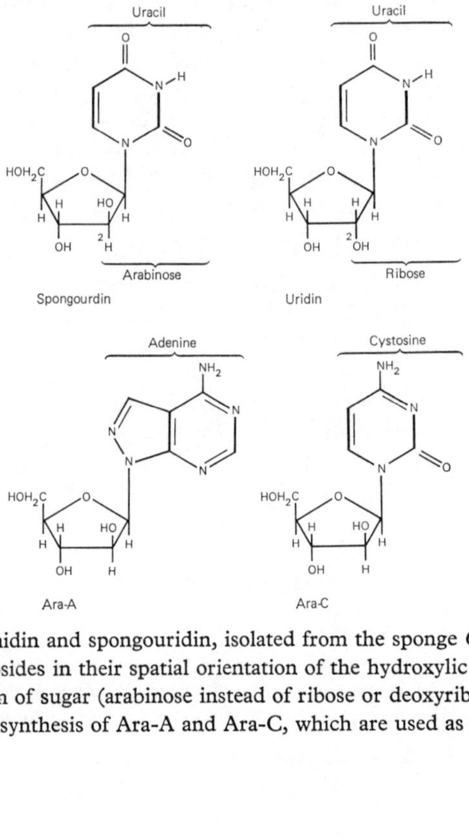

FIG. 2. Spongotimidin and spongouridin, isolated from the sponge *Cryptothetya crypta*, differ from nucleosides in their spatial orientation of the hydroxylic function in the No. 2 carbon atom of sugar (arabinose instead of ribose or deoxyribose). They are used as models for the synthesis of Ara-A and Ara-C, which are used as antileukaemic substances.

Another important field of research in the chemistry of marine natural products is that of chemical ecology, the object of which is to study chemical interactions between organisms. This field of research, too, which is intrinsically of great scientific interest, has had practical applications—to the harmful effects of pollution on marine life for example. Many interactions between marine organisms occur through the medium of chemical agents rather than as a result of physical forces or contact stimulu,[9] and the presence of pollutants which affect these delicate processes can have dramatic effects on the behaviour and physiology of organisms. It is therefore of the utmost importance that we should know as much as possible about the phenomena of chemical interaction and chemoreception between marine organisms.

Marine toxins

Tetrodotoxin: puffer-fish toxin

Reference to the toxicity of the 'puffer fish' (of the Tetraodontidae family) and descriptions of poisoning caused by the ingestion of fish belonging to the tetradontidae family occur in ancient writings from various parts of the world, especially China and Japan. A Chinese book, *Sankaikyo,* written about 2,000 years ago, describes the death of a man who had eaten the intestines of the puffer fish. The initial symptom of poisoning is paralysis of the tongue and lips, which occurs at an early stage (between twenty minutes and three hours after ingestion) and is progressive. Death occurs between six and eight hours after absorption of the toxin. The toxin is present in a large number of species of Tetraodontidae (twenty-nine in Japan), approximately half of which are edible. *Fugu* is an expensive Japanese delicacy. As the toxin is concentrated in the intestines and skin, the art of preparing *fugu* lies in removing these without contaminating the rest of the fish. Stringent regulations for the handling of *fugu* were introduced during the Edo period (1603-1868), and in Japan today all cooks must pass a searching examination on its preparation before they are qualified to serve *fugu*.

The toxicity of these fish is caused by tetrodotoxin, one of the most poisonous non-protein toxins known to man (LD_{50} *ca.* 10 $\mu g/Kg$ in rats). Attempts to isolate the toxin in its pure state date back to 1909, and it was not until 1964/65 that its structure was established—by three research teams, working independently of each other (Woodward, 1964; Goto, 1965 and Tsuda, 1965).[10] The long wait was not in vain; tetrodotoxin (Fig. 3) proved to be extremely interesting chemically, in that it was found to have previously unknown structural characteristics. It is a neurotoxin, similar to curare, for example, which inhibits the transmission of nerve impulses through the selective blockage of the transfer of sodium ions across the membrane of nerve cells (blockage of the sodium channel). It has been widely used in neurophysiological research, the findings of which have provided information on basic aspects of the mechanisms which regulate the transmission of nerve

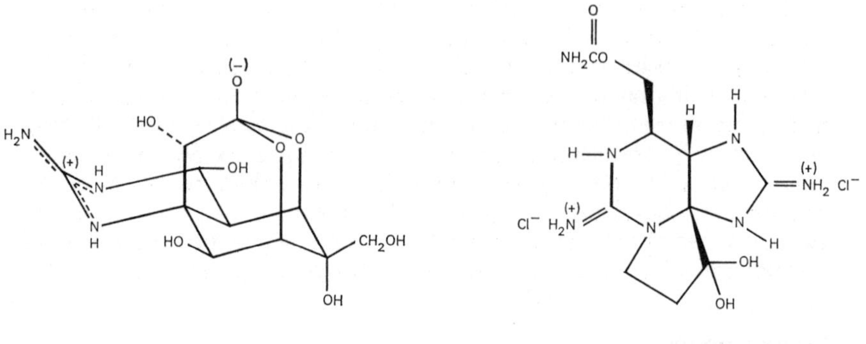

Tetrodotoxin   Saxitoxin

FIG. 3. Tetrodotoxin, isolated from the puffer fish, and saxitoxin, isolated from the microscopic red algae which cause 'red tides', are extremely toxic and block the transmission of nerve impulses. They have been widely used in neurophysiological research.

impulses. Tetrodotoxin has also been used to a lesser extent in clinical medicine; in Japan it is used as a muscle relaxant and as an analgesic in the treatment of certain tumours.

Saxitoxins: paralytic shellfish poison

Another cause of food poisoning in humans is to be found in mussels and shellfish, which are rendered poisonous by the contamination and ingestion of toxic red microalgae (dinoflagellates). Poisoning occurs when the toxic dinoflagellates suddenly increase in numbers; filtering organisms, particularly mussels and molluscs, concentrate the toxin in their bodies, thus becoming in turn toxic and inedible. Paralytic shellfish poison poses a very serious problem in the United States and Canada, as the puffer fish does in Japan. There are similarities between the poisons—they have both caused many deaths, they produce very similar symptoms and in both cases the causative agents are highly toxic and act in much the same way—by blockage of the sodium channel. The first of these toxins to be isolated in the pure state was saxitoxin; it was originally obtained from an edible bivalve, the *Saxidomus giganteus,* which is rendered toxic by the ingestion of dinoflagellates. Its structure (Fig. 3) was established in 1975 by Shantz and his collaborators, after more than twenty years of research.[11] Various other toxins closely resembling saxitoxin were subsequently isolated from dinoflagellates of the genus Gonyaulax, mainly as a result of research carried out by Shimizu at the University of Rhode Island.[12] Their identification, together with their use in neurophysiological research, has made it possible to obtain information about the structure of the receptors of excitable membranes.

Palitoxin: the most toxic marine product

Palitoxin, which is isolated from a zoantharian (phylum celenterati), *palythoa toxica,* also has a long history; it is mentioned in an ancient Hawaiian legend, which describes a potent poison found in seaweed known as *limu-make-O-Hana* (the deadly seaweed of Hana) with which the islanders of old smeared the tips of their lances. It was not until 1961 that the habitat of the organism described in the legend was discovered and scientists were able to collect the poisonous seaweed, as it was called which closer investigation showed to be a zoantharian. Several research teams set to work to isolate the toxic agent and establish its structure (Scheuer and Moore in Hawaii, Hashimoto and Irata in Japan). Moore and Scheuer isolated the toxin in 1971,[13] and its structure was definitively established ten years later.[14] Polytoxin has an unusual structure; despite its great molecular weight (around 3,300 daltons), it is not formed by the repetition of simple structural units, as in the case of polypeptides and polysaccharides. Palitoxin is extremely poisonous (LD_{50} 0.5 μg/Kg in rats) and, *in vivo*, possesses antitumoral properties. A dose of around one-tenth of the minimum lethal dose completely cures Ehrlich's tumour in rats. It has recently been used as a local anaesthetic in maxillofacial surgery, allowing surgeons to operate for several hours at a time. It is a powerful vasoconstrictor and is potentially useful in the study of angina in animals.

Nereistoxin

Nereistoxin, a powerful insecticide, was discovered by Japanese fisherman who noticed that the worm *Lumbriconereis heteropoda* which they used as bait killed all insects with which it came into contact. In 1934 Nitta isolated the toxin, and its structure was established by Hashimoto and Okaichi in 1960.[15] Its structure is simple (Fig. 4), though unusual compared with that of other natural products, and it acts as a highly powerful agent against various insects, especially the larvae of the rice-stem borer and the American

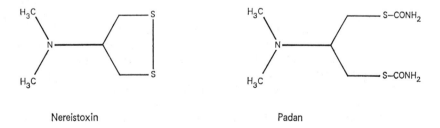

Nereistoxin   Padan

FIG. 4. Nereistoxin was isolated in the worm *Lumbriconereis heteropoda,* and has been used as a model for the synthesis of Padan, which is employed as an insecticide in Japan.

cockroach. The above-mentioned characteristics prompted research workers to synthesize a number of compounds with a similar structure to that of the toxin, the most effective of which, 1,3-bis-(carbomolicio)-2-N, N-dimethylamine propane, was patented, and has been used for agricultural purposes since 1967. It has been estimated that in 1972 the total production of this compound was 1,500 tonnes. It has been applied to many crops besides rice.

Antitumoral marine products

There are references in scientific works to many other compounds which can be classified as marine toxins, including some whose antitumoral properties have aroused considerable interest. Dibromo-aplysiatoxin, first isolated from the digestive gland of the sea hare, *Stylocheilus longicauda*[16] and subsequently discovered in a blue-green alga, the *Lyngbya majuscula,* on which the mollusc feeds, possesses antineoplastic properties. As part of a research project on blue-green algae, R. E. Moore, of the University of Hawaii, isolated another antineoplastic compound, oscillatoxin A, which is effective in the treatment of leukaemia P-388.[17] Both products are toxic (minimum lethal dose in rats is about 0.2 mg/Kg), and their effectiveness against tumours requires concentrations approaching those of chronic toxicity, so that it does not appear possible to use them as antitumoral drugs. Didemnins, cyclic peptides, which have been isolated from a tunicate of the genus *Trididemnum* by a research team under Rinehart at the University of Illinois,[18] seem to be much more promising. Didemnin B is particularly effective in the treatment of leukaemia P-388 and of melanoma B.16 in rats, and is therefore now under clinical investigation. Another group of cyclic peptides, the dolastatins, has been isolated from a mollusc by a research unit under Pettit at the University of Arizona.[19]

Marine antibiotics

More has been written about the antibiotic effects of substances extracted from marine organisms (especially algae and sponges), either as fractions or as pure compounds, than about their other effects. It is certainly true that most of the pharmacologically active products that have been isolated from the sea (especially from algae and sponges) are antibiotics.[20] Nevertheless, no marine antibiotic has yet been used therapeutically, with the exception of cephalosporin C. I am not sure whether cephalosporins should be classified as marine natural products. The fungus, *Cephalosporium acremonium*, was isolated in 1952 near a sewer outlet on the Sardinian coast, so that it is doubtful whether it is of marine or terrestrial origin. The active principle of the fungus was identified as cephalosporin C, which has a structure similar to that of penicillin and is effective against very many bacteria, including those which are penicillin-resistant. This important characteristic of cephalosporins is attributable to their resistance to penicillase, a bacterial enzyme which destroys penicillin.

The constant need to find new antibiotics is largely the result of an increased resistance in pathogenic bacteria caused by the continued use of these drugs. To date, approximately 4,000 antibiotics have been identified, mainly from terrestrial micro-organisms. There is every reason to believe that marine micro-organisms could be a valuable source of new antibiotics. The initial findings in this field, obtained by Okami in Japan, appear encouraging.[21] An active principle which inhibits *Plasmodium berghei*, the cause of malaria, was isolated in a marine actinomyces. In view of its properties, the substance was given the name of aplasmomycine. Two antibiotic aminoglycosides, histamines, have been isolated from a new species of marine streptomyces. They greatly inhibit Gram-positive and Gram-negative bacteria, including those resistant to known antibiotic aminoglycosides.

Other marine products of use to man

More pharmacological properties have been discovered amongst secondary metabolites of marine organisms than those briefly described above. Moreover, applications other than pharmacological have been found for several marine products. It is worthy of note that one of the oldest dyes used by man, Tyrian purple dye, is obtained from a marine mullusc, the *Murex brandaris*. The structure of the dye was established in 1968 in Australia by Baker and Sutherland as 6.6´-dibromo-indigo.[22] It has also been demonstrated that the dye does not exist as such in the living animal, but is activated by light on a monomeric precursor.

Applications have likewise been found in fields that than pharmacology for polysaccharides of algae, compounds formed by the linking of many monosaccharide molecules. Agar-agar is widely used in bacteriology in the preparation of culture media, while alginates are used in the food industry. Agar-agar, carrageenin and laminaria, a sulphated polysaccharide, have anti-coagulant, antiseptic and anti-ulcer properties. In many cases the use of carrageenin has proved to be extremely effective therapeutically in the treatment of ulcers. In addition, alginic acid, which is commercially processed from *Macrocystis pyrifera* and other algae, is used as a protective agent against intoxication by radioactive strontium, because of its ability to form salts with metallic ions.

The anti-parasitic properties of red algae, *Digenea simplex*, have been known for at least a thousand years. The active agent has been identified as K-acid, an *a*-amino acid whose structure is related to that of proline.

Many extracts from marine organisms have an effect on the cardiovascular system, largely due to the presence of catecholamines, which have been found to occur in many sponges. However, some unusual and highly effective compounds which affect the action of the heart are being extracted from the sea—for example automium, a brominated derivative of phenylethylamine, isolated from a sponge, cardiotonic peptides isolated from the sea anemone, *Anemonia sulcata,* and doridosine, a nucleoside isolated from one of the nudibranchia (shell-less molluscs) and subsequently from a sponge which has a powerful hypotensive effect and slows cardiac rhythm in rats.[23]

Conclusions

It is clear from what has been said that marine organisms are an invaluable source of secondary metabolites which possess a range of structures and of biological and pharmacological properties as great as those identified in terrestrial organisms in the course of a century of extensive research. Yet very few marine products are used in clinical therapy, which is disappointing. However, to be just we must bear in mind at least two important points. First, the development of a new drug takes a very long time. It is estimated that, on average, ten years elapse from the indentification of a new active molecule to its commercialization. In this connection it should be noted that more than twenty years elapsed between Bergmann's discovery in the 1950s of spongo-timidin and spongouridin and the time when the antiviral drugs Ara-A and Ara-C were entered in the pharmacopoeia. Second, it is only in relatively recent times that a sustained and vigorous effort has been made in the field of research on new drugs of marine origin, and few chemists and pharma-cologists are engaged in this field compared with those who have been investigating terrestrial sources for at least a hundred years.

I think, therefore, that systematic research into marine products, algae and invertebrates, including micro-organisms, should be pursued. Only the future will tell if 'medicines from the sea' were a pipe dream of the 1970s or if the efforts made so far will produce really useful results.   ■

Notes

1. P.J. Scheuer, *Chemistry of Marine Natural Products,* New York, Academic Press, 1973.
2. P.J. Scheuer (ed.), *Marine Natural Products—Chemical and Biological Perspectives,* Vols. 1-4, New York, Academic Press, 1978-81.
3. M.H. Baslow, *Marine Pharmacology,* Huntington, NY, Krieger, 1977.
4. J.T. Baker and V. Murphy, *Compounds from Marine Organisms,* Vols. 1-2, CRC Press, 1976-81.
5. B.W. Halstead, *Poionous and Venomous Marine Animals of the World,* Princeton, NJ, Darwin Press, 1978.
6. Y. Hashimoto, *Marine Toxins and Other Bioactive Marine Metabolities,* Tokyo, Japan Scientific Societies Press, 1979.
7. A. J. Weinheimer and R.L. Spraggins, *Tetrahedron Letters,* 1969, p. 5185.
8. W. Bergmann and D.C. Burke, *J. Org. Chem.,* Vol. 20, 1955, p. 1501; Vol. 21, 1956, p. 226.

9. P.T. Grant and M.A. Makie (eds.), *Chemoreception in Marine Organisms,* London/ New York, Academic Press, 1974.

10. R.B. Woodward, *Pure Appl. Chem.,* Vol. 9, 1964, p. 49.

11. E.J. Schantz, V.E. Ghazarossian, H.K. Schnoes, F.M. Strong, J.P. Springer, J.O. Perranite and J. Clardy, *J. Am. Chem. Soc.,* Vol. 97, 1975, p. 1238.

12. Y. Shimizu, 'Dinoflagellate Toxins', in J.P. Scheuer (ed.), *Marine Natural Products–Chemical and Biological Perspectives,* New York, Academic Press, Vol. 1, 1978, p. 1.

13. R.E. Moore and P.J. Scheuer, *Science,* Vol. 172, 1971. p. 495.

14. R.E. Moore and G. Bartolini, *J. Am. Chem. Soc.,* Vol 103, 1981, p. 2491.

15. Y. Hashimoto and T. Okaichi, *Ann. N.Y. Acad. Sci.,* Vol. 90, 1960, p. 667.

16. Y. Kato, P.J. Scheuer, *Pure Appl. Chem.,* Vol. 48, 1976, p. 29.

17. R.E. Moore, op. cit., Vol. 54, 1982, p. 1919.

18. K.L. Rinehart, J.B. Gloer, J.C. Cooks, S.A. Mizsakae and T.A. Scahill, *J. Am. Chem. Soc.,* Vol. 103, 1981, p. 1857.

19. G.R. Pettit, Y. Kamano, P. Brown, D. Gust, M. Inoue and C.L. Herald, op. cit., Vol. 104, 1982, p. 905.

20. D.J. Faulkner, in P. Sammer (ed.), *Topics in Antibiotic Chemistry,* Vol. 2, p. 13, New York, Wiley, 1978.

21. Y. Okami, *Pure Appl. Chem.,* Vol. 54, 1982, p. 1951.

22. T.T. Baker and M.D. Sutherland, *Tetrahedron Letters,* 1968, p. 43.

23. P.N. Kaul, *Pure Appl. Chem.,* Vol. 54, 1982, p. 1963.

At Prince Edward Island, a Canadian province in the marine environment of the north-western Atlantic Ocean, a private group concerned with the ecological and cultural heritage of national parks has developed a public cartoon display of one of nature's most abundant materials. This resource is sand, the metamorphosed particulate rubble, averaging two millimetres in diameter, of the mineral familiarly called quartz. Quartz is silicon dioxide (SiO_2) and it belongs to the sedimentary sandstones known as arenaceous rocks. The substance is most often transparent or yellowish in colour, although brown, green and purple sands exist too; on Prince Edward Island, there is much red sand. Sand has many industrial uses.

Chapter 9

The great wind
and dune story

'Mr Sand'

This picture story was developed under the sponsorship of Parks & People Association, Inc., Box 1506, Charlottetown C1A 7N3, Canada. Art director, Lee Sackett (Halifax); drawings by Michael Fog (Montreal); captions by Gerry L'Orange (Montreal); production by Ken Shelton, with Paul DeMone and Philip Michael as technical advisers. A project of Parks Canada/Prince Edward Island National Park.

Come let Mr Sand guide you
through the life of a dune.
Learn how to protect the sand
dunes, Mr Sand and his
friends who live with him.

BE A FRIEND TO MR SAND!

This story is portrayed in a full
colour outdoor exhibit to be
found at Cavendish main beach
in Prince Edward Island
National Park.

Once upon a time, bedrock was formed. In this part of the world, red sandstone was formed by deposition, underwater, grain by grain, layer by layer, some 250 million years ago. Some of it later rose out of the gulf to become an island.

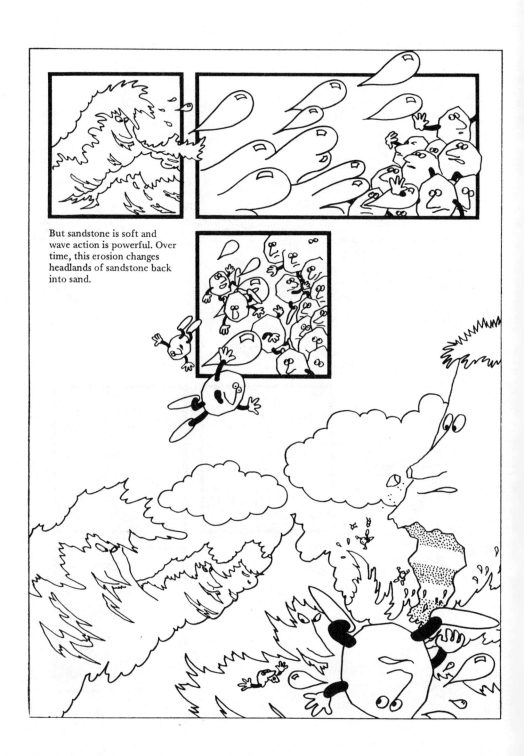

But sandstone is soft and wave action is powerful. Over time, this erosion changes headlands of sandstone back into sand.

Currents carry the now loose grains of sand away from the headlands.

A salty bath changes the red sandstone of Prince Edward Island into the off-white sand of our park.

Offshore bars—underwater accumulations of sand—grow, grain by grain, up to and above the water level until they become barrier beaches.

Now dry, the sand flies with the wind. Taking off in winds
of 20 km/h or more, landing as windspeed drops, it collects
on the lee side of obstacles along the flight path.

These accumulations grow into dunes. And as they grow, they move with the wind.

Dunes continue to shift until they become colonized by marram grass, whose roots help stabilize and anchor them, and encourage further dune growth, which in turn encourages further growth of marram grass.

As it stabilizes, a dune becomes a more hospitable home for other plants. Bayberry, wild rose, hudsonia, and eventually white spruce move in and take root, adding moisture and shelter and promoting additional colonization. In the low wet areas cranberry and crowberry grow.

With plant growth, insects come to the dunes, followed by the small plant- and insect-eating shrews, jumping mice, and white-footed mice. These small creatures attract mink, owls, hawks and red fox.

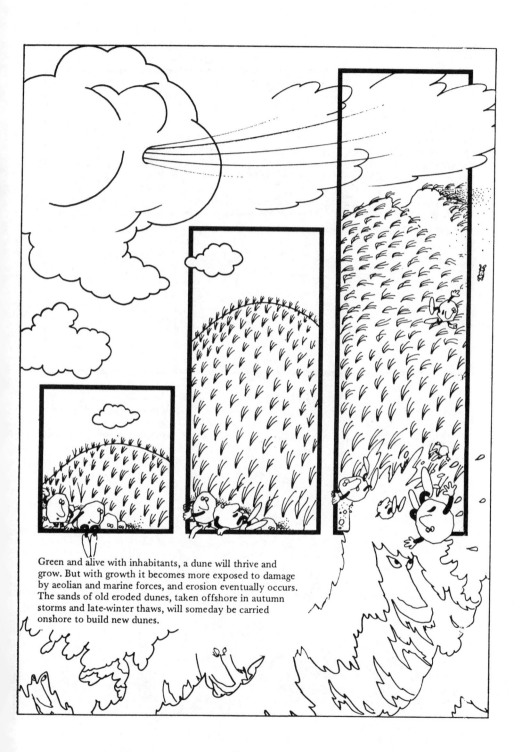

Green and alive with inhabitants, a dune will thrive and
grow. But with growth it becomes more exposed to damage
by aeolian and marine forces, and erosion eventually occurs.
The sands of old eroded dunes, taken offshore in autumn
storms and late-winter thaws, will someday be carried
onshore to build new dunes.

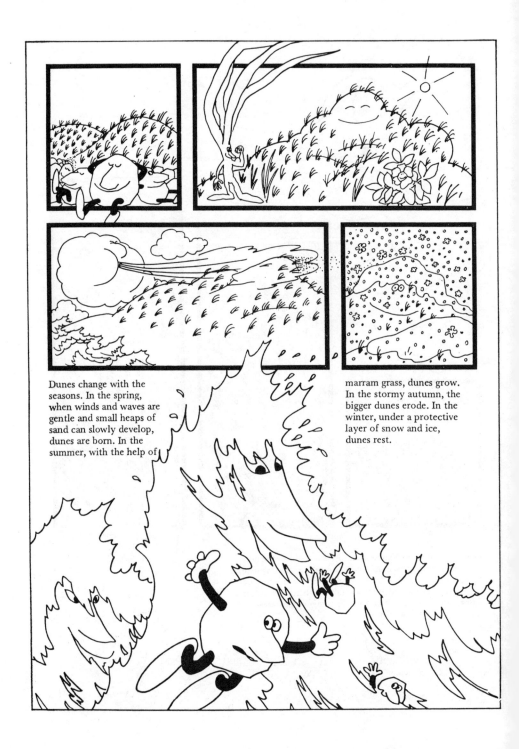

Dunes change with the seasons. In the spring, when winds and waves are gentle and small heaps of sand can slowly develop, dunes are born. In the summer, with the help of marram grass, dunes grow. In the stormy autumn, the bigger dunes erode. In the winter, under a protective layer of snow and ice, dunes rest.

Because sand is easily displaced by strong onshore winds, a dune has a natural tendency to migrate inland, and cover everything along its route.

Humankind is a potential foe for a dune. With its ability to inadvertently destroy marram grass, the human element in the park can have a severely damaging effect on dune stability.

This dune land is valuable and unique. Various conservation practices are undertaken to protect it: Snow-fences serve as obstructions behind which dune formations can begin; boardwalks protect the dunes from pedestrian traffic; and a programme of transplanting encourages regrowth in exposed areas and along pathways.

These dunes and beaches have been brought to you by the wind and the waves. The conservation practices of Prince Edward Island National Park are our attempt to preserve this area. They will be effective if all visitors co-operate. Let's take care of our park . . . future visitors thank you.

Most of the world's water is saline. The small fraction of fresh water remaining cannot be used, for the most part, because it is frozen in polar or glacier ice. Once it has evaporated, however, this form of water takes an active part in the world's water cycle. Here we pay a brief visit to an alpine glacier.

Chapter 10

High seas in the uplands

Jacques G. Richardson

The author acknowledges, with gratitude, the supply of data concerning the Morteratsch Glacier by Karl E. Scherler of the World Glacier Inventory and the Department of Geology at the Eidgenossische Technische Hochschule (ETH), Zurich.

The water in the sea, although making up 97 per cent of all the water available on the planet, needs partial replenishment through the process of transpiration that returns water—through the atmosphere—from the terrestrial land-masses to the ocean. Of the 3 percent of non-saline water existing, two-thirds are locked in the ice of the polar caps, snow fields and glaciers. The glaciers are thus the highest 'bodies of water' on the continents.

A typical glacier is the Morteratsch Vadret (or Morteratsch Glacier, in the local Romansh language) near the Bernina Pass in the Engadine region of Switzerland. Draining successively into the Inn and Danube rivers and finally the Black Sea, the glacier currently has an exposed surface of less than 16 km^2. Between 1900 and 1970, the glacier's (downward) tip receded by 1,318 m. The surface shrank about 23 per cent between 1943 and 1983. The recession of glaciers is worldwide and suggests that the globe's average temperature is rising.

The Morteratsch has a mean length of 6.8 km, a mean width of 800 m and an average elevation of 3,000 m above sea-level. This and other mountain glaciers—sometimes removed 1,000 km or more from the nearest ocean shore—are part of the total water cycle; originating both on land and in the sea. Glaciers were more widespread during the ice ages of the Pleistocene (which began about 2 million years ago).

The skiing and snowshoeing at Morteratsch are good, although fairly difficult going uphill since there are neither ski-lifts nor towropes. In the spring, as the snow cover loosens and aggregates, alpenstocks and avalanche rockets are required.   ■

*What do we know about the world's 20,000 glaciers, how they evolve
and what their effects are on the environment? The world's hydrologists
and climatologists have organized the modern science of glaciology,
inventorying this unusual earthly feature, helping to attenuate its
disaster potential, and working closely with other specialists to use our
knowledge of glaciers to mankind's optimal advantage.*

Chapter 11

Glaciers: new interest in ancient ice

David Spurgeon

*Mr Spurgeon is Unesco's science journalist and the author of a booklet
on the Organization's science programmes,* Why the 'S' in Unesco?
Formerly editor of the Canadian publication Science Forum *and science
reporter on* The Globe and Mail *(Toronto), the author has also worked
in Kenya (with the International Council for Research in Agroforestry
and the United Nations Environment Programme), Australia and
Ethiopia (with international agricultural organizations). He wrote the
original version of the present chapter, here adapted, for* Unesco Features.

Ice and climate changes

During the past twenty years, many of the world's climatologists have become concerned about the relationship between climatic change and an increase in the carbon dioxide content of the atmosphere. A well known example is a 1979 study by the U. S. National Academy of Sciences suggesting that, if mankind went on burning fossil fuels at the then-current rate, the world might find itself facing—within a few decades—an increase in global temperatures of 2-3 degrees centigrade, with dire effects.

The temperature rise would result from the so-called greenhouse effect, with the carbon dioxide increase from the fossil fuels trapping heat from the sun much in the same way the glass walls and roof of a greenhouse do. Like the greenhouse, the earth's atmosphere is largely transparent to solar radiation, but it strongly absorbs longer-wave radiation from the ground. The amount of such radiation absorbed is greatly increased when its carbon dioxide content rises; much of this radiation is re-emitted towards the ground, thus warming the earth's surface.

The effects of the warming would include drastic changes in the earth's ice cover and in the water cycle in densely populated chains of mountains at low altitudes. Regions in developing countries and at the poles would also be seriously affected. Some specialists have even speculated that water might rise so high that it would inundate the world's ports.

One result of such speculation was a renewed interest in changes in the world's glaciers. At the Second Planning Meeting on World Climate Programming/Water held in Paris in 1982, scientists decided that a new programme of world glacier surveillance should be developed before 1985. The *World Glacier Inventory* is now being completed: it will contain information on variations in about 1,000 glaciers in more than twenty countries. A World Glacier Monitoring Service is also expected to be under way by 1986 and, in the Soviet Union, a *World Atlas of Snow and Ice Resources* is close to completion. This *Atlas* evaluates natural ice as a potential source of fresh water and reveals the globally climatic significance of glaciers and snow cover.

Glaciers and the evolutionary process

Interest in glaciers did not arise, of course, only as recently as 1979. Glacier observations have been under way in some Alpine countries for many decades, and a regular observation programme was suggested as early as 1773. The International Commission on Glaciers was founded, in 1894, with this as its objective. Since 1960, glacier observations have been standardized internationally and published every five years by the Permanent Service on the Fluctuations of Glaciers.

Such interest is well founded because glaciers have had a profound effect on the earth and its peoples for centuries. Not only have they produced climatic changes, altered the levels of the seas and sculptured the earth's landscape, glaciers have also deeply affected the course of early human evolution, development and migration through climatic changes related to their advances and retreats.

More than half the earth's land surface shows landforms resulting from climatic conditions associated with the Ice Age that ended 10,000 years ago, such as lakes, coastal features and river courses. (More than fifty lakes caused by glacier movement have appeared during the past 150 years.) Depression and upheavals of the earth's crust resulting from variations in glacier loading still occur today: the Scandinavian Peninsula north of the Gulf of Bothnia is rising at the rate of about 88 centimetres per century, for example.

Glaciers have been responsible for many of the earth's most precious natural resources. For instance, the debris transported across the surface of the globe and deposited in various locations led to the development of fertile soils. And where climatic conditions coincided with these—for example, in northern Europe— highly productive agriculture emerged.

Glacial terrains are generally associated with the abundant surface and sub-surface water essential to human settlement and industrial development, while modern rivers of glacial origin provide some countries with hydroelectric power. Thousands of lakes in Scandinavia, Canada and the United States, which are of great value for recreation and commerce, have their origins in glaciers. Groundwaters in glaciated areas are contained in aquifer consisting of sands and gravels which, in turn, are used to produce concrete, while the clays deposited in ancient glacial lakes are used to make bricks and tile.

Glacier stock-taking and disaster avoidance

So glacier observations are necessary for a variety of reasons. They help increase man's knowledge of local, regional and global water cycles. They provide data for studying climatic change. And they serve as data bases for fresh-water resource planning, irrigation, hydroelectric power schemes, recreational facilities and disaster prevention.

Disaster can be caused by glacier surges—sudden catastrophic advances or the raising and lowering of ice surfaces, which are relatively common in some areas. In the Yukon between 1965 and 1969, glacier advances were observed with surge velocities of up to 18 metres per day, while the Brasvalsbreen Glacier on Spitsbergen (Norway) slid forward in the late 1940s by eight kilometres in less than five months.

In 1953, the Kuthean Glacier (Tyan-chan in Chinese) moved along a valley in East Asia at an average speed of 113 metres per day, covering a distance of 12 kilometres during a two-month period. Destroying buildings and trees in its path, it finally formed a dam in a river, creating a huge lake. When the glacier began moving again, the dam of ice broke, producing disastrous flooding.

Such cataclysms may even result in death: in 1965 the break-up of the lower part of the Allalin Glacier in the Swiss Alps—1 million cubic metres in volume— killed 88 workmen engaged in dam construction.

Today, glacier inventories contain information about more than 20,000 glaciers the world over. The *World Glacier Inventory* will provide a snapshot of ice conditions on earth taken during the second half of this century. But more comprehensive efforts will be needed in the future. Some of these may be provided through the use of new observational techniques such as satellite imagery.

During the past decade a special branch of glaciology has been developed which should help in forecasting global interactions between climate and glaciers, in learning how ancient glacier sheets arose during the planet's evolution, and in evaluating the influence on glaciers of human activity. Such research could be significant in our understanding not just of the evolution of the earth but even, in future, more widely throughout the solar system—since many other planets also have glaciers and huge ice sheets, such as the polar ice caps of Mars and the ice cover of Pluto.

Forecasting and Unesco's role

The proposal for a World Glacier Monitoring Service came from a working group of the International Commission on Snow and Ice (ICSI), an offshoot of the International Association of Hydrological Sciences. At an ICSI bureau meeting held at Unesco in Paris, December 1984, an ad hoc advisory panel was appointed to examine the proposal before its submission to the United Nations Environment Programme. Unesco will contribute to both the publication of the *World Glacier Inventory* and the support of the World Glacier Monitoring Service; it has already provided financial assistance to some aspects of the *World Atlas of Snow and Ice Resources.*

The proposed World Glacier Monitoring Service is designed to merge the existing Permanent Service on the Fluctuation of Glaciers with the Temporary Technical Secretariat of the World Glacier Inventory. The new Service would continue and improve the work of these two bodies, but it could also enhance the collection and publication of data from regions about which little or nothing is currently known.

In this way, the new group should also promote further research and analysis in an area of great potential importance, increasing man's knowledge and ability to control the effects of one of the world's great natural features—the living rivers of ice that history has bequeathed us.   ■

PART 3.

RESEARCH

*The tendency of man's increasing activities to contaminate or
destroy the ocean environment needs the countervailing pressure of
accumulated scientific knowledge. The world's coastal regions badly
need such technical attention via a multidisciplinary approach. In
China, anticipated finds of new petroleum sources off-shore may
exacerbate a problem of already polluted coastal waters. In both
cases, science can help enormously.*

Chapter 12

Amassing scientific knowledge
to preserve the marine environment

Luo Yuru

*Dr Luo occupies the posts of director of the Council of the Chinese
Society of Oceanography as well as director of his country's National
Bureau of Oceanography, in Beijing, publishers of* Acta Oceanologica.

1

The tremendous achievements of modern science and technology are turning man's ideal of developing total use of the oceans into reality, in both breadth and depth. Meanwhile, however, sea-born activity combined with other industrial and agricultural pursuits have led to deterioration of the marine environment through damage to its resources. Thus have we hindered or even excluded the exploitation of certain marine resources, thereby offsetting some of the socio-economic benefits possible through development of the seas.

In turn, we are under the obligation to rely on advanced science and technology if we are to prevent and cure marine pollution, or to conserve and replenish the living resources of the ocean. Without advanced technology, for example, it is out of the question that we monitor the large-scale features of the pollution of the sea rapidly or that we treat the 'three wastes' (waste gas, water, and industrial residues) efficiently. The key to conservation in the ocean, therefore, with the aim of its full but reasonable development, is to give to technology its full play as the appropriate lever.

China has attached considerable importance to basic investigations and other studies related to the marine environment and its resources. Beginning in the early 1970s and in a systematic way, we have conducted basic research related to the prevention and control of marine pollution. Having completed the base-line studies lasting more than a decade concerning pollution in most of our off-shore areas (covering a surface of about 450 000 km^2), we now have, by and large, a clear picture of the level of pollution in the areas investigated.

At the same time, the pollution monitoring maintained in selected zones has provided important, fundamental data necessary in order to keep abreast of the pollution dynamics—analysing their trends and evaluating environmental quality. On the basis of these marine studies and monitoring, we have conducted 'basic applied' research on the technology required for marine environmental protection and on the processes of the ocean's self-purification. These have suggested the scientific bases, methodology and technical means required for the future preservation of the ocean environment.

2

Here I would like to deal particularly with the need to enhance marine ecological studies in order to preserve the marine environment and conserve its resources. As one of the main branches of holistic ecology, modern marine ecology is multidisciplinary—comprising the natural sciences, mathematics, systemics, and the social sciences; it stresses research on the interrelationships between human activities and marine biological populations and their habitats. The fundamental motive force in the rapid progress of marine ecology over the past ten years or so have obviously come from the practical needs of nations in the development, protection and overall 'management of the ocean.' The most practical significance of improving studies in marine ecology has four major aspects.

First, improved knowledge in marine ecology contributes to environmental protection. In normal conditions, the processes representing material cycles and energy flow proceed continuously throughout marine biota and between these biota and the sea's environment, thus maintaining an ecologial balance. Once the quantity present of a given material outweighs the ocean's capacity for self-purification, it may cause pollution harmful to certain organisms, thus changing their population structure and function to such an extent as to destroy ecological equilibrium.

By making thorough environmental studies, we should be able to assess correctly the chronic effects of pollutants on the structures and functions involved, and thus understand counteractions of the ecosystem on pollutants as well as biological adaptive capacity. The assimilative function of an ecosystem as regards pollutants and its tolerance to inroads made upon it, moreover, then serve as the basis for formulating standards for discharge into the environment. Only with such knowledge can we conscientiously try to adjust the ecosystem's balance and—once an existing natural system loses its equilibrium as a consequence of marine exploitation—rebuild dynamic equilibrium at a higher level, with interference by man.

Secondly, better knowledge helps raise the management level concerning fishery resources. Being renewable, the ocean's living resources are constrained by ecological laws. If fishery production is not in conformity with these laws, resources may be wasted (as in underfishing) or depleted (the case in overfishing), thus destroying fish productivity. It is therefore necessary to reinforce fishery management, defining optimal catch limits, before we can develop the sea's living resources rationally. This makes it imperative to conduct environmental studies designed to help us understand productivity levels and the law of energy conversion at different trophic, or food-chain, levels. Equipped with such knowledge, we should be able to understand the interrelations among diverse fishery-resource populations and between these and the marine environment, as well as the reactions of fishery resources to (a) fishing activities and (b) environmental variation. Because of poor knowledge of the interactions among the many marine species, the process of single-species management has been, until now, common practice in many countries—including the developed nations.

So how to manage fishery production, comprehensively based on the nature of the ecosystem, is another major question which ecological research must answer. In order to define the optimal yield of a given fishery, for example, we need to predict the natural 'recruitment' to some important populations; this necessitates a clear explanation of air-sea interactions, density structures and levels of fertilization in sea-water, propagation patterns of natural variability through food webs, and the like. We can thus conclude that the amelioration of marine ecological studies will benefit our capability to reinforce fishery management.

3

Thirdly, new knowledge is helpful in the development and rational use of coastal and off-shore resources. Coastal-zone ecosystems are the most fragile in the whole marine environment, the most vulnerable to disturbance or destruction by man's activities. With population increase and economic growth, there have emerged new activities contending for development and use of the coastal regions: manufacture, agriculture and animal husbandry, commerce, communication and transport, tourism, petroleum and gas exploitation, waste discharge, residential development, fishery at sea, and so on. In China, there are more than 170 trades directly pertinent to her coastal zones. Earlier, the lack of unified management by orderly methods not only reduced the socio-economic benefits possible in development and use; it also led to serious damage to the environment and natural resources of quite a number of regions, resulting in further deterioration of seaboard ecosystems.

There is little doubt that ecological research can provide the scientific basis for rational development and exploitation of coastal areas and adjacent seas. Some developed countries have attached particular importance to such effort. During studies made for an eventual tidal-power station in the Bay of Fundy (Nova Scotia), for instance, the Canadian government organized marine ecologists to study the impact of such a station on the ecosystem thereabout. In the case of China, the large-scale off-shore petroleum development about to take place will become an enormous source of pollution. So the off-shore activities planned to develop oil resources on the continental shelf will have to be examined with the aim of maintaining ecological balance; corresponding standards, regulations and procedural measures will have than to be elaborated.

4

Fourthly and lastly, better knowledge helps increase the capability to make forecasts related to the marine environment. Changes in this environment can cause ecosystem variation, and sudden variability may even cause destruction to the ecosystem. Since biological and non-biological factors interact and condition each other within the same ecosystem, changes in the ocean environment are often reflected by biological variation in community structures, species of organisms and their quantities, biological morphology and physiological and biochemical processes. From these, we can infer changes in the environment. In the past, on the basis of this approach, methods to determine diversity index and species abundance have successfully been used to monitor the marine environment.

So long as we go deep into the causalities between internal and external variations—as well as between preceding and subsequent variations, we should be able to make environmental forecasts. With these, we can maximize the advantages while avoiding the disadvantages, thereby reducing the adverse effects producted by environmental variability.

* * *

All of the four aspects mentioned are essential to ocean conservation. Whether these problems are to be solved properly is closely dependent upon our understanding of marine ecosystems. We can thus conclude that improvement of marine ecological studies is of vital importance to the preservation of the ocean's environment and its considerable natural resources. ■

Biogeochemical cycles in marine ecosystems are discussed with special reference to the behavior of nitrogen. Some research recommendations are presented which should be considered in the framework of the proposed International Geosphere and Biosphere Programme (IGBP).

Chapter 13

The biogeochemical cycle in marine systems

Akihiko Hattori

Director of the Ocean Research Institute of the University of Tokyo, Dr Hattori can be reached at his Institute's offices at 1-15-1 Minamidai, Nakano-ku, Tokyo 164 (Japan), telephone (03) 376-1251, Telex J25607 ORIUT. This article appeared originally in Global Change, *a volume published in 1985 by the International Council of Scientific Unions and Cambridge University Press and by whose kind permission the text is included in* Managing The Ocean. *The author expresses his thanks to Drs I. Koike, T. Saino and D. D. Swinbanks for their valuable suggestions and Ms M. Ohtsu for her assistance in the preparation of the manuscript.*

Introduction

Life on the earth is sustained by an intricate interaction with complex physical and chemical systems. The availability of various elements, or building blocks, and energy for the formation of cellular constituents has been decisive for the development and maintenance of the present global ecosystem. Furthermore, the development and function of living organisms have modified the patterns of distribution and circulation of these elements, and in turn contributed to changes in the habitats of these organisms and their environments. Some controlling mechanism must be operating so as to make the survival of organisms on earth possible, and a subtle balance has been maintained between the geosphere and biosphere throughout geological time. It has been argued that the impact caused by human activities may perturb this balance.

The biosphere extends from the top of the earth's highest mountains to the abyss of the oceans. The occurrence of benthic animals and other organisms in the deepest oceanic trenches has been documented. Water is essential for life. Life is believed to have originated from the sea. Although the supply of water often regulates the growth of terrestrial organisms, this is not the case for marine life.

Besides oxygen and hydrogen, carbon, nitrogen, phosphorus and silica are of primary importance for marine life and form major components of the biosphere, but with the exception of carbon they are relatively less abundant in the hydrosphere. Here I shall try, from the point of view of marine ecology, to summarize problems and topics which should be included within the framework of IGBP. I will focus on the problems associated with biogeochemical cycling of nitrogen in marine ecosystems, because nitrogen in sea water is present in various chemical forms; some are volatile while others are soluble or solids (see Figure 1). Their conversion within marine ecosystems is closely linked to the cycles of carbon, phosphorus, and other essential biophilic elements. Some nitrogenous compounds are produced and decomposed either chemically or photochemically, but these processes largely occur in the atmosphere. Only biological processes are of quantitative importance with respect to nitrogen transformations in the aquatic system. Information on nitrogen behavior provides a basis for the understanding of short-term and long-term variations in the biogeochemical cycles as well as information on the interaction between the hydrosphere and the atmosphere and between the hydrosphere and the lithosphere. The subject which I discuss here should be considered as a part of the comprehensive and interdisciplinary study of global biogeochemical cycles.

Nitrogen biogeochemical cycle in marine ecosystems

The ocean contains 16 mg of N_2 per kilo of sea water. The amounts of combined nitrogen consisting of nitrate, nitrite, ammonium (an ion of

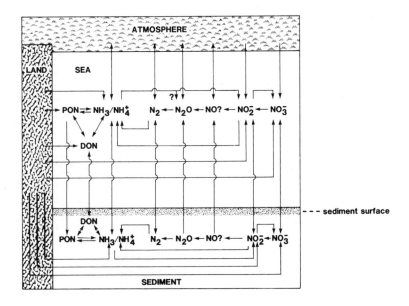

Figure 1. The nitrogen cycle in the sea, after A. Hattori (1982). PON represents particulate organic nitrogen, DON dissolved organic nitrogen.

ammonia) and amides are relatively small in sea waters, about 700 mg per kilo of sea water in total. Nevertheless, combined nitrogen plays a crucial role in plant, animal and bacterial life, and marine biological productivity is often limited by its supply. Except for a few species of bacteria (including cyano-bacteria), marine organisms cannot directly utilize N_2 for their growth.

Combined nitrogen is introduced into the ocean by river runoff and by precipitation and dry deposition over the sea surface. According to current estimates of the oceanic nitrogen budget (Table 1), 20 to 110 Tg [teragrammes or 1×10^{12} g] of combined nitrogen is transported annually to the sea. Another 20 Tg/year are added by pelagic and benthic biological nitrogen-fixation. The pool of marine biomass contains about 300 Tg of organic nitrogen on a global scale. Recent data of marine primary productivity[1] suggests that about 8,000 Tg of nitrogen are taken up by phytoplankton every year.[2] If this estimate is valid, the turnover time of phytoplankton nitrogen is calculated to be about half a month, and more than 97 per cent of the nitrogen is recycled within the marine system.

The budget (Table 1) further suggests that the annual supply of combined nitrogen exceeds the loss by burial in sediments by 40-90 Tg N. Although some nitrogen can be lost from the sea by volatilization or sea spray, the majority of this excess of nitrogen must be released to the atmosphere in gaseous form if a steady state is to be maintained in the sea.[3,4,5] Denitrification appears to be the mechanism by which this is accomplished.

TABLE 1. Oceanic budget of combined nitrogen (Tg N/year).

	Flux	Reference *
Gains		
River runoff	19	Emery *et al.* (1955)
	10	Tsunogai (1971)
	10	Holland (1973)
	13-24	Note 7
Precipitation on sea surface	59	Emery *et al.* (1955)
	23	Tsunogai (1971)
	10	Holland (1973)
	30-83	Note 7
Nitrogen fixation		
Pelagic	5	Carpenter (1983)
Benthic	15	Capone (1983)
Losses		
Burial in sediments	8.6	Emery *et al.* (1955)
	8	Tsunogai (1971)
	10	Holland (1973)
	38	Note 7
Denitrification		
Pelagic North Pacific	19-23	Note 3
South Pacific	19-25	Codispoti and Packard (1980)
Benthic	300	Note 4
	44	Note 5

* See Notes and To Delve More Deeply, at the end of this chapter.

   Some marine bacteria can produce or consume N_2O. Although its concentration in sea water is far less compared with the other forms of nitrogen, understanding of the fate of N_2O in the marine system is critically important because it may affect the mixing ratio of N_2O in the atmosphere. The importance of N_2O in the chemistry of atmosphere ozone has been discussed[6]; therefore, minor components such as N_2O should not be overlooked. It is further stressed that minor components can often be used as tracers to monitor temporal and spatial variations in biogeochemical cycles.

Knowledge of the qualitative aspects of the global nitrogen cycle has advanced during the past decade.[7,8] However, it is still difficult to present a quantitative picture. There is considerable uncertainty with respect to most data on nitrogen fluxes. The reported annual fluxes of nitrogen compounds have been estimated, based on data obtained in limited areas and over limited periods of time. A difference of an order of magnitude is not uncommon between lowest and highest values. In order to make any prediction about global change, either natural or anthropogenic, more precise and extensive information is undoubtedly needed. This is specially the case in estuarine and coastal areas where the effects of man's intervention are expected to have profound significance. Extrapolation of limited local data to a global scale is not warranted.

Research recommendations

Summarized below are some of the research themes which should be considered in the IGBP.

Identification of biochemical processes responsible for nitrogen transformation and their regional characteristics

A great deal of information is available on the biochemistry of nitrogen metabolism. However, various kinds of organisms participate in nitrogen transformation in the sea. Some processes, e.g., nitrification, dentrification and nitrogen fixation, are mediated only by some special groups of bacteria, but seaweeds, phytoplankton and bacteria share the capacity for producing organic nitrogen from inorganic nitrogen. We only poorly understand individual processes occurring in the marine ecosystem. Products of these reactions vary depending upon the conditions to which organisms are exposed. Nitrous oxide has traditionally been viewed to originate from anaerobic reduction of nitrate by denitrifying bacteria. Recent studies, however, suggest that N_2O is also produced in association with the oxidation of ammonium [an ion of ammonia] by nitrifying bacteria[9] and that this process contributes significantly to the production of N_2O in marine systems.[10,11] Date on ^{15}N abundance in dissolved N_2O were recently reported that support this inference.[12]

Another point that should be considered under this heading is primary production in the open ocean. It is regulated by the supply of nitrogen, although phosphorous may control phytoplankton growth in some tropical areas where nitrogen-fixing cyanobacteria prevail. Variation of primary production might effect CO_2 exchange across the air-sea interface and, in turn, the mixing ratio of CO_2 in the atmosphere. A thorough understanding of biology and biochemistry in marine primary production is yet to be attained.[13] Some chemolithotrophic bacteria* can grow without a supply of solar energy or energy reserved in the form of organics. Evidence was recently

---

* A chemolithotrophic organism is one that oxidizes inorganic substances for its 'nourishment.'

presented for the occurrence of CO_2 fixation mediated by nitrifying bacteria in deep water.[14]

Marine cyanobacteria with a nitrogen-fixing capacity produce H_2.[15] The presence of H_2 in sea water and its diurnal variation have been reported.[16] Hydorgen gas can be used as a tracer to indentify the sites of oceanic nitrogen fixation.

Identification of chemical species and their functions

The biosphere is ultimately supported by the sun's energy. The life of animals and other heterotrophs [organisms that obtain their nourishment from outside sources] depends on organic substances produced by the photosynthesis of plants, and its function is regulated by their supply. Nutrients in the euphotic layer [penetrated by light] are extracted by phytoplankton and replenished from depth or by remineralization in situ. Their distribution is also affected by water movement and circulation.

We can distinguish the fertility of the sea on the basis of nutrient data. Unfortunately, the amounts and chemical forms of organics present in sea water are not sufficiently well understood. For example, circumstantial evidence suggests that the rate of nitrate reduction and denitrification in oxygen- depleted oceanic water is limited by the supply of organics. However, we do not know which organic compound forms the key in this process. There is a controversy as to whether N_2O in the atmosphere is currently increasing and as to whether the sea acts as a sink or source of the atmospheric N_2O.[17-20] Although much effort has been devoted to determination of the amounts of N_2O in the atmosphere and in sea water, this problem is still left unresolved.

Identification of biological components responsible for nitrogen transformation and transportation and their regional characteristics

Identification of plankton, bacteria and other organisms is time consuming and requires specialized knowledge. Without detailed information about biomass and species composition, however, it is difficult to assess the structure and function of the ecosystem and the rates of processes within it with certainty. Quantification is another problem. The currently available techniques are not precise or dependable. Intensive studies applying modern techniques such as immunofluorometry and other biochemical techniques, and automated image analysis should be encouraged.

Although most organic matter once produced by phytoplankton is consumed by zooplankton and fishes, organic debris and fecal pellets formed in association with grazing accelerate vertical transport of nutrients and other chemical species in particulate form. Since heavy metals and man-made chemicals such as DDT and PCB are often concentrated in biota, this process of vertical transport may play an important role in their cycling. Natural abundances of ^{13}C and ^{15}N have been used for characterization of particulate organic matter.[21,22] Isotope fractionations of nitrogen and carbon are associated with decomposition of particulate organic matter. The decomposition of organic nitrogen is much faster than that of organic carbon. Recent data show that vertical distributions of heavy metals such as cadmium, zinc, nickel and copper are similar to those of nitrate and phosphate,[23] suggesting a close link between organic matter and heavy metals.

The fate of nitrogen and other biophilic elements in estuaries and embayments

Available data on Tokyo Bay indicate that nitrogen and phosphorus that had been introduced from land are quickly taken up by phytoplankton but that the bulk of these nutrients, once trapped, are exported after recycling through food webs within the water column.[24,25] Only less than 10 per cent of the nitrogen and phosphorus load settled on the Bay's floor. However, the situation might vary from place to place. The factors that regulate their transformations should be identified on a worldwide basis at representative locations; otherwise, it is difficult to make any global assessment. Heavy metals such as copper, lead and zinc are buried in the sediments. We can easily trace the history of metal pollution from the records in core samples of coastal sediments.[26] Emphasis should also be placed on interactions of biochemical processes with physical and abiotic chemical processes.

Episodic events as a model of events on geological time-scale

Climatic change and changes in the marine ecosystem associated with El Niño are well documented. Such changes in the marine ecosystems often lead to irreversible biological transitions (e.g., changes in species composition). If the changes in ecosystems associated with physical episodic events such as the El Niño are extensively investigated, this can provide useful information about the interaction between the geosphere and biosphere, and this knowledge in turn contributes to understanding of ecosystem change linked to change in the geosphere over geological time scales.

Geological records of biogeochemical events

Records of past activities in sediments provide important information for understanding the interactions between the hydrosphere, atmosphere and biosphere. In this respect, collection of biochemical information together with physical, chemical and biological data undoubtedly contributes to the advance of the studies along this line. Many deep-sea sediment samples necessary to do this are available, if IGBP is conducted in close co-ordination with the on-going Ocean Drilling Project and other related projects.

Nitrogen cycles in polar ecosystems and thermal vents

In polar regions, we can probably see most significantly the effects of variations of solar radiation on the earth. Nitrate is relatively abundant in the soils of Antarctica, and its ^{15}N content is extremely depleted.[27] The possibility has been suggested that this nitrate originates from NO_x, produced by auroral activity. Biological activity around geothermal vents on the ocean floor is another subject that can provide novel information about the possible origins of life. Some data on the abundance of ^{15}N near vent areas have been presented.

Concluding remarks

The research themes discussed above are closely related to each other; none can be achieved in isolation. Quantitative understanding of the biosphere and its interaction with the geosphere is of fundamental importance and is indispensable in formulating any mathematical models for predicting global change. In this respect, the development of new techniques is one essential task, and the extension of geographical coverage the other. Currently available techniques for determination of biomass and biochemical rate processes in the marine system are not sufficiently accurate, and instrumentation is still far behind compared to the other physical and chemical disciplines. The information collected by direct measurements can be complemented with that obtained by remote sensing and other sophisticated modern techniques. In order to fulfil the goals of IGBP the continuation of international and interdisciplinary research efforts over at least a period of ten years is necessary.   ■

Notes

1. C.C. DeVooys, Primary Production in Aquatic Environments, In: B. Bolin, E.T. Degens, S. Kempe, P. Ketner (eds.), *The Global Carbon Cycle,* New York, John Wiley, 1979.
2. A. Hattori, The Nitrogen Cycle in the Sea with Special Reference to Biogeochemical Processes, *J. Oceanogr. Soc. Japan,* Vol. 38, 1982, p. 245.
3. L.A. Codispoti, F.A. Richards, An Analysis of the Horizontal Sequence of Denitrification in the Eastern Tropical North Pacific, *Limnol. Oceanogr.,* Vol. 21, 1976, p. 379.
4. J.J. Goering, Denitrification in Marine Systems, In: D. Schlessinger (ed.), *Microbiology–1978,* Washington, American Society of Microbiology, 1978.
5. A. Hattori, Denitrification and Dissimilatory Nitrate Reduction, In: E.J. Carpenter, D.G. Capone (eds.), *Nitrogen in the Marine Environment,* New York, Academic Press, 1983.

6. P.J. Crutzen, Atmospheric Chemical Processes of the Oxides of Nitrogen, including Nitrous Oxide, In: C.C. Delwiche (ed.), *Denitrification, Nitrification and Atmospheric Nitrous Oxide,* New York, John Wiley, 1981.
7. R. Söderlund, B.H. Svensson, The Global Nitrogen Cycle, *Ecol. Bull.* (Stockholm), Vol. 22, 1976, p. 23.
8. C.C. Delwiche, *Denitrification, Nitrification and Atmospheric Nitrous Oxide,* New York, John Wiley, 1981.
9. J.M. Bremner, K.L. Blackmer, Terrestrial Nitrification as a Source of Atmospheric Nitrous Oxide, In: C.C. Delwiche, *op. cit.*
10. J.W. Elkins, S.C. Wofsy, M.B. McElroy, C.E. Kolb, W.A. Kaplan, Aquatic Sources and Sinks for Nitrous Oxide, *Nature,* Vol. 275, 1978, p. 602.
11. Y. Cohen, L.I. Gordon, Nitrous Oxide in the Oxygen Minimum of the Eastern Tropical North Pacific: Evidence for Its Consumption during Denitrification, and Possible Mechanism for Its Production, *Deep-Sea Res.,* Vol. 25, 1978, p. 509.
12. N. Yoshida, A. Hattori, T. Saino, S. Matsuo, E. Wada, $^{15}N/^{14}N$ Ratio of Dissolved N_2O in the Eastern Tropical Pacific Ocean, *Nature,* Vol. 307, 1984, p. 442.
13. P.G. Falkowski, *Primary Productivity in the Sea,* New York, Plenum Press, 1980.
14. D.M. Karl, G.A. Knauer, J.H. Martin, B.B. Ward, Bacterial Chemolithotrophy in the Ocean is Associated with Sinking Particles, *Nature,* Vol. 309, 1984, p. 54.
15. T. Saino, A. Hattori, Aerobic Nitrogen Fixation by the Marine Non-heterocystous Cyanobacterium *Trichodesmium (Oscillatoria):* Its Protective Mechanism against Oxygen, *Mar. Biol.,* Vol. 70, 1982, p. 251.
16. F.L. Herr, E.C. Frank, G.M. Leone, M.C. Kennicutt, Diurnal Variability of Dissolved Molecular Hydrogen in the Tropical South Atlantic Ocean, *Deep-Sea Res.,* Vol 31, 1984, p. 13.
17. H.B. Singh, L.J. Salas, H. Shigeishi, The Distribution of Nitrous Oxide, (N_2O) in the Global Atmosphere and the Pacific Ocean, *Tellus,* Vol. 31, 1979, p. 313.
18. D. Pierotti, R.A. Rasmussen, Nitrous Oxide Measurements in the Eastern Tropical Pacific Ocean, *Tellus,* Vol. 32, 1980, p. 56.
19. R.F. Weiss, The Temporal and Spatial Distribution of Tropospheric Nitrous Oxide, *J. Geophys. Res.,* Vol. 86, 1981, p. 7185.
20. J. Hahn, Nitrous Oxide in the Oceans, In: C.C. Delwiche, *op. cit.*
21. P.M. Williams, L.I. Gordon, Carbon-13:Carbon-12 Ratios in Dissolved and Particulate Organic Matter in the Sea, *Deep-Sea Res.,* Vol. 17, 1970, p. 19.
22. T. Saino, A. Hattori, ^{15}N Natural Abundance in Oceanic Suspended Particulate Matter, *Nature,* Vol. 283, 1980, p. 752.
23. K.W. Bruland, Oceanographic Distributions of Cadmium, Zinc, Nickel and Copper in the North Pacific, *Earth Planet. Sci. Lett.,* Vol. 47, 1980, p. 176.
24. A. Hattori, E. Matsumoto, N. Handa, Behavior of Pollutants in Coastal Areas, In: T. Hirano (ed.), *Sciences of Marine Environment,* Tokyo, Koseisha-Koseikaku, 1983 (in Japanese).
25. A. Hattori, M. Ohtsu, I. Koike, Distribution, Metabolism and Budgets of Nitrogen in Tokyo Bay, *Chikyu Kagaku* (Geochemistry), Vol. 17, 1983, p. 33 (in Japanese).
26. E. Matsumoto, The Sedimentary Environment in Tokyo Bay, *Chikyu Kagaku,* Vol. 17, 1983, p. 27 (in Japanese).
27. E. Wada, R. Shibata, T. Torii, ^{15}N Abundance in Antarctica: Origin of Soil Nitrogen and Ecological Implications, *Nature,* Vol. 292, 1981, p. 327.

To delve more deeply

CAPONE' D.G., Benthic Nitrogen Fixation, In: E.J. Carpenter, D.G. Capone (eds.), *Nitrogen in the Marine Environment,* New York, Academic Press, 1983.
CARPENTER, E.J., Nitrogen Fixation by Marine *Oscillatoria (Trichodesmium)* in the World Oceans, In: Carpenter and Capone, *op. cit.*

CODISPOTI, L.A., PACKARD, T.T., Denitrification Rates in the Eastern Tropical
    South Pacific, *J. Mar. Res.*, Vol. 38, 1980, p. 453.
EMERY, K.O., ORR, W.L., RITTENBERG, S.C., Nutrient Budgets in the Ocean,
    In: *Essays in the Natural Sciences in Honor of Captain Allan Hancock,*
    Los Angeles, University of Southern California Press, 1955.
HOLLAND, H.D., Ocean Water, Nutrients and Atmospheric Oxygen, *Pro. Symp.
    Hydrogeochem. Biogeochem. (Vol. 1),* Washington, Clarke Co., 1973.
TSUNOGAI, S., Ammonium in the Oceanic Atmosphere and the Hydrosphere,
    Geochem. J., Vol. 5, 1971, p. 57.

Future population trends in India, with their associated implications and demands on natural resources and industrial development, are surveyed. The author considers the degree to which India's ocean environment could contribute to these future needs and reviews the ways in which the natural resources of the sea are now used. He then evaluates the potential of such resources to meet India's needs by the end of this millennium.

Chapter 14

A technological forecast of ocean research and development in India

S. Z. Qasim

The author is secretary to the Government of India's Department of Ocean Development, South Block, New Delhi 110011. He is a former director of the National Institute of Oceanography, Dona Paula, Goa, India. He has been deeply involved in the development of marine sciences in India for more than thirty years and his present interest is Antarctic research.

Introduction

There is a general agreement among most demographers that the population of India by the end of this century will approach the 1,000 million mark. Projections generally range from a low estimate of 875 million to a high of 1,020 million. Figure 1 shows the likely trend of population growth towards A.D. 2001. By the year 2000, the population will be around 998 million. Of this, 508 million will probably be males and 490 million females.[1] These estimates clearly indicate the immensity of the population problem to be tackled in India (see Table 1).

Table 2 gives the existing population of India and some of the data related to land use. As can be seen from the table, the total land area of India is 3.276×10^6 km² which is only 0.64 per cent of the earth's surface. This would also mean that about one-sixth of the world population would

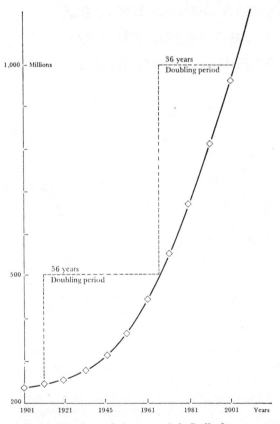

FIG. 1. Projection of population growth in India for 100 years, from 1901 to 2001.

TABLE 1. Some socio-technological forecasts

Forecast	Year of application	Probability (%)
World population would be 5,000 million. If this is not reached it would mean mass application of an effective birth-control system, which would require a drastic change in social education	1990	90
Simple and inexpensive fertility control	1987	99
Most effective and reversible birth-control measure	2000	99
Complete understanding of hormonal control of fertility	1995	90
Intra-cellular surgical techniques applicable to birth control	1987	90
Zero population growth for the entire world	2000	90
Capability to choose sex of the unborn	2000	?

become congregated in a 150th part of the world's surface area and this population will have to be sustained from the agricultural land of about 1.65×10^6 km².

During the next two decades, one of the areas in which the most spectacular advances are expected to be made is the ocean sector. Coastal and offshore activities are rapidly on the increase in India. The establishment of the new Department of Ocean Development shows that the Government of India is giving a high priority to the frontier areas of ocean development. Very large investments are being planned for the exploration and exploitation of various ocean resources such as food, chemicals, minerals, transportation, recreation and energy to improve the socio-economic conditions of the vast population. The offshore construction and operational activities have been rapidly increasing largely because of the exploration and exploitation of oil and gas.

India has a coastline of about 6,000 kilometres. Assuming that 25 per cent of the present population of 680 million live in the coastal areas (within a few hundred miles from the sea shore), it can be estimated that about 170 million people are directly dependent on the sea for their living or livelihood. Even for those who live in landlocked regions, the influence of the sea on their daily lives is significant. Along the coastline, the many rivers discharge about 1,645 km³ of fresh water into the sea. Of this quantity nearly 75 per cent is discharged into the Bay of Bengal (Table 2). Fourteen major rivers in India carry about 85 per cent of the total runoff, covering 83 per cent of the drainage basins, while about 80 per cent of the population live in these

TABLE 2. Present population and related data for India

Population	680 million
Coastal population	170 million
Land area	3.276×10^6 km^2
Agricultural area	1.65×10^6 km^2
Exclusive Economic Zone	2.015×10^6 km^2
River runoff (annual mean)	1,645 km^3
Annual rainfall (on land)	3.4×10^{12} m^3
Annual rainfall over Bay of Bengal	6.5×10^{12} m^3
Annual rainfall over Arabian Sea	6.1×10^{12} m^3

basin areas. The average rainfall over the Indian sub-continent is 3.4×10^{12} m^3. Thus it is clear that the seas around India receive about four times as much rainfall as does the land.

Ocean and climate

Large-scale oceanic circulations are driven by atmospheric winds and by an exchange of heat with the overlying atmosphere. In addition, there are other transfer mechanisms, such as evaporation, precipitation and river discharges, which couple atmospheric systems with the oceans.

Recent years have witnessed two important developments in the global climate programme, and these are likely to play an important role in the future. The first includes space-based platforms, which have a continuous link with weather satellites, and the second is related to large computer capabilities for simulating ocean–atmosphere exchanges.

The response time of the ocean is generally much slower than that of the atmosphere, hence the ocean is often referred to as being the memory of the atmosphere. This is an important feature of climatic changes. One of the difficulties that has been encountered is the lack of precise knowledge about how anomalies in the characteristics of the ocean affect the atmosphere.

Numerous experiments are now in progress to determine the atmospheric response to fluctuations in sea-surface temperature, the reflective power of land and sea (albedo), ground hydrology and many other features. The principal difficulty lies in the lack of understanding of how the momentum of the wind is transferred to the ocean.

Perhaps the most striking example of oceanic response to atmospheric changes lies in the Somali current. This is of special relevance to India because of its close association with the summer monsoon. The Somali current represents a narrow current of water which flows northward during the northern summer along the coast of Kenya and Somalia from Mombasa

(4° S.) to Sacotra near 12° N. During the northern winter, the current reverses its direction and flows southward. Several features of the Somali current are as yet poorly understood.

Observations indicate that the Somali current is accompanied by strong coastal upwelling. This is strengthened by the existence of a low-level air jet.

Of the oceanic anomalies that are believed to have an impact on the atmospheric circulation, sea-surface temperature anomalies have received considerable attention in the recent past. This is because there is a tendency for sea-surface temperature (SST) anomalies to persist longer than atmospheric anomalies. For example, it has been found that the northern Pacific SST anomalies in the summer are significantly correlated with the following season's pressure, especially over the Aleutian islands. The correlation was found to be —0.76 using twenty-seven years' data.

Using a general circulation model (GCM), it has been found that a cold SST anomaly leads to a substantial reduction in model-generated rainfall. This is later reinforced by a statistically significant correlation between SST over the Arabian Sea and rainfall in India a month later, from June to August. Another experiment with a general circulation model revealed precipitation changes near the region of SST anomaly, but there is no conclusive evidence of its impact on rainfall at larger distances away from the anomaly.

In the early years of this century, Sir Gilbert Walker embarked on a search for predictors which could be associated in a statistical sense with the summer monsoon rainfall over India. During his search, he found three important oscillations in the world weather. By far the most important of these was the oscillation located over the Indian Ocean, which is now known as the Southern Oscillation. It refers to a tendency for the accumulation (high pressure) of air over the Indian Ocean to be associated with the removal of air (low pressure) from the Pacific and vice versa. The intensity of the Southern Oscillation shows marked changes from one year to the next. The oceanic response to the Southern Oscillation appears to be linked with the El Niño phenomenon off the coast of Peru.

Very little is yet known of the reverse process, namely the atmospheric response to oceanic anomalies. Clearly, there are global, long time-scale atmosphere–ocean systems our understanding of which is still not clear.

The next two decades will see the most exciting revelations of our knowledge of the ocean and climate, and by the year 2000 our understanding of the monsoons, on which the livelihood of millions of people depends, will become far greater.

Living resources

Capture fishery

The oceans occupy 70.8 per cent of the earth's surface with an average depth of 3,730 m. The zone of maximum importance to mankind today for the exploitation of natural living resources is the uppermost layer of 100 m. This is the zone where most of the photosynthetic production of organic matter occurs. It forms about 1.8 per cent of the world oceans. From this zone, more than 50 per cent of the world's fish catch is obtained at present. The regions occupying this zone are either fairly close to the coast or in very fertile areas of the coastal or offshore upwelling regions called 'oases' in the sea. Since such areas constitute only 25 per cent of the total oceanic area, it can be assumed that 75 per cent of the ocean area may be termed as oligotrophic with moderate to low production rates.

In India, marine fish production consists largely of capture fisheries and for these the intensively exploited areas are found in the narrow coastal belt. In 1947, fish production in India was about 0.4 million tonnes. In 1979/80 it had increased to 1.4 million tonnes. In 1970, it passed the million mark and thereafter its increase has been somewhat unsteady. In certain years there was a decrease in the total catch from the previous year and hence some fishery biologists have referred to the period 1973–80 as stagnating years of Indian fisheries.

The estimates of potential fish yield from the Indian Ocean vary from about 7 million to 17 million tonnes. Of this potential, India's contribution is expected to be of the order of 5–9 million tonnes. Thus a four- to sixfold increase over the existing production is envisaged.

From the present stage of our knowledge about the potential and sustainable yield, the fish production from the Indian Exclusive Economic Zone (EEZ) is expected to be about 3 million tonnes. The projected fish requirements of India's population by the year 2000 are estimated to be 11.4 million tonnes. Of this, 60–75 per cent is expected to come from the sea and the rest from fresh-water sources. The increase in fish production during the last three decades has been attributed to the increase in mechanized fishing vessels, of which there are now more than 15,000 in the country. However, the operating costs of these vessels has increased so much that unless a substantial quantity of high-quality fish is caught with 20 per cent or more of the catch containing prawns, fishing becomes uneconomical. It is, therefore, very unlikely that the expected targets for the year 2000 will be met from mechanized fishing alone.

Mariculture

Another important and productive sector, namely mariculture, is not properly organized in India. It is being practised on a small scale in the enclosed backwater and estuarine areas of Kerala, Karnataka and West Bengal. The culture is largely based on traditional methods of trapping the juveniles of prawns and fishes brought in by the tidal currents into the enclosed areas provided with sluice gates, where they are allowed to grow from three to nine months before harvesting. The total production of fish and prawns from aquaculture practices is about 10,000 tonnes. Table 3 shows the brackish-water areas under cultivation in four states of India, and the total areas available for cultivation in nine maritime states and the existing annual production from mariculture. It is clear from Table 3 that only a small percentage of the total area is being cultivated at present. The potential in this sector appears to be most promising and by the year 2000, the annual production of 4 million tonnes (2.5 from coastal and 1.5 from fresh-water areas) appears to be within reach. World aquaculture production is expected to reach 70 million tonnes by the year 2000 and this is supposed to come largely from the developing countries. Intensive efforts are being made in Asian and South-East Asian countries to attain a substantial increase in fish production by the year 1990 so as to achieve the targets laid down by the turn of the century.

TABLE 3. Total brackish-water areas available in different maritime states of India together with the areas under cultivation and annual production

State	Potential area available for culture (hectares)	Area currently under culture (hectares)	Annual production (tonnes)
Gujarat	376 000	88	Not available
Maharashtra	81 000	—	
Goa	19 000	—	
Karnataka	8 000	4 800	1 000
Kerala	243 000	5 700	6 000[1]
Tamilnadu	80 000	—	
Andhra Pradesh	200 000	—	
Orissa	299 000	—	
West Bengal	405 000	20 000	3 000
	1 711 000	30 588	10 000

1. Includes paddy-fields.

Seaweeds

Seaweeds are one of the important living resources exploited by man for food, animal feed, fertilizers and for chemicals and pharmaceutical products. The total marine algal yield of the world has been estimated as 172,000 tonnes per year. Of this, India contributes only 1.09 per cent of the total (Table 4). The demand for agarophytes and alginophytes by industry in India and

abroad is increasing very rapidly. Unfortunately, India has not yet fully
utilized its seaweed resources (Table 5). Seaweed-based industries in India

TABLE 4. Annual yield of marine algae from different countries as a percentage
of total world production

Country	Annual yield (%)
Japan	62.26
Republic of Korea	10.63
Norway	9.27
Canada	4.37
Mexico	3.52
Argentina	3.09
United Kingdom	3.09
India	1.09
France	1.09
Spain	1.08
West African countries	0.45
Madagascar	0.04
United States	0.02
	100.00

TABLE 5. Total seaweed landings in India

Area	Annual yield in tonnes (fresh weight)
Gujarat	7 000
Maharashtra	6 000
Goa	1 000
Karnataka	Negligible
Kerala	Not known
Tamilnadu	5 000
Andhra Pradesh	Not known
Orissa (Chilka Lake)	5
West Bengal	Not known
Andaman and Nicobar Islands	Not known
Lakshadweep Islands	Not known

came into existence only during the last twenty years and, because of the
shortage of natural resources, many industries are confronted with serious

problems. Before the advent of an indigenous seaweed industry, about 200 tonnes of dried seaweed used to be exported from India annually. Shortage of seaweed is faced by most of the industries because of the depletion of natural seaweed resources after harvesting. The natural regeneration in the beds is not fast enough to meet the demand, so that the only way to generate extra resources is by cultivation of seaweed on ropes and wooden frames. Production by cultivation is likely to increase substantially during the next two decades, but intensive efforts combined with research and development are required to achieve the targets.

Mangroves

In India, mangrove ecosystems, as in several other countries, have been severely depleted during the last two decades. In the past they have been treated as unwanted plants and were largely used as a source of timber and charcoal. It is only in recent years that they have been recognized as ecologically vital areas. Mangroves play a very important role in protecting the shoreline from major erosion damage. The ecosystem forms an ideal nursery for juvenile forms of many economically important species such as mullets, sea trout and shrimps. A large percentage of detrital food that supports a variety of young fishes and shrimps is generated from mangroves.

Awareness of conservation issues and the need of protection for mangroves has been developing rapidly in most of the tropical and subtropical regions and by the year 2000, while mangroves will be denuded from most of the shorelines of India because of the population pressure on land, some mangrove forests will be protected as biosphere reserves. It is only in these areas that future studies on the mangrove ecosystem will be undertaken.

Coral reefs

Coral reefs are among the most biologically productive, taxonomically diverse and aesthetically important living communities. While their massive occurrence provides much needed protection for the coastline from waves, their biological productivity yields a multitude of fauna and flora dependent on the coral-reef ecosystem. These communities also form the main attraction for skin diving, underwater photography, sport fishing and shell collecting. They thus provide a vital stimulus to the tourist industry.

Due to population pressure, most of the coral reefs have become extremely vulnerable to pollution and industrial development along the coastline. Hence, unless protection is offered to coral reefs in the future, most of them will shrink in size and will ultimately die. The forecast is that most of the fringing reefs in India along the main coastline will not survive by the year 2000, unless extensive protection is offered to them in the form of coastal marine parks. The only reefs that would probably survive would be on the atolls of Laccadives and on some of the islands of Andaman and Nicobar.

Non-living resources

Fresh water from sea-water

In terms of population growth, the world supply of fresh water is dwindling very rapidly every year, and therefore measures are being undertaken to obtain drinking water at least from all possible sources. There are many areas in India where potable water is in short supply and thus people even resort to drinking saline water, very often containing objectionable chemicals. For example, the presence of fluorine in drinking water causes what is commonly known as fluorosis—a disease leading to painful symptoms of bone deformity. The following desalination technologies are being employed to generate fresh water from sea-water.

Solar stills. These are well suited for small and isolated communities where water is limited and where power is either not available or is in short supply and the transport of large quantities of water from neighbouring places is not practicable. Solar stills are ideal for small coastal villages as they run on a non-expendable energy source. They are simple to construct and their operating and maintenance costs are minimal. A solar still of 5,000 litres per day capacity has been installed in Avnia village in Gujarat where 500 families obtain their drinking water from this source.

Flash distillation. In this process, heated saline water is allowed to flow through a series of chambers which are maintained at different pressures below atmospheric, and progressively decrease towards the end of the series. Saline water thus evaporates in each section of the chamber, the vapour is released and then condensed over a bundle of tubes cooled by circulating sea-water inside them. Distillate of fresh water produced at each stage is gathered either separately or collectively to be used as fresh water.

Electrodialysis. This technique employs ion-selective membranes for the desalination of brackish water. Electrodialysis is more economical for salinities below 5,000 p.p.m. The energy cost of the process is directly proportional to the salinity, thus beyond 5,000 p.p.m., the process is no longer economical.

Reverse osmosis. This is the most widely used desalination technique. In this process, suitable osmotic membranes are used which reject salts and allow the water to pass through. Several plants with capacities of 10,000 to 15,000 litres have been set up in Indian villages to supply potable water to the villagers.

In the future, desalination technology of different types will play a distinct role particularly in India's rural development programme for the supply of potable water. However, it is not certain that desalination technology can produce enough water to meet the demand of a growing population. It can only supplement other technologies but will not provide a substitute.

Drugs from the sea

In India the utilization of marine plants and animals as a raw material for effective and safe drugs and pharmaceuticals is of recent origin. Of the 100 or more organisms that have been screened so far, forty-two have given promising results. The most remarkable feature is the antifertility properties of ten marine organisms. Recently prostaglandins, which play a major role in controlling biological reproduction, have been isolated from the seaweed *Gracilaria* sp. by Australian researchers. Studies in India indicate that all those species of marine algae that exhibit antifertility properties may also contain prostaglandins.

Researches in this field and also on the cultures of marine bacteria, fungi, yeast, etc., will advance considerably during the next two decades for the production of bioactive substances.

Marine chemicals

Of the sixty elements present in sea-water, only six are recovered commercially. These are sodium and chlorine in the form of common salt, magnesium in the form of its compound, bromine, calcium and sulphur in the form of calcium sulphate (gypsum). Owing to its low concentration, the recovery of potassium directly from sea-water is not considered economical. However, it is possible to recover potassium from bitterns (the mother liquor from salt extraction). Efforts are being made to recover many useful elements commercially, namely iodine, uranium and gold from sea-water. So far, owing to the availability of cheaper extraction methods from land deposits, the technology of obtaining some of these valuable elements either from sea-water, sea-brine or seaweed is not economical. Recently, a commercial process for the recovery of uranium from sea-water has been reported. It is almost certain that research in this field will accelerate considerably and a very large number of elements will be recovered from sea-water in commercial quantities by the year 2000.

Placer deposits

Chemically stable minerals are not decomposed by weathering processes, and as the rocks surrounding such minerals become dissolved and disintegrate, the heavy particles settle to the bottom in layers and become continuously enriched as heavy, chemically stable minerals. All such concentrates are called placers. Mineral placers along the seashores, usually known as black sand, occur in many localities along the Indian coast. Deposits on the west coast are largely concentrated as high-grade beach and low-grade dune deposits, extending from Kanyakumari to the Maharashtra coast with interruptions in between. These deposits mainly contain ilmenite, rutile, zircon and monazite with varying proportions of magnetite and garnet.

Mineral placers are also reported on the coastal tracks of many countries in the world. However, economically exploitable deposits are known only from the beaches and the shelf of Australia and India (rutile, zircon), India and Brazil (monazite and ilmenite), Norway and Japan (magnetite), Malaysia (wolframite and cassiterite), South Africa (chromite and precious diamonds), Alaska, Canada, USSR and South Africa (native gold and platinum), the USSR and United States (ilmenite and rutile). Apart from these major occurrences of placer deposits, minor deposits have also been found off Mozambique, Senegal, Indonesia, the Korean peninsula, Thailand and Sri Lanka. As time passes, their importance and value will go on increasing and by the turn of the century they will probably be exploited extensively.

Offshore mining

Mining from the sea is either by tunnelling, pumping or dredging. Mining companies are extracting coal from the sea beneath the shelf in several countries by tunnelling under the sea for more than five kilometres. Iron ore is also extracted by tunnelling. However, undersea tunnelling for minerals is expensive and sometimes hazardous. Special engineering problems of undersea tunnelling add to the extraction costs, making competition with land producers totally uneconomical. Pumping of many minerals from the sea bed is also an increasing activity, as in the case of sulphur and potash in many countries. Thus, many offshore and placer deposits are exploited by pumping the material from the sea bed to the shore. Dredging in the sea is becoming increasingly important for mining of heavy minerals such as aragonite and biogenic material (corals, shells, etc.).

By the turn of the century, the demand for almost all minerals will increase two- or threefold. Thus their production from land sources alone will be difficult to meet the rising demand. In India, efforts are underway to collect all possible information on the nature of marine mineral resources to work out the most efficient extraction method of both placer and offshore deposits so that the withdrawal of minerals from the sea bed can become economical. It is to be hoped that environmental protection will not be ignored in the search for profitability.

Deep-sea mining

The past two decades have witnessed the development of technology to mine polymetallic nodules from a depth range of 4,000–5,000 metres and to extract the economically important metals from them. They have also witnessed the development in the Third United Nations Conference on the Law of the Sea, of the legal and institutional framework in which the exploitation of the nodules will take place. Who will exploit these resources, how will this be done and who will benefit from this exploitation? The crucial factor in the development of resources such as polymetallic nodules is the economics of their recovery. Due to uncertainties in the scientific explanation regarding the nature of the resources, their relative inaccessibility and the lack of adequate technologies, even the most basic questions about the magnitude and distribution of polymetallic nodules are still imperfectly answered.

It was only in the early 1960s that the nodules were recognized as the largest resource on the deep-sea bed. They exhibit varied physical and chemical properties and occur in different sizes (from 0.5 to 10 centimetres in diameter, though most of them are in the size range 3–4 centimetres), and they are essentially porous. Generally they have been described as dark potato-shaped lumps. Their chemical composition also varies widely, as is evident from Table 6.

The information on their occurrence and distribution is largely derived by two methods—sediment samples and deep-sea photographs. From these two methods, their percentage occurrence seems to vary from 3 to 18 per cent (Table 7). The most promising area in the world for commercial exploitation is the Clarion-Clipperton Zone (7–15° N. and 115–160° W.) in the North Pacific, where their occurrence is as high as 64 per cent. The total areas covered by polymetallic nodules in the world oceans is given in Table 8. From the Indian Ocean about 120 published and 700 unpublished analyses are available. These data do not cover the potential mining sites.

It is now accepted that the first-generation polymetallic-nodule sites should have the following characteristics: an area of about 30,000 km²; nodule abundance 5 kg/m², preferably more than 10 kg/m²; nodules to contain 2.47 per cent nickel, copper and cobalt; reserves should be about 60 million tonnes.

The sea bed at the site should not have large topographic variations and should have good weather for 200–250 days in a year.

Based on the analysis of all the available data, two target areas in the Indian Ocean have been identified: (a) para-marginal area (2.47 per cent Ni+Cu+Co), Central Indian Basin; (b) sub-marginal area (1.63–2.47 per cent Ni+Cu+Co), Wharton Basin, south-western Australian Basin, Seychelles–Somali Basin, west of Laccadives.

TABLE 6. Metal content of polymetallic nodules (percentage weight)

Sample No.	Mn	Fe	Co	Ni	Cn
I	9.50	17.00	0.16	0.13	0.07
2	25.60	4.10	0.16	1.57	1.28
3	18.63	19.64	0.48	0.31	0.13
4	42.30	2.47	0.17	0.26	0.15

TABLE 7. Percentage occurrence of polymetallic nodules
in four different regions of the world oceans

Area	Total stations	Occurrence of nodules	%
North Pacific Ocean (0–50° N., 90° W.–110° E.)	16 546	3 004	18
South Pacific Ocean (0–70° S., 70° W.–110° E.)	9 077	1 118	12
Indian Ocean (70° S.–20° N., 110–30° E.)	5 406	486	9
Atlantic Ocean (70° S.–70° N., 30° E.–70° W.)	19 178	654	3
All oceans	50 207	5 262	10

TABLE 8. Areas covered by nodules in the three oceans (million km^2)

Oceans	Area covered	Prime areas
Pacific	23	5.20
Atlantic	8	0.85
Indian	15	0.50

India has launched a massive exploration programme to identify the mining sites. A systematic and detailed study has been initiated and extensive nodule samples have been collected from different locations for analysis. Intensive chemical analysis of the nodules is in progress. For the selection of candidate sites, exploration at a spacing of fifty kilometres has been completed. However, to demarcate the site with absolute certainty, exploration will have to be carried out at a spacing of ten kilometres which will be further narrowed down to one or two kilometres in specific areas. It is estimated that

nearly one-eighth of the Central Indian Basin may have
demarcate the candidate sites for mining.

On 30 April 1982, the Third United Nations Conferenc
Sea adopted a Convention and a Resolution on preparat
pioneer activities relating to polymetallic nodules. According to this resolution, four countries, namely France, India, Japan and the USSR, and four
multinational consortia (Kennecott Group with five mining companies;
Ocean Mining Associates with three companies; Ocean Management Inc.
with four companies and Ocean Mineral Company with five companies) were
recognized as 'pioneer investors'. Each pioneer investor by definition is
expected to sign the convention stating that prior to 1 January 1983, the
investor has expended an amount no less than $30 million in pioneering
activities of deep-sea mineral extraction and has expended no less than
10 per cent of that amount in the location, survey and evaluation of a specific
mining site. This site should be large enough to be later divided into two
pioneer areas of equal estimated commercial value. The size of one pioneer
area should not exceed 150,000 km^2.

This development will accelerate the preparatory work connected with
deep-sea mining. However, the technology for carrying out deep-sea mining
is today only available in a very few countries. No country or any mining
company has yet demonstrated a properly evaluated and economically viable
system for the mining of polymetallic nodules. However, it is expected that
by 1990, commercial exploitation of polymetallic nodules will start in
several countries including India.

Ocean engineering

Engineering tasks associated with the ocean are many times more expensive
than similar activities on land. Moreover, the hazards of work are greater and
the management of operations more demanding. The farther a structure
lies from shore or the greater the distance from the sources of supply, the
more prodigious will be the cost of any marine operation. Special problems
pertaining to ocean engineering include designs and constructions to withstand very high pressure and hydro-elastic forces due to flow and wave action
especially on slender structures, estimation of wave and current forces in a
particular location, transport of polyphase mixtures via submarine pipes,
design and analysis of submarine pipelines and offshore structures, pipe-
laying techniques under different sea conditions, control of underwater
machinery, underwater surveying, oceanographic measurements, analysis of
data, etc. The problems can broadly be divided as follows.

Coastal zone management

The vast coastal zone of India is being utilized for the development of ports and harbours, fisheries, beach resorts, land reclamation, location of shore-based industrial complexes, human settlement, agriculture, disposal of wastes, etc. The coastal states of India are presently confronted with the problems of coastal erosion. Pollution in estuaries and nearshore waters is becoming a serious problem requiring scientific solution. Due to the increase in population and industrialization and steadily increasing tourist traffic, there is an ever growing demand for more and more recreational beaches, tourist resorts and scenic spots along India's shoreline.

Moreover, the increase in offshore activities for oil and gas extraction, fisheries, mining of sea-bed minerals, extraction of renewable energy from the sea, shipping and marine transportation, etc., poses special problems in the coastal zone since it would be necessary to provide special onshore facilities in harmony with the natural environment in order to cope with the new demands on the coastal zone.

Thus the coastal zone is subjected to multiple use leading thereby to conflicting demands for the exploitation of the various coastal resources by different interest groups and user agencies. It has therefore become very vulnerable to the destructive forces caused by pollution and several other man-made changes. There is consequently a need for a comprehensive national policy and guidelines with requisite enforcing powers for managing the various coastal developmental activities in India.

Ports and harbours

Increase in the volume of seaborne traffic in petroleum and petroleum products, mineral ore, coal, fertilizers, food grain, etc., has necessitated the development of new harbours and the expansion of the existing harbour and port facilities in the country. Sea trade has become very competitive in the world. It is, therefore, necessary for India to reduce the unit cost of transportation by introducing large ships, cargo containerization and adopting new methods of handling cargo at high speed and in large bulk. In almost all harbours there is a need for regular dredging to maintain required depths at navigable approaches. Congestion is becoming a serious problem in India, particularly at large ports. It has, therefore, become necessary to divert ships to uncongested ports and develop others to receive the increased traffic.

Control of coastal erosion

Various parts of the east and west coasts of India are confronted with severe erosion problems. Erosion is caused by the interaction of natural forces, which, very often, are aggravated by man-made changes along the shoreline.

The most important factors affecting the stability of a beach or a shoreline are waves, tides, currents, beach-material characteristics, geomorphology and man-made changes. Engineering measures designed to control erosion and stabilize the eroding shore can be put under two general categories. First, the construction of structures such as seawalls, revetments, groynes, breakwaters and jetties to reduce or prevent the wave energy from reaching the erodible material on the shore. Second, the artificial nourishment of an eroding shoreline to make up the deficiency in the supply of sand with or without structures like groynes, etc., to reduce the rate of the loss of entrapped material.

Control of coastal pollution

Due to the increase in population and industrial activities, the estuaries and nearshore waters are being polluted as a result of the disposal of waste materials. Problems of thermal and oil pollution are also becoming very common in many parts of the world.

To develop suitable solutions to pollution problems, investigations are to be undertaken to study the diffusion and dispersion characteristics of coastal waters in which waste disposal is planned. To deal with oil-pollution problems, research and development on chemicals and oil-cleaning equipment will have to be intensified. Suitable oil booms and skimmers are to be developed or obtained.

Production of offshore oil and gas

India has mounted a major effort to increase its oil exploration and development capabilities. Over the next few years this effort is likely to increase rapidly. Though the present activities related to offshore production of oil and gas are confined to Bombay High areas, the exploration activities are intensively going on both on the east and west coasts of India. To support the efforts in offshore exploration it would be necessary for India to develop various support services indigenously. A large number of supply vessels and other multi-purpose support and inspection vessels, helicopters, crew boats, etc., are required. A large number of drilling and process platforms, including submarine pipelines, have already been installed and more of them will be constructed in the future.

Extraction of energy from the sea

It is well known that the vast seas around India have great potential for renewable sources of energy available in the form of ocean thermal energy, waves, tides and salinity gradients. The time may not be far off when the extraction of energy from these sources may become economical because of the increasing cost of power generation from the depletable sources of energy such as coal, oil and gas.

A major research and developmental effort has to be undertaken in the country in order to develop the necessary capabilities in this frontier area. Considerable research and developmental efforts are required in connection with data collection, analysis, environmental modeling, selection of materials, design of various components, fabrication, sea trials, environmental impact studies, etc.

Conclusion

India's future oceanographic programmes are being carefully planned to suit many important and urgent needs. During the next twenty years, India will have to make a much greater effort towards accomplishing each of the above plans of work so that a new era of exploration and exploitation of our vast ocean resources for the economic and social development of the country is opened.

Our growing need for the resources from the sea demands increased knowledge of ocean characteristics and predictions of its future state. A wide range of environmental services including various types of ocean industries are required in India in the next twenty years to support increasing marine activities: (a) forecasting the weather, storms, sea state, and storm surge; (b) surveying and charting coastal waters and continental shelves; (c) keeping archives of marine data; (d) conducting geophysical investigations; and (e) developing the necessary expertise and capabilities for solving the various ocean-related problems. The new age we are entering is not only going to be the age of the atom, electrons and space, it is also going to be the age of the sea.   ■

Note

1. The official United Nations population estimate for India as of 1 July 1980, is 663,576,000, comprising 343,328,000 males and 320,248,000 females; both the author and the United Nations agree that the population comprises a majority of 51.7 per cent of males—Ed.

In this chapter, the author summarizes the geological history of the Baltic Sea and the development of pollution problems arising from local industrial and agricultural practices. He goes on to review the early creation of organizations from many countries whose primary concern was the protection of the marine environment in the Baltic; he concludes with a comprehensive survey of current research, scientific surveys of current research, sponsored surveys, and international agreements sponsored and monitored by international organizations.

Chapter 15

The Baltic Sea — pollution problems and natural environmental changes

Klaus Voigt

Professor Voigt has been director of the Marine Research Institute in Rostock-Warnemünde (German Democratic Republic) since 1965. A corresponding Member of the Academy of Sciences of the German Democratic Republic, he has worked as a seagoing oceanographer since 1955, on the Atlantic equatorial system and in the Baltic. In 1968 he was elected secretary and, from 1972 to 1976, vice-president of the Scientific Committee on Oceanic Research. He served the Inter-governmental Oceanographic Commission in its Long-term and Expanded Programme of Oceanic Research (LEPOR), from 1976 to 1980. In 1982 the IOC General Assembly unaminously elected Professor Voigt as Vice-Chairman (Ocean Science). His address is: Direktor, Institute für Meereskunde der Akademie der Wissenschaften der DDR, 2530 Rostock-Warnemünde (German Democratic Republic).

Evolutionary history

The Baltic Sea (471,000 km²) is one of the largest brackish-water areas in the world. Extremely shallow (mean depth 55 m, maximum 459 m) and geologically very young (formed 12,000 years ago), this semi-enclosed body of water (22,000 km³) provides an ideal experimental test-bed for the proper understanding of marine environmental changes under the influence of man and nature.

The evolution of the Baltic Sea went through several marine and lacustrine phases after the retreat of the last continental ice sheet from Europe (see Fig. 1). The rise in world sea-levels and the present recorded uplift (up to 10 mm/yr in the north) or submergence (1–2 mm/yr in the transition to oceanic areas) of tectonic plates in the Baltic region have greatly influenced marine environmental features. Varying amounts of river runoff (440–480 km³ or about 2 per cent of the volume of the Baltic Sea), precipitation input from the atmosphere, releases of phosphorus from the sea bed (induced by changing sea-bed chemistry) and flow between the open oceanic areas and the Baltic all contribute to the natural variability of the marine environment in the Baltic Sea.

Pollution

Pollution problems in the Baltic stem from sewage discharges from 17.5 million people and from assorted industrial wastes, such as pulp and paper effluents, chemical, steel and metal industries. There are inputs from diffuse land-based sources such as agriculture and airborne pollution from a much wider region than the immediate drainage area. Deepening of shipping lanes, river regulations, offshore dredging of sand and gravel, shipborne pollution and forthcoming offshore oil/gas drilling are additional activities, which though regulated and controlled by established water and environmental management services, constitute a risk to the human and marine environment. Man-made deterioration of the beach zone during one summer season can cause as much damage as a storm surge. The occasional occurrence of sea-ice enhances natural hazards to ships and platforms and also diminishes exposure of effluents or spills to effective mixing, diffusion or solution processes.

The Baltic Sea is connected with the North Sea and its general pollution load and climatic change. This is documented by harmless but detectable gradients of artificial radionuclides, in particular of caesium-137, which stem from the discharge of reprocessing plants and reach the Baltic Sea on marine pathways in two to five years. With an average residence time for the water in the Baltic at about thirty-five to forty years, the global fall-out from

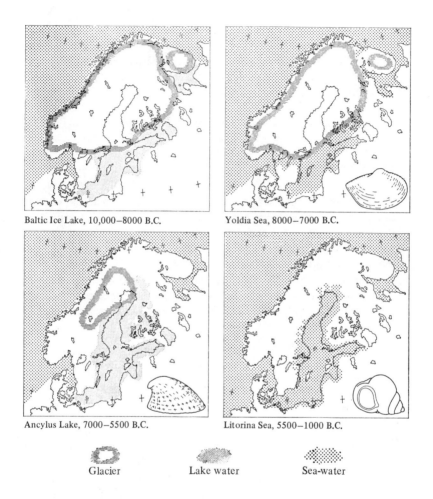

FIG. 1. Postglacial geological and lacustrine (lake) development of the Baltic Sea. Some corresponding sea snails are shown. (After M. Sauramo and *The Gulf of Finland—A Joint Concern*, p. 4, Helsinki, Finnish–Soviet Commission for Scientific–Technical Co-operation, 1981.) (Publication No. 10.)

nuclear weapon tests in the atmosphere in the 1950s and early 1960s is still detectable.

Survey, research and legislation

Endeavours to understand this area and its resources including man's impact by means of observations, fishery statistics and appropriate surveys date back to the last two decades of the nineteenth century. At the Sixth International Geographical Congress held in London in 1895 Professor Otto Petterson of Sweden put forward a scheme for international marine science co-operation with special emphasis on the Baltic. The resolution of the Congress recognized

The scientific and economic importance of the results of recent research in the Baltic, the North Sea and in the North Atlantic especially with regard to fishing interests, and [the Congress recorded] its opinion that the survey of the areas should be continued and extended by the co-operation of the different nationalities.

The statement that the physical, chemical, biological and geological processes governing phenomena occurring in the Baltic Sea can only be identified as a result of continuous, large-scale and long-term co-ordinated observation programmes of the highest scientific standard is confirmed today.

First in the field was the International Council for the Study (then Exploration) of the Sea (ICES) founded in 1902, followed by the 1957 Conference of Baltic Oceanographers (CBO-I) and more recently the Baltic Monitoring Programme (BMP)[1] of the Baltic Marine Environment Protection

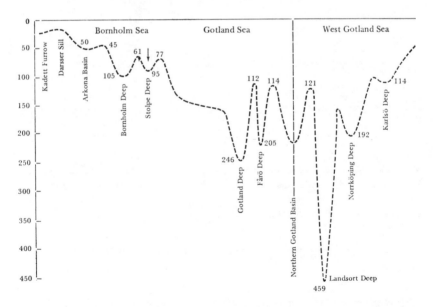

FIG. 2. Schematic display of the bottom profile along a longitudinal section in the Baltic Sea. (After Melvasalo et al. (eds.), *Assessment of the Effects of Pollution on the Natural Resources of the Baltic Sea*, Helsinki, 1981.)

Commission (HELCOM)—which came into force in May 1980—which adjusted the scheme of agreed marine studies from the approaches to the Atlantic into the fjord-like Baltic Sea (see Fig. 2).

However, observations of marine climatic changes in the Baltic had already been undertaken since the twelfth and thirteenth centuries. In the chronicles of their cities on the Baltic Sea and upstream of the rivers, the guilds of merchants made recordings of sea-ice, storm surges, fish (in particular herring catches), all of which had considerable impact on the fleet-based commerce between the medieval towns and caused rivalries in the feudal societies developing in central and northern Europe at this time. In the middle of the nineteenth century, man in the Baltic could still fall mortal victim to the failure of crops, fishing and trade because of natural climatic changes.[2]

The Second World War inflicted great damage to both the environment and to human populations in the Baltic (the worst example of the latter being Leningrad where the people suffered death *en masse*). The most significant threats to the environment were those due to oil spills from sunken vessels, extended mine operations, bunker building along beaches and the random disposal of ammunition and chemicals in the Baltic basin.

The beginning of international accord

The 1970s brought into being the Convention on Fishing and Conservation of Living Resources in the Baltic and the Belts (Gdansk Convention), signed on 13 September 1973, with its secretariat in Warsaw, Poland, and the already mentioned Convention on the Protection of the Marine Environment of the Baltic Sea Area (Helsinki Convention) signed on 22 March 1974 with the secretariat in Helsinki, Finland (see page 419). Both governmental agreements for the Baltic Sea consider that a rational protection and proper management and exploitation of its resources should rest as far as possible on scientific inquiry and agree that international co-operation is the optimal way of achieving satisfactory results in this direction. A motto already laid down in the preamble to the rather precise plans of the hydrographical and biological programme adopted at the second international conference in May 1901 to prepare for ICES, held in Oslo under the presidency of Nansen, with Hjort as secretary-general, Knipovich, Knudsen, Heincke, D'Arcy Thompson, Garstang and others.

The Helsinki Convention was also the first regional treaty to cover land-based pollution sources and, in particular, to combat marine pollution from oil in the whole drainage basin which is 4.3 times as large as the area of the Baltic Sea itself. Certainly the inspiration to initiate conventions for similar sized semi-closed marine areas like the Red Sea, the Persian Gulf, the

Mediterranean, and others, was drawn from the success of the intergovernmental action in protecting the Baltic Sea. This success was partly due to the well-developed organizational pattern of Baltic marine science[3] with almost twenty different scientific or administrative bodies dealing with the Baltic Sea on a regular basis. Such a pattern encourages new ideas and project proposals. There are also opportunities for young marine scientists and new groups and schools to influence the course of events in marine scientific co-operation.

These two conventions and their contracting parties request scientific advice and guidance, and facilitate the implementation of advanced projects requiring the consent of bordering countries.

Based on this complex advisory machinery, HELCOM and ICES presented an extensive assessment[4] of the effects of pollution on the natural resources of the Baltic Sea, with co-ordinated contributions from twenty-eight principal authors from seven countries.

Assessing the effects of pollution

The report states that large variations, including long-term changes in oxygen and nutrient concentrations, have occurred during the present century, and at present it is not really possible to determine how much man's influence has contributed to these changes.

As high-quality data have been available since the 1880s, an increase in temperature and salinity in the deep water of the Baltic can be detected, from 0.6 to 2.7 °C and 0.8 to 1.7 parts per thousand. The depth of the primary halocline has been displaced slightly upwards by about 5 to 10 m in the Gotland basin. Marine species are now found further north in the Baltic.

The deep water of the Baltic proper is periodically renewed by short-term inflows of water with a sufficiently high density to force out old stagnant bottom water. During the period between such strong inflows from the transition area to the North Sea (which may last several years), the oxygen is continuously consumed by the oxidation of natural organic material in the surface sediments and organic material sinking from above, as well as by respiration by organisms. If the period between bottom-water renewals is long enough, all the dissolved oxygen will be consumed and anaerobic conditions will ensue, with the subsequent formation of hydrogen sulphide. This in turn is accompanied by a deterioration of the benthic community and even the disappearance of benthic fauna if anaerobic conditions develop and prevail. Anaerobic conditions also cause, among other things, a rapid release of bound phosphorus from the sediment.

This assessment, written in 1980 when, after a long period of stagnation, hydrogen sulphide occurred in most of the Baltic basins, proved to be correct.

Two years later, in November 1982, the current metering station at the Darsser Sill (established by the Institute for Marine Research in Rostock-Warnemünde, German Democratic Republic) signalled the next climatic change: an irregular intrusion of water with higher salinity, temperature and oxygen. This was confirmed by the regular observation programme of research vessels during the following months. From these data, it is predicted that the HS_2 content and oxygen-depleted layers in most of the Baltic basins are going to diminish considerably or disappear entirely in the near future.

Integration of sampling and analysis

The HELCOM report also recorded declines in DDT, PCBs and mercury in Baltic biota during the last few years following restrictions on the use and release of these substances. Trace-metal levels in sediments, on the other hand, have exhibited an increase during the last century and there seems to be no sign of their diminishing. All these phenomena must be taken into account when comparing the results of sediment samples from different parts of the Baltic Sea.

HELCOM research activity is subjected to continuous review by a number of groups, who have recommended that efforts should be made to obtain a more realistic measure of the spread of chlorinated hydrocarbons (CHCs) in the Baltic. CHCs are being increasingly applied in agriculture as a replacement for DDT, and it is essential that we should have sufficient data to enable us to predict the effects on the marine environment of such pollutants—which also include commercial forms of polychlorinated terpenes.

The ICES/SCOR Working Group on Pollution of the Baltic Sea is currently initiating a thorough integration of sampling and analysis techniques to achieve comparable results for further assessment. This exercise is based on the collection of relevant samples from representative stations by marine scientists, particularly from Finland and the German Democratic Republic, and their distribution to interested marine scientists worldwide for analysis and comparison. All parties involved in the Baltic co-operation have so far stressed the importance of this approach and that the monitoring of Baltic sediments will not be effective until a thorough integration of results is achieved and a full agreement is reached on which methods shall be used. Without such basic scientific studies it is not possible to assess the fate of pollution by trace metals entering and leaving Baltic waters from their accumulation in recent sediments.

The list of problems in understanding the strongly layered water body of the Baltic and its living and non-living resource potential still looks insurmountable, even after a century of co-operative international research.

Other unresolved issues awaiting further joint studies are the nutrient supply to the Baltic, eutrophication probability and increase in primary production—natural or man-made.

Groups associated with the Helsinki Convention (1974)

The objectives of the Helsinki Convention are implemented by a great number of organizations such as ICES, the Scientific Committee of Oceanic Research of ICSU, the International Hydrological Programme of Unesco, the Nordic Council, the Council for Mutual Economic Assistance and others. In addition, several bi- and multilateral bodies (governmental and non-governmental) have been formed with the sole objective of intensifying co-operation in various Baltic regions or projects of mutual interest. Among these are the Sound Commission (Denmark and Sweden), the Odra Haff Commission (German Democratic Republic and Poland), the Committee for the Gulf of Bothnia (Finland and Sweden) and the Working Group on the Protection of the Gulf of Finland (established in 1968 and associated with the Finnish–USSR Committee for Scientific and Technical Co-operation). A series of Baltic Sea Symposia with subsequent joint field studies has also been organized under the Swedish–USSR Committee on Economic, Scientific and Technical Co-operation.

There are also several academic marine-science groups of multilateral character. The Conference of Baltic Oceanographers (the thirteenth conference took place in 1982)[5] has met since 1957 on a rotating basis, and is reviewing the progress in the physical, chemical and geological sciences of the Baltic Sea at biennial meetings in one of the seven countries bordering on the Baltic—the next to come will take place in Poland in 1984. In between, the Baltic marine biologists have been meeting since 1968; their last session was held in the University of Rostock, German Democratic Republic, in 1981, with the latest meeting taking place this year in Lund, Sweden.

Recommendations of both groups are implemented with the support of the above organizations or through national institutions and joint field exercises, particularly if coastal and estuarine investigations are required. Mention should also be made of an ongoing series of experiments to study hydrodynamical processes[6] supported by the Academies of Sciences in the USSR, Poland and the German Democratic Republic.

Conclusion

In closing this personal review, I am sure that Baltic marine scientists will continue to advise and guide the organizations responsible for the protection of the Baltic Sea and its resources in the best tradition of international marine-science co-operation. O. Krümmel, in 1904, after having analysed (on behalf of ICES) the first co-ordinated international oceanographic exercise in the Baltic sea, wrote:

The fragments of the oceanography of our local seas presented here have shown that many problems still await solution. We hope that our successors will at least acknowledge that we, by the international organization of observations spanning the whole region simultaneously and working with identical methods, have honestly tried to do our part. ∎

Notes

1. *Guidelines for the Baltic Monitoring Programme (BMP) for the First Stage* (updated version), Helsinki, Baltic Marine Environment Protection Commission, June 1982.
2. C. Weikinn, *Quellentexte zur Witterungsgeschichte Europas von der Zeitenwende bis zum Jahre 1850 (Hydrographie)* (4 volumes), Berlin, Akademie-Verlag, 1958–63.
3. B. Dybern, The Organizational Pattern of Baltic Marine Science, *Ambio*, Vol. IX, No. 3/4, 1980, pp. 187–93.
4. T. Melvasalo et al. (eds.). *Assessment of the Effects of Pollution on the Natural Resources of the Baltic Sea*, Baltic Sea Environment Proceedings, Nos. 5A and 5B, Helsinki, 1981.
5. *Proceedings of the XIII Conference of Baltic Oceanographers*, Helsinki, 24–27 August 1982 (2 volumes), 1982.
6. K. Voigt et al., Theoretische und praktische Arbeiten zur Untersuchung der Ausbreitung von Beimengungen in Meer, *Beiträge Meereskunde*, No. 30/31, Berlin, 1972; C. Druet et al., The Interaction of the Sea and the Atmosphere in the Nearshore Zone [of the Baltic Sea], *Proceedings of the EKAM 73, International Field Study in the Baltic Sea Nearshore Zone*, Gdynia, 1975 (Raporty MIR, Seria R, No. 2a); C. Druet, Properties and Transformation of Hydrodynamical Processes in the Coastal Zone of Nontidal Sea, *Results of the International Projet Lubiatowo 74 in the Coastal Zone of the Baltic Sea*, Gdynia, 1976. (Raporty MIR, Seria R, No. 1a.)

To delve more deeply

VOIPIO, A. (ed.) *The Baltic Sea.* Amsterdam/Oxford/New York, Elsevier, 1981. (Oceanography Series No. 30.)

The author introduces us to changing technology in the oceanic medium, explains the climatic and global cycles of the environment including the ocean, describes and specifies our new technological prowess in viewing the planet globally, and finally projects where these tools should lead us as we master understanding of oceanic processes through satellite sensing.

Chapter 16

Satellite measurements of the oceans — a new global view

D. James Baker

Dr Baker is Dean of the College of Ocean and Fishery Sciences and professor of oceanography at the University of Washington, Seattle. He is a member of the Board on Ocean Science and Policy and the Climate Research Committee of the United States National Academy of Sciences as well as of the Earth and Space Sciences Advisory Committee of the National Aeronautics and Space Administration; he has written more than fifty articles on the general circulation of the ocean, the ocean's role in climate variation, and ocean technology. His address is: College of Ocean and Fishery Sciences, HG-35, University of Washington, Seattle, WN 98195 (United States of America).

Introduction: a changing technology for the oceans

When I entered oceanography in the early 1960s, the field was making rapid strides towards describing and predicting the ocean's behaviour, but with a technology and with research ships that had serious limitations. My first expedition was to spend six months in the Indian Ocean helping to measure the currents and winds in the equatorial regions there. The ship we used was the *Argo*, a submarine rescue ship deeded from the United States Navy to the Scripps Institution of Oceanography, and the equipment we had was largely devoid of electronics. Our officers used celestial navigation,[1] dead-reckoning,[2] and local measurements. The limitations in obtaining any kind of global view or photograph of conditions at any one time over a large area of the Indian Ocean were painfully obvious.

Today, our instruments are much more sophisticated largely because of the revolution in microelectronics and computers, but the limitations of measurements from ships are still with us. Measurements of the ocean from satellites promise to liberate us from many of those shipboard limitations. In fact, navigation by satellite, a technique that uses satellites and on-board receivers and computers, is now commonplace for research ships and commercial vessels. Accurate positioning is available in most regions of the world and in any weather conditions every two hours. We are also just beginning to use other satellite data now to get photographs of the behaviour of the ocean—its temperature, the shape of its surface, its waves and its chemical and biological properties.

Climate and global cycles of the environment

The ocean is a subject of interest to the applied scientist for a variety of reasons: the coupled flow of ocean and atmosphere determines the global climate, and the circulation of the ocean affects chemical and biological cycles, pollutants, marine transportation and national defence. The description of the ocean that we can eventually expect to obtain from a mixture of satellites, ships and computers will help us develop understanding of the ocean environment. This description and understanding will help us to elaborate a scientific strategy for managing our global environment or, if not managing in the sense of controlling it, at least understanding the choices that we could make as we affect climate and the global cycles that are important to life on earth.

In the past few years, there has been a growing interest among specialists in many disciplines that the new global views offered by satellites could help us to understand climate and the global chemical cycles. The climate, the productivity of the earth and the quality of the air we breathe and the water that we drink are the basis for the habitability of the earth. An international programme has been proposed called Global Habitability, which would have as its central elements satellite measurements by many nations, and shipboard and land-based measurements and modelling of the ocean, the earth, and the biosphere. The goal of the programme is to understand the climate and the global cycles.

The concept of global habitability arises from the fact that, in the past, increasing energy and food requirements could be met by expansion of frontiers, use of chemicals for fertilizers and pesticides, and by using available reserves of oil and gas. Today, however, humanity is being confronted by the finite dimensions of its world. We are at the point where we affect both regional and global climate. We are a factor in the global cycles of carbon, nitrogen, phosphorus and sulphur because we produce significant amounts of these chemicals and dump them into the air and the sea.

It is essential that we understand how the overall, coupled ocean–atmosphere system works if we are to live successfully with global change. The problem is global, and becomes even more severe when we look to the developing countries as they improve their standards of living. Such improvement invariably brings increased energy use and dissemination of various chemicals into the environment through farming and industry. We have to understand the cycling of energy, waste and essential chemicals through the atmosphere, land and oceans.

Climate, global cycles and the ocean

The oceans regulate the climate and make it more predictable. The slow variations of ocean currents which affect the transfer of heat to the atmosphere, could, in the views of many experts, provide a mechanism for predicting the climatic changes in the atmosphere. But in order to understand these variations of ocean currents, we need to understand the general circulation of the ocean, how the currents are driven, and how and why they change. Another largely unknown factor in the climatic system involves the role of sea-snow and sea-ice. Both of these are quantities that can be measured and monitored by satellite.

The distribution and cycling of nutrients is also important to habitability. The nutrients are related to the biological productivity cycle through the atmosphere, rivers, coastal zones and the deep ocean. The continental shelf is a critical area. Most of the world's fisheries occur here, much of the world's petroleum production may be in these areas in the future, and it is here where most of man's pollution enters the ocean. The shelf is a direct source of food for humanity, and the pressures on this food supply are increasing rapidly. We must know what is happening here, and we must be able to understand the mechanisms that control the transport of these nutrients from terrestrial sources to oceanic sinks. To do this, we must be able to understand the interactions between physical and biological processes in the oceans. Large-scale satellite views of nutrients in the ocean are necessary for this understanding.

Changes in the amount and timing of global rainfall are critical for farmers. The green revolution has shown dramatically how genetic engineering can provide farmers with exactly the right kind of crops for a maximum of productivity if the water resources are known. Thus rainfall is critical, but we must know it globally. The problem is not yet solved at all; in fact, the natural rainfall patterns over the globe are only marginally predictable by the use of climate models now available. Global rainfall is another parameter that we can measure by satellite.

Technology for global views

For the oceanographer, ships are essential. There is no other way to get samples of water, chemicals or biological organisms from the sea. But ships alone cannot give us an overall view of the ocean. A typical research ship travels at ten knots, or about five metres per second. At that speed it takes about ten days to make a single crossing of the North Atlantic; it can take months to complete a careful survey of any large region, and in that time, the state of the ocean can change drastically.

From ships, and from the relatively few buoys that are moored in the ocean for weather measurements and for research purposes, we get an average view of the ocean's properties (for example, its temperature and salinity) and of near surface weather. This average view is important, but we have to recognize that the ocean is continually changing. It does not change as fast as the atmosphere, but it does behave in a similar and turbulent way. The atmosphere changes its weather in a few days and its climate varies from year to year. The ocean's 'weather', that is, the currents and temperatures, change over periods of a few weeks instead of a few days. The ocean's 'climate' also changes from year to year.

The anomalous weather of the winter of 1982–83 around the world is a good example of year-to-year climate changes. In early 1983 the west coast of the United States suffered severe rainfall, much in excess of the average. The 1982–83 winter is also a good example of how the ocean works together with the atmosphere to cause climatic variation. The changes observed are believed to be attributable to a large pool of warm water appearing off Peru late in 1982 (the so-called 'El Niño'), which, in turn, is believed to be caused by changes in the trade winds in the Western Pacific Ocean.

But how do we get an overall view of the ocean? Our instruments have to be placed in a position away from the earth in order to see it all at once. Satellites are the only way. Typically satellites receive their power from the

MEAN SURFACE TEMPERATURE FOR JANUARY 1979

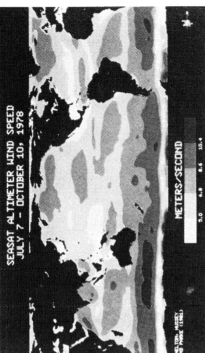

SEASAT SMMR WATER VAPOR
JULY 7 – OCTOBER 10, 1978

GRAMS/CENTIMETER²

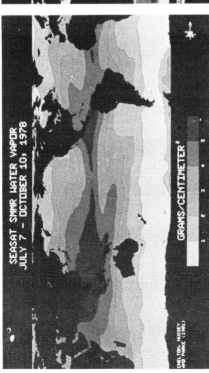

SEASAT ALTIMETER WIND SPEED
JULY 7 – OCTOBER 10, 1978

METERS/SECOND

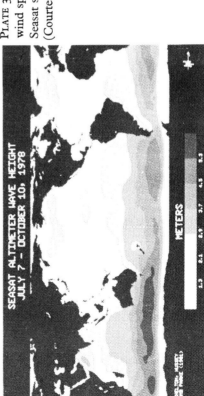

SEASAT ALTIMETER WAVE HEIGHT
JULY 7 – OCTOBER 10, 1978

METERS

PLATE 3. Global measurements of water vapour, wind speed and wave height, made by Seasat satellite during the period 7 July to 10 October 1978. (Courtesy of Dr D. Chelton, NASA/Jet Propulsion Laboratory.)

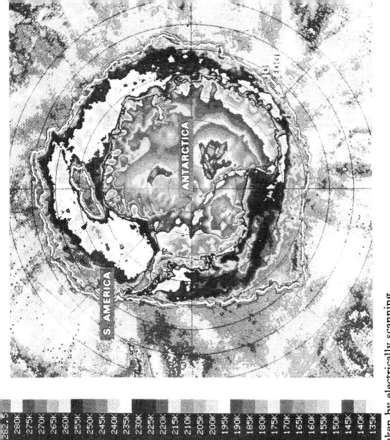

PLATE 4. Wintertime cover of sea ice, taken by electrically scanning microwave radiometer aboard Nimbus 5 satellite (8 March to 16 September 1974). (Courtesy of NASA, Washington.)

sun, and can be classified as either passive or active. The passive satellites receive the radiation that comes from the earth and attempt to interpret it; the active satellites illuminate the sea with specified radiation and observe the reflected pulses. We now have satellites that can measure the shape of the ocean's surface, the surface temperature of the ocean, the wind speed near the surface, the extent of polar ice, the surface chlorophyll, and a variety of properties in the atmosphere.

From these instruments we get a global view: either an image of almost half the earth at one time from a satellite in geostationary orbit, or from a series of tracks of satellites closer to the earth that repeat their orbits about every ninety minutes. Such a 'low earth orbiter' also gives coverage of large areas of the earth in less than a day. From these satellite measurements, we have a new and remarkable series of pictures of the ocean and the land. This new imagery has dramatically changed the way in which we view the ocean, allowing us to plan for new research programmes. We now see the possibility of describing and understanding some of the important cycles of climate, of nutrients and of other environmental factors that affect global habitability.

Some particular and spectacular examples

The data that have been collected since the early 1960s, when satellites were first available, are enormous. But it is only in recent years that satellite pictures that can be used for real scientific understanding of the ocean have become available. I have included in this article four examples of these pictures, because the images themselves are the best demonstration of the power of this new technique. These pictures have all been made available through the courtesy of the United States National Aeronautics and Space Administration (NASA).

Plate 1 shows the first global map ever put together of the surface temperature of the earth. The temperatures shown on the map correspond to those of January 1979. Since the cold polar regions cover only a small part of the globe relative to the warm equatorial regions, the mean surface temperature of the earth is clearly dominated by the temperatures in the tropics. The map shows several cold regions such as Siberia and northern Canada, and a hot Australian continent during the southern hemisphere's summer. The effect of the Gulf Stream is clearly seen off the eastern coast of the United States: the temperatures there are cooler than at comparable latitudes on the eastern side of the Atlantic Ocean. Maps such as these are used to establish the basic parameters of global air/sea interaction for understanding climate.

Plate 2 shows a comparison of sea surface temperature measurements and surface chlorophyll measurements made by satellite; it is one of the most spectacular pictures made of the ocean because of what it reveals in the rich and complex structure of these fields. The upper figure shows the variations in chlorophyll concentration in the north-west Atlantic Ocean that were derived from a Nimbus 7 satellite on 14 June 1979. Lower productivity is shown in areas like the Gulf Stream (in the lower portion of the picture), whereas higher productivity is seen in areas approaching the Grand Banks of Newfoundland and the continental shelf. The chlorophyll patchiness and wavelike patterns along the Gulf Stream show clearly the difficulties that researchers have had in attempting to measure biological productivity from ships. Note the eddy or ring south of Cape Cod (A) which became separated when a meander of the Gulf Stream was pinched off. The ring contains low-chlorophyll Gulf Stream water.

The lower part of Plate 2 shows the variations in sea surface temperature that correspond to the chlorophyll patterns in the upper figure. The water over the very productive Grand Banks region is cold and nutrient-rich, having been mixed by tides upwards from depths of 70 to 100 metres in the Gulf of Maine. This tidal mixing is primarily responsible for the sustained high productivity of this important fisheries region.

Winds, waves, ice and snow

Plate 3 illustrates global measurements of water vapour, wind speed, and wave height made by the SEASAT satellite between 7 July and 10 October 1978. The water vapour in the atmosphere affects the earth's surface radiation balance and, together with wind speed, is important for studies of climate. As expected, most water vapour is found in the tropics, where evaporation is high. The wind speed is measured by inference from the shape of the waves, which in turn is measured by the altimeter. The modern altimeter is a radar device that sends down a pulse from the satellite, which is reflected back to the satellite. The time it takes for the pulse to go from satellite to ocean and back again tells how far the sea surface is from the satellite, and hence the shape of the surface under the satellite track. From the surface's shape, the waves and the general topography of the surface can be determined.

The wind speed and wave height show the strongest intensity in the southern oceans, particularly south of Australia. These measurements illustrate the importance of satellite measurements for such data, since this is a region generally devoid of ships to be used for direct measurements. For the first time, we now have measurements of wind and waves in this region.

The altimeter can measure more than waves, since variations in the sea's surface height are also caused by ocean currents. For example, the Gulf Stream has a variation of height across its width of about one metre. Such variations can be measured by the satellite altimeter, and promise to provide global pictures of ocean circulation as well as waves. This is one of the most exciting possibilities of the new satellite data.

Plate 4 shows how the changing patterns in the world's cover of sea-ice can be readily monitored by using satellites. Here data from passive microwave emissions show the difference between open ocean and sea-ice. Microwave imagery has the capability of permitting global, all-weather, day or night observations of sea-ice. The images in Plate 4 show the extent of sea-ice at close to the annual maximum. In the northern hemisphere, the ice cover extends well out into the Bering Sea, the Sea of Okhotsk, and Baffin Bay, and along the east coasts of Greenland and Newfoundland. In the southern hemisphere, the winter ice surrounds the entire Antarctic continent. A large *polynya* or open ice region can be seen in the Weddell Sea near the top of the picture. In both hemispheres, significant retreat of the ice occurs in spring and summer, although in the northern hemisphere much of the central Arctic remains ice-covered throughout the year.

The variations in sea-ice and snow cover are essential ingredients in the variation of climate, because they can cause large changes in reflected radiation. If the ice cover grows, more radiation is reflected. If it diminishes, then less radiation is reflected. In either case, the climate is affected, and it is clearly essential to monitor how the ice cover is changing. The ice cover in the southern hemisphere seems to be decreasing over the past twenty years, a possible indication of the global warming that is predicted because of the increased carbon dioxide in the atmosphere caused by burning of fossil fuels. Thus, sea-ice cover is a potential indicator of important, long-term climatic changes.

What next?

Proponents of the Global Habitability programme have noted that man has the ability to manage his resources, to plan intelligently for his future and to preserve the necessary elements of his habitat. If we are to be successful in this endeavour, then we must take steps to develop the body of knowledge required to permit wise policy choices in the future.

Using satellites, ships and computers, environmental scientists have the technology to carry out the required research. In addition, their theoretical tools have advanced to the point where the integrated study of the global system is becoming feasible. With scientists working together internationally, it is possible that a rational management of the earth and its resources may be achieved.   ■

Notes

1. The calculation of a vessel's position from astronomical observations.
2. The estimation of a vessel's position using ship's log, ocean currents, wind, and the last known position.

The world has become increasingly involved with man's activities on the global scale. The seas are not only a promising source of food, mineral and energy supply and an important transport waterway; being an integral part of the biosphere, the world ocean also determines in many respects the result of human activities—thereby regulating climatic and weather conditions. The following is a Soviet view concerning forecasts and control of ocean dynamics.

Chapter 17

Satellite monitoring of the ocean climate

B. A. Nelepo and G. K. Korotaev

Gennadiy Konstantinovich Korotaev (born 1946) is a principal research worker and department head at the Marine Hydrophysical Institute, Academy of Sciences of the Ukrainian S.S.R., with more than eighty publications to his credit. Dr Korotaev's co-author is Dr B.A. Nelepo, who heads the same institute, situated in Sevastopol on the Crimean peninsula. He prepared the following article specifically for this book.

Developing a system of control and prediction of the state of the world ocean may have most promise in the remote measurements possible by earth satellites. These are unique means to provide systematic information on a global scale of all the seas as well as the observational base necessary to construct a physical model of the ocean—a model taking into account the interrelational processes operating on different space-time scales.

It is theoretically possible to measure remotely from satellites oceanic processes reflecting a wide range of space-times scales.[1] But it is quite normal, first of all, to state a problem in ocean-climate monitoring in terms of the processes related to perennial, seasonal and synoptic variabilities of the ocean 'fields' (hereafter called the large-scale variability). It is precisely such scales that involve the major energy of variability in the oceanic fields, scales that are responsible for the conditions in which smaller scale processes are developed.

Control of large-scale oceanic variability on a global scale is also important from a practical point of view. Large-scale interaction between atmosphere and ocean determines the earth's climatic and weather conditions.[2] It is large-scale variability in the ocean's physical fields that is responsible for variations in the concentration density and unit-count of fish 'herds'; these parameters condition the efficiency of commercial fishery.[3] At the same time, a more concrete statement of the problem, i.e. how, where and when to arrange fishery in a region with potentially high numerical strength of the fish herds, must be based on the analysis and prediction of smaller scale processes. Information on large-scale variability of oceanic fields is necessary to meet the requirements of navigation, programmes for the development of mineral resources and for the prevention of ocean pollution.

We have much experience in remote sensing of oceanic characteristics from satellites. The USSR and the USA have launched specialized satellites to conduct marine research from space.[4,5] In the Soviet Union, specialized oceanographic experiments were conducted from abroad the Kosmos 1076 and 1151 satellites between 1979 and 1981.

Methods of approaching the main problems

Experiments in oceanic research from space have shown that it is possible to perform regular satellite measurements of the global distributions of ocean-surface temperature, velocity vectors of near-water wind, ocean surface-radiation balance, slopes of the 'level' surface and, in perspective, ocean surface-heat balance. And there are further possibilities of measuring characteristics of the ocean's surface. Now a question arises in this connection. What should be the minimal set of measured parameters, with what accuracy, in order to provide effective control of large-scale variability in the ocean medium?

Taking into account the fact that parameters measured from satellites refer to the ocean's surface while large-scale variability covers the upper, active ocean layer (or all depths), quite a reliable diagnosis of the global ocean based on remote measurements is possible. But, in order to do this, satellite measurements must be combined with calculations derived from process models characterizing large-scale variability in the ocean processes themselves. This is why, in determining the composition of parameters and the formulation of requirements for accuracy of their measurements, it is quite natural to base the possible applications of hydrophysical models—as well as the description of large-scale variability processes—on satellite measurements.

The variability of the ocean manifests itself differently in the open seas, the equatorial ocean and the regions of intensive, Gulf-Stream type currents; variability is also evident in the active layers 200-300 metres deep as well as within the main aquatic body of the sea. Below, we shall discuss these regions separately.

A few complicating factors

First, let us think about the feasibility of remote measurements as applied to the dynamic control of the upper, quasi-uniform layer of the open sea having the most pronounced seasonal variability. Recent investigations enable us to describe these dynamics by a vertically integral model.[6,7,8] Using such an integral model, calculations of depth and temperature can be made on condition that tangential wind stress, radiation balance, heat loss because of evaporation, and contact heat-exchange at the ocean-atmosphere interface as measured. Furthemore, integral transfer of mass fluid by drift currents, surface geostrophic current velocity (relating to deflection caused by the earth's rotation) and temperature—just below the quasi-uniform layer—must be known.

The modulus, or degree of property possessed, of tangential wind stress and the integral transfer of mass by drift currents are calculated from known relations via the wind-velocity vector in the near-water layer. The velocity of surface geostrophic current is found from the dynamic slopes of the ocean-level surface. Temperature below the upper quasi-uniform layer, as a first approximation, may be taken as equal to the mean climatic temperature for the given month. Calculations of depths and temperature (and hence, heat content) of the upper quasi-uniform layer can be made by the remote-sensing of radiation and heat balance at the ocean's surface, of wind velocity in the near-water layer, and of deviations of the ocean-level surface from the equilibrium parameter.

The most difficult to measure are deviations of the ocean-level surface from its undisturbed state and the heat balance of the ocean's surface. Since open-ocean geostrophic currents are rather weak, in the absence of level measurements it is possible to replace real geostrophic currents by mean climatic ones: for instance, by those calculated by the diagnostic method.[9] Dynamic models of a quasi-uniform layer also permit use of ocean-surface temperatures instead of data based on heat balance. In this application, the ocean's surface heat-balance can be found along with other characteristics of the oceanic quasi-uniform layer.

Estimates have shown that one can calculate monthly mean values of the characteristics of the quasi-uniform layer to at least 20 per cent accuracy—on condition that the near-water velocity and the direction of the wind, as averaged over one month in a latitudinal-longitudinal area measuring 5° by 5°, can be determined with an accuracy of (respectively) 5 per cent and 5°. Temperature at the ocean's surface can be measured with an accuracy of 0.1° C and the level-surface slopes with an accuracy of 1 cm/100 km.

The effects of atmospheric wind

Variability in the regions of jet currents manifests itself as variations in a current's axis, width and density. Of greatest interest appear to be the disturbances synchronously evident at all depths. One can then calculate distributions of temperatures and current velocities over depth from their known distributions at the ocean's surface, using empirical relationships. The most reliable quantitative characteristic of current velocity can be obtained on the basis of remote measurements of the slopes of the ocean-level surface.

Since currents in a jet are gradient currents, calculations based on geostrophic relations will give the value of the current-velocity component along the direction of its axis. The level-surface slopes can thus be measured with an accuracy of 10-15 cm/km. And since ocean-surface temperature drops across a current by 1.5-3° C or more,[10] the acceptable accuracy of its determination is 0.5-0.7° C. A week may be taken as the time interval over which the averaging of measured parametres is permissible.[11] Averaging over space in order to resolve current structure should be performed over an area whose diameter does not exceed 50 km.

Variability in the ocean's equatorial zone results mainly from the action of atmospheric wind. Wind oscillations with a period of less than ten days induce inertial gravity waves (trapped at the equator) which do not cause current variations.[12] Wind variability having a 10-50 day period results in variations within the ocean's subsurface layer which do not involve the equatorial countercurrent. Wind variability of 50-150 days leads to fluctuations in a zonal pressure gradient sufficient to transform the equatorial countercurrent. The process of adjustment of the equatorial zone to the 'wind field' in this case is determined by Kelvin waves*—waves radiated from the western coasts of oceans.

Other synoptic considerations

Calculation of variability in hydrophysical characteristics at various depths is most expediently performed by using a complete hydrodynamical model of circulation off the equator. The wind-velocity vector should be the main

---

* A Kelvin wave is a tidal phenomenon occurring in fairly confined bodies of water in which the range between high and low tides increases to one side of the tide's direction of travel. —Ed.

measured parameter in making such calculations. In addition, ocean-surface temperatures and those of level-surface slopes are also desirable. Ten days may be taken as an acceptable scale for averaging the measured fields in time. The equivalent permissible scale for spatial averaging is 100 km.

Theoretical analysis has shown that the equatorial ocean's response to external action is close to a linear projection. Assuming a linear response of the ocean to atmospheric disturbances, one can easily estimate that the calculated accuracy for temperature variations and those of current velocity within the ocean's mass via the hydrodynamical model can be raised by computing (a) wind velocity averaged over 10 days and over an area 100 by 100 km^2 to within 10 per cent, and (b) wind direction to within approximately 7°. Ocean-surface temperature averaged over the space-times scales indicated above is required with an accuracy of some 0.2° C.[13] Level-surface slopes (associated with variations in wind conditions) to estimate the variability in a zonal pressure gradient at the equator should be calculated to within about 1 cm/km.[14]

The synoptic variability of the open ocean shows up intensive eddies and the so-called Rossby waves. The most typical feature of both wave and eddy motions on a synoptic scale is the universal character of their wave structure.[15,16] Because of this characteristic, it is possible to calculate their distribution with sufficiently high accuracy. For the 'young' rings of jet currents, their vertical structure can be calculated from the distribution of ocean-surface temperatures. Computing the vertical structure of open-ocean synoptic eddies, however, requires measurement of the distribution of current velocity in the surface layers, i.e. measurement of the sea's surface-level slopes.

Estimates show that measurements of the surface-level slopes to within 5 cm/100 km allow calculation of current velocities at each level accurate to 5 cm per second and of temperature deviation accurate to 0.3° C. As synoptic eddies displace over distances equal to their radius in a period of 15-30 days, their time-averaging should be 7-10 days. The corresponding space-averaging is 30-40 km.

Space-time errors and general accuracy

The amplitudes of synoptic-scale motions over the major oceanic 'territories' are relatively small,[17] so that such motion must have a wave character. Wave motions induce surface currents, but recording these is a doubtful procedure at present. It is practicable, however, to determine the spectral characteristics of synoptic wave motions from measurements of the ocean-surface temperature—provided that its mean gradient is expressed.[18] The basic idea here is that wave motion induces forced currents in the upper ocean layer, thereby causing fluctuations in temperature at the ocean's surface.

The space-time spectrum of Rossby waves can be computed when ocean-surface temperatures, averaged weekly, are found as accurate as 0.2° C and the mean temperature gradient is of the order of 10^{-7} degree per centimetre. Analyses performed have shown that, in order to check large-scale variability in the ocean's state, one needs to measure surface temperature, velocity of the near-water wind vector, and level-surface slopes.

From the point of view of remote-sensing methods, it is most attractive that calculations should involve using characteristics averaged over a 7-30 day time period and a spatial interval of 30-500 km, but precise computation of the efficiency of spatial and temporal averaging of remote measurements needs much study. Errors still arise from instrument noise and that associated with interfering atmospheric effects. Space-time averaging can, however, overcome instrument noise. And the most effective method to overcome interference by atmospheric effects is to use multichannel measurements.

Other interfering factors not taken into account produce error conditioned variability in the state of the atmosphere. Assuming that this variability is determined basically by the atmospheric processes, we can conclude that averaging remote measurements over space will not reduce error because the spatial scale of the atmosphere's synoptic processes is rather large. Still, because of time-averaging, the calculation error of oceanic parameters taken from satellite measurements (caused by atmospheric variability within a synoptic period) can be reduced.

Table 1 shows that the achieved level of remote measuring accuracy is acceptable for solving problems related to control of large-scale variability in the ocean's state. It is worth noting that realization of a system of control is possible provided that—as a minimum—synchronous measurement of the ocean-surface temperature, level-surface slopes, and near-water wind velocity vector are made and applied to models of hydrodynamic processes. ■

Table I. Consistency of required (and obtained) accuracies in measuring characteristics of the atmosphere and ocean.

Nature of accuracy / Parameter	Required accuracy in averaging characteristics	Required accuracy of single measurements	Obtained accuracy of single measurements
Ocean-surface temperature	$0.1\text{-}0.5^{\circ}C$	$0.3\text{-}1.0^{\circ}C$	$0.6^{\circ}C$
Wind velocity vector	5-10 per cent	10-20 per cent	10-20 per cent
Wind direction	$5\text{-}7^{\circ}$	10-20 per cent	16°
Level-surface slopes	1-15 cm/100 km	5-20 cm	8 cm

Notes

1. B. Nelepo, Yu. Terekhin, V. Kosnyrev, B. Khmyrov, *Sputnikovaya Gidrofizika* (Satellite Hydrophysics), Moscow, Nauka, 1983.
2. E. Kraus, *Atmosphere-Ocean Interaction,* Oxford, Clarendon Press, 1972.
3. T. Levastu, I. Khala, *Promyslovaya Okeanografiya* (Fisheries Oceanography), Moscow, Gidrometeoizdat, 1974.
4. B. Nelepo, et al., Eksperiment 'Okean' na Iskusstvennykh Sputnikakh Zemli iz Kosmos 1076, 1151 (The 'Ocean' Experiment aboard Earth Satellites Kosmos 1076, 1151), *Issledovaniye Zemli iz Kosmosa,* No. 3, 1982.
5. G. Born, et al., Seasat Mission Overview, *Science,* Vol. 204, No. 4400, 1979.
6. V. Kalatskii, *Modelirovanie Vertikal'noy Termicheskoy Struktury Deyatel'nogo Sloya Okeana* (Modelling of the Vertical Thermal Structure of the Active Ocean Layer), Moscow, Gidrometeoizdat, 1978.
7. R. Davis, et al., Variability in the Upper Ocean during MILE, Pt. II, Modelling of the Mixed Layer Response, *Deep-Sea Res.,* Vol. 28, No. 12A, 1981.
8. G. Prangsma, et al., Development of the Temperature and Salinity Structure of the Upper Ocean over Two Months in an Area 150 km x 150 km, *Results of the Royal Society's Joint Air-Sea Interaction Project (JASIN),* London, The Royal Society, 1983.
9. A. Sarkisyan, *Chislennyi Analiz i prognoz Morskikh Techenii* (Numerical Analysis and Prediction of Sea Currents), Moscow, Gidrometeoizdat, 1977.
10. E. Baranov, *Srednemesyachnye Polozheniya Gidroopticheskikh Frontov v Severnoy Chasti Atlanticheskogo Okeana* (Monthly Mean Positions of Hydro-optical Fronts in the Northern Atlantic Ocean), *Okeanologiya,* Vol. 12, No. 2, 1972.
11. G. Halliwell, C. Mooers, The Space-Time Structure and Variability of the Shelf Water-Slope Water and Gulf Stream Surface, *J. Geophys. Res.,* Vol. 84, No. C12, 1979.
12. S. Philander, R. Pacanowski, Response of Equatorial Oceans to Periodic Forcing, *J. Geophys. Res.,* Vol. 86, No. C3, 1981.
13. J.Merle, Seasonal Heat Budget in the Equatorial Atlantic Ocean, *J. Phys. Ocean.,* Vol. 10, 1980.
14. E. Katz, et al., Zonal Pressure Gradient along the Equatorial Atlantic, *J. Mar. Res.,* Vol. 35, 1977.
15. G. Korotaev, et al., Ob Ustoichivosti Vertikal'noy Struktury Vozmushchenii Polya Temperatury Sinopticheskikh Vikhrei (On the Stability of the Vertical Structure of Temperature Field Disturbances in Synoptic Eddies), *Eksperimental'nye Issledovaniya po Mezhdunarodnoy Programme Polimode* (Sevastopol), 1978.
16. E. Mikhailova, N. Shapiro, Parametrizatsiya Polei Plotnosti i Skorosti Techenii na Sinopticheskikh Masshtabakh (Parameterization of Fields of Density and Current Velocities on Synoptic Scales), *Morskiye Gidrofizicheskiye Issledovaniya* (Sevastopol), No. 2, 1978.
17. V. Kamenkovich, M. Koshlyakov, A. Monin, *Sinopticheskiye Vikhri v Okeane* (Synoptic Eddies in the Ocean), Moscow, Gidrometeoizdat, 1982.
18. V. Kosnyrev, Otsenka Kharakteristik Barotropnykh Voln Rossbi po Izmereniyam Temperatury v Kvaziodnorodnom Sloye Okeana (Estimation of Characteristics of Barotropic Rossby Waves from Temperature Measurements in the Quasi-Uniform Oceanic Layer, *Izvestiya AN SSSR,* (FAO), Vol. 18, No. 12, 1982.

*There has been a tremendous increase in the scientific and
economic importance of the world's seas during the last few decades. In
geology, this has taken the form of measurement of global geophysical
fields and development of global tectonics—making it possible, in
particular, to understand the formation of mineral deposits. Economically,
this importance has taken the form of drilling for oil at sea and
recovering ferromanganese nodules. In the future, the extraction of
sulphide ores and metalliferous deposits can be anticipated. To accomplish
all these things, man's organism needs to adapt itself physiologically
and psychologically.*

Chapter 18

Some problems concerning the exploitation of the sea's treasure in petroleum and minerals

A. S. Monin

*Andrei Sergeyevich Monin, doctor of physical and mathematical sciences,
is professor and Corresponding Member of the Academy of Sciences of the
USSR, director of the Oceanological Institute of that Academy, honorary
member of the National Academy of Sciences (Washington) and the American
Academy of Arts and Sciences (Boston). The author has written some four-
hundred scientific works, including* Oceanic Turbulence, The Variability
of the World's Seas *and* An Introduction to Climate Theory.

Our changing knowledge

Where the earth sciences and the science of the universe are concerned, the importance of the world's seas obviously lies in the fact that they occupy 70.8 per cent of the earth's surface and consequently cover more than two-thirds of all geophysical fields: those of gravitation and magnetism, geo-thermal heat flux, reflected solar radiation, terrestrial heat radiation, and so on.

The need to know all geophysical fields as a whole led scientists to organize the measurement of their characteristics by means of research ships in all the world's sea areas. An example of this is the measurement of the geomagnetic field (known, since the Middle Ages, in its simplest form as 'magnetic deviation') carried out, over a number of years, by the Soviet non-magnetic schooner, *Zarya*. Such measurement is now done by a large number of research vessels, using towed magnetometers.

While many people know about and understand such investigations, they may be unaware of the fact that the information gathered by oceanographers over the last fifteen years on the geology and geophysics of the ocean floor has completely revolutionized geological science; these data have led to the creation of a new global tectonics (or plate tectonics) in which is included a revised version of Wegener's continential drift theory—unjustly rejected in the years before the Second World War.*

According to contemporary data, the upper hard crust of the earth—the lithosphere—consists of individual plates that move in relation to each other under the action of convection currents in the earth's thick lower shell, or mantle. Where the plates move apart, i.e. at the lines of spreading or the axes of mid-oceanic ridges, a new oceanic lithosphere is built up through the volcanic eruption of basalt. When there is 'subduction' by the oceanic plate under the continental plate (as with the deep oceanic trenches), this gives rise to deep-focus earthquakes and 'granistoid magmatism' that builds up the continental lithosphere, as in the case of the chain formed by Japan, the Kurile Islands, Kamchatka, the Aleutian Islands, the Cordillera and the Andes. When the continental plates collide, this results in the formation of huge mountain ranges: the Himalayas, the Caucasus, the Alps.

The distribution of the globe's continents and ocean basins has changed radically with time; see, for example, L.P. Zonenshain's and A.M. Gorodnitskiy's 1977 models for the Phanerozoic period, or the last 570 million years. At some time, South America, India, Antarctica and Australia formed Laurasia which, afterwards, merged with Gondwandaland. This was Wegener's Pangaea. At the beginning of the Jurassic period, 150 million years ago, Pangaea broke up, and its fragments moved apart. They formed the continents as we know them.

---

* Alfred L. Wegener (1880-1930) developed his ideas more than seventy years ago.—Ed.

Exploitation of the sea's treasure in petroleum and minerals

Rewriting geology's history

Such models are built up by using three kinds of data. First, data on the bands of magnetic anomalies found at the ocean floor, making it possible to reconstruct the drift of the continents as the oceans spread; this information covers 150-160 million years of the earth's history. Secondly, data on the magnetization of rocks of various ages on continents, making it possible to reconstruct the directions of 'palaeomeridians' and 'palaeolatitudes'; a great deal of information is already available for periods dating 600 million years, and some findings have been made for periods dating 2,850 million years. Thirdly, there are data on rocks indicating seams in the continental crust resulting from old zones of spreading, subduction, the shrinking of the oceans, and the collision of continental masses.

Oceanographic data show that the earth's historical geology, written from the point of view of a fixed distribution of continents and oceans, needs to be rewritten. Work on this has already begun, and it is clearly leading to a radical adaptation of our ideas about the origin and distribution of mineral deposits in the earth's crust. The presence of oil-bearing strata, for example, has been predicted in the Rocky Mountains of North America.

Attention has been focused until now, however, far more on continents than on the world's seas, and it is clear that the progress of mankind calls for the broad development of oceanic research. It will be possible in the future to carry out some of this research with the help of satellites; but, in the decades immediately ahead, research vessels will continue to play the primary role. It would be desirable, therefore, for all coastal nations to offer these ships maximal assistance on their cruises, particularly by simplifying procedures for entry into port, obtaining fuel, provisioning and rest.

Role of the petroleum economy

The enormous role played by oil in the world's economy as the most convenient and widely used fuel in both power production and automotive transport has invested with particular importance the intercontinental shipments of petroleum from oil-producing to oil-consuming regions, especially from the Gulf to Europe and the United States. It has proved advantageous in this connection to use bigger and bigger tankers; quite clearly, the sudden lengthening of sea routes when the Suez Canal was closed encouraged the development of supertankers displacing more than 200,000 t. Of the 4,200 tankers in existence today, some 700 are supertankers. An idea of their size can be judged from one of the most recent French supertankers: more than 400 m long, 70 m wide and 40 m high. Plans have been made for supertankers displacing between 700,000 and 1 million t.

The growth of merchant shipping and, in particular, the appearance of supertankers are making greater demands regarding the provision of safe navigation; they also raise the question of the effective forecasting of the surest and, at the same time, most economical routes for each voyage of every large tanker. Supertanker disasters are no more frequent than those befalling other ships, yet they cause extremely serious ecological damage because of the tremendous amount of oil spilled. French resort areas along the English Channel suffered considerable damage in 1967 when the *Torrey Canyon* went down, spilling 117,000 t of oil, and again in 1978 when the supertanker *Amoco Cadiz* sank and dumped another 229,000 t of petroleum.

Of the economic problems dealt with by marine geology, foremost is undoubtedly the extraction of oil and gas through the sea and ocean floor. Its magnitude can be understood from the fact that one-fourth of all the petroleum extracted today comes from wells on the sea floor—and economists predict that by the year 2000 the proportion will reach 40 per cent.

Oil has been extracted for many years from the bottom of the Caspian Sea. Today, many countries are dependent on Arabian oil drilled from the bottom of the Gulf. A large number of marine oil-drilling platforms, some 7,000 so-called Texas towers, are in operation along the Louisiana coast in the Gulf of Mexico. The north-south axis of the North Sea, now divided into zones among its coastal countries, has proved to be an area of continuous oil fields; the region extends even into the Norwegian Sea, whence the Norwegians pump oil to the shores of Scotland. Petroleum has been found off the coasts of California, Venezuela, Indonesia, in the Gulf of Suez, in the littoral area of the Mediterranean near the Sinai peninsula, and—to everyone's surprise—even in the basalt volcanoes of the Tonga islands. There, following one of the ideas of plate tectonics, oil may form from marine sediments in the area where the eastern oceanic plate subducts beneath the western plate, along the Tonga trench.

There are now about 10,000 marine oil-drilling platforms in operation throughout the world. The largest of these, operating at depths of 300-400 m, are multi-level constructions manned by as many as 400-500 specialists who, working from one or two rigs, drill from 25 to 35 oblique wells (each from 10 to 12 km in length) in order to extract oil and gas. Such platforms require a capital investment of around a thousand million dollars each.

Even if the average cost of a platform is one-tenth of this figure, it still must be recognized that a new marine industry has emerged, representing a capital investment of a million million dollars. It is a growing industry, operating at increasingly greater depths—posing and solving new scientific and engineering problems. Here we shall consider two problems of general oceanographic importance, namely geophysical prospecting for undersea oil and the servicing by divers of deep-water oil fields.

The geophysical challenge of finding undersea oil

Geophysical prospecting for underwater petroleum requires the application of the latest oceanological methods such as satellite navigation, permitting the fixing of a ship's position correct to a few score metres; area-by-area measurement of sea-bed relief by means of side-scanning sonar; shooting of the ocean floor's permanent abyssal seismic profile (several kilometres thick) by means of seismic acoustic waves derived from ship-towed 'air guns.' Here, the reflected waves are received by the hydrophones of a multichannel, towed 'seismic array'; future plans call for the use of harmonic vibration emitters and seismic holography, the signals so received being subsequently processed by the ship's computers. Prospecting also needs magnetometry and gravimetry; measurement of geothermal heat flux; probing by gas samplers, and so on. All these activities are of great importance not only for oil prospecting but for marine geology as a whole; they are followed by exploratory drilling of promising structures carried out from surface or semi-submerged vessels equipped with derricks having an automatic device for running drill pipes in, as well as stabilizing tanks and systems for dynamic positioning and satellite navigation.

During both exploratory and production drilling for oil and gas at sea and in the operation of marine oil-drilling platforms, many diverse operations (notably, the screwing and welding of pipes), at depths at present of between 300 and 400 m, can be performed only by divers. At one time, the peculiarities of man's hyperbaric physiology seemed to be an insurmountable obstacle to work of this kind. Even a short period—ten minutes or so—spent under high pressure (i.e. several atmospheres, and each ten metres' depth is equivalent to 1 atmosphere) causes the diver's breathing mixture to be dissolved in the blood and tissues. When rapid decompression takes place, as when a diver comes up, this gas then leaves the bloodstream and tissues in the form of bubbles, causing injury (caisson disease) or even death.

The only way of avoiding the disease is to decompress very slowly (even over a period of a few days), but to do this would make undersea diving work wholly uneconomical. It should also be mentioned that, under pressures higher than a few atmospheres, the nitrogen of the air becomes toxic and has a narcotic and incapacitating effect on a person similar to alcholic intoxication (the 'laughing gas' effect). For this reason, helium is used in the breathing mixtures for deep-sea divers, being much less toxic than nitrogen.

How many atmospheres will man withstand?

The remarkable discovery of man's hyperbaric physiology made it possible to overcome this difficulty. It turned out that, by using mixtures of helium and oxygen, man can live and work for many days and weeks under pressures of at least 30-45 atmospheres. Nowadays, divers on drilling vessels and platforms live under high pressure in compression chambers and are transferred (in diving bells) at the same pressure to the required depth. There, they enter the water equipped with aqualungs, carry out their work, and are then returned in the diving bells—with no change of pressure—to the compression chambers. They undergo decompression only after completing a shift of several days. A 'saturation regime' of this kind makes deep-sea diving work fully profitable.

When divers were subjected to rapid compression at pressures exceeding 27-30 atmospheres, a new difficulty arose, that of high-pressure nervous syndrome (HPNS). The signs of HPNS are spasms ('helium tremor'), nausea and vomiting, intermittent somnolence, and a total incapacity for work. Initially HPNS made it seem that the maximal pressure man could withstand was 35 atmospheres. Subsequently, this threshold was increased to at least 40-60 atmospheres by changing over to extremely slow compression.

In 1980-81, Peter Bennet of Duke University (North Carolina) suggested that HPNS could be overcome, even under conditions of rapid compression, by adding a small amount of nitrogen to the helium-oxygen breathing mixture. He raised the record for the prolonged submersion of divers under high-pressure conditions to almost 70 atmospheres (24 hours' submersion is maximal pressure) which is equivalent to an ocean depth of 700 metres. Judging by experiments carried out on higher animals, even this is probably not the ultimate limit. Clearly, the possibility of man making direct descents to the ocean depths is of enormous interest not only to the petroleum industry but also in connection with many other problems related to the study and exploitation of the world's seas.

The challenge of mining the sea's floor

The second most important task of marine geology today, from the economic point of view, is concerned with much greater depths. This is the recovery on an industrial scale of the sea's ferromanganese nodules that, in some places, like cobblestone pavements, cover vast areas of the central regions of the ocean floor; they can extend over thousands of kilometres and reach to a depth of 4-6 kilometres. According to some estimates, the sea contains as much as 2-3 million million t of such nodules. It is likely that a certain percentage—thousands of millions of tonnes—may become important as ore.

These nodules have been studied since the Second World War by a number of research institutions, particularly extensive studies having been made by the Academy of Sciences of the USSR's Oceanological Institute which, in 1976, published a special monograph, Ferromanganese Nodules of the Pacific (*Transactions*, Vol. 109). The richest area, as regards both quantity and precious metal content, has proved to be a region in the North Pacific basin between the Clarion and Clipperton fracture zones. The requests made by various countries for the recovery of nodules on an industrial scale will probably be met by granting concessions precisely in this area [France is the legal owner—Ed.].

As iron ore, marine nodules have not great value. They contain on average only 12 per cent iron (and a maximum of 33 per cent). As a rule, they are poor in maganese ore: the average content is 18 per cent, while the maximum is 42 per cent. When the manganese is extracted in connection with the processing of other metals, however, it can prove profitable and should be of considerable value to countries lacking manganese resources on land.

The chief value of marine nodules lies, in fact, in their non-ferrous metal content—in which respect they are on a par with medium- and high-grade ores found on land. The three metals mainly concerned are copper, nickel and cobalt. The nodules' average content of these is 0.38, 0.59 and 0.33 per cent respectively, with maximal values of 1.9, 2.48 and 2.53 per cent. On land, ores with concentrations of 0.5, 0.3-1.0 and 0.3-0.5 per cent (respectively) are regarded as up to standard. The nodules are also valuable for the titanium, lead, zinc, molybdenum and other useful elements they contain. It should be added that, where the technology of their processing is concerned, marine nodules have shown themselves to be very convenient ores.

Undersea thermal currents transport metals

The scientific back-up for the recovery of ferromanganese nodules and the process of recovery itself are complicated matters owing to the great depth of the ocean floor. Scientific back-up must consist primarily of making the best possible choice of areas (the richest, as regards quantity and quality of the nodules); these should offer the most favourable conditions for their recovery. This calls for both (a) extensive exploration and mapping of large areas of the sea floor and (b) study of the small-scale variability of nodules in specially selected zones.

As yet, the means do not exist for carrying out such investigations directly from a ship; at present, point-to-point photography is used, and samples are collected by means of bottom trawls. The first unmanned underwater craft, however, have now become available, equipped with cameras and television, side-scanning locators, small seismic profilographs and other apparatus. These are towed close to the bottom, enabling continuous monitoring of the sea floor, all along a ship's course.

As for the recovery of nodules, dredging systems using continuous conveyor-belts are being developed and tested. (This has been done, for example, on the Japanese vessel, *Chiyoda-Maru*.) Hydraulic suction systems (used by the American vessels, *Deep Sea Miner* and *Prospector*) and even self-contained diving apparatus are being developed by French specialists. The experimental recovery of nodules is already under way.

Of the minerals obtained from the sea bed, mention should be made—after oil, gas and ferromanganese nodules—of metal-bearing sediments. These were first discovered many years ago in the 'deeps' containing hot brine in the Red Sea; currently, experiments are being conducted there on the industrial recovery of such deposits from the Atlantis II Deep. In recent years Alexander P. Lisitsyn of our Oceanological Institute established, during expeditions in the vicinity of the East Pacific Rise, that the concentration of metals in sedimentary deposits rose as one approached the axis of the mid-oceanic ridge. Thus, the presence of these metals should be explained by their having been carried via thermal waters from the new oceanic crust forming on the line of spreading as a result of the extrusion of basalts from the earth's mantle. In short, metals come directly from the mantle and form deposits in spreading areas.

Lisitsyn's conclusion is of major importance for theoretical geology, and it may very well have immediate practical significance. Divers operating from the manned submersibles *Alvin, Ciana* and *Pisces* explored the axial areas of the mid-oceanic ridges and discovered metalliferous sediments there in the form of extremely rich sulphide-polymetallic ores of hydrothermal origin. This is the same probable origin of the pyrite ores found on land, in Cyprus, for example.

Potential riches beyond man's immediate needs

Ores of this kind have been found in the Galapagos fracture zone, in a number of places on the East Pacific Rise, in the Gulf of California, in the Juan de Fuca and Gorda seamount chains, as well as in the rift valley along the Red Sea's axis. In places, ore bodies exist in massive amounts: it is estimated, for example, that the Galapagos fracture zone contains 8 million cubic metres of ore and the Gulf of California 1.2 million m^3. On the average, the ores contain 35 per cent iron, as much as 10 per cent copper (in the Galapagos zone) and 30-55 per cent zinc (in Juan de Fuca, Gorda and the Gulf of California), as well as lead, and silver and other precious metals.

Similar ores were found in the Red Sea, although in small quantities, as a result of persistent searches using the *Pisces* manned submersible carried out during an expedition that I led. An ore specimen picked up by *Pisces'* mechanical arm and taken on board was soft and black in colour, and it was difficult to wash off the hands—even with soap.

Comparable ores are likely to be found on the axes of all mid-oceanic ridges, which have a total length of around 60,000 kilometres. Their extremely high concentrations of metals and the medium depths at which they lie (2,500-3,000 m) make large ore beds of this kind extremely profitable fields for recovery on an industrial scale. An extensive industrial prospection effort is therefore, naturally, to be expected here.

Among other important underwater minerals, mention should be made finally of phosphorites and of placer deposits of diamonds, gold, titanium, tin and other metals lying in the sands of some coastal-zone areas.

Predicting, finding, studying and making it possible to recover all these minerals from the sea floor—such can be the contribution of geological and oceanological research to the peaceful progress of mankind. ∎

An Egyptian specialist recounts the roles of the sea, as both destroyer and preservative, in the archaeological past of the bimillennial city of Alexandria, its lighthouse and monuments, and nearby Abu Qir (Aboukir)— where the naval forces of Napoleon Bonaparte and Horatio Nelson engaged in one of history's most decisive struggles on an ocean battlefield. Future underwater exploration will help us better understand our total cultural past.

Chapter 19

Submarine archaeology and its future potential: Alexandria casebook

Selim A. Morcos

The author, a physical and chemical oceanographer, earned the B Sc at the University of Cairo, the M Sc at the University of Alexandria and the Ph D at the University of Kiel. He taught at the University of Alexandria where he became professor, taking time out for research aboard the Scripps vessel Argo *(on a Unesco fellowship) during the International Indian Ocean Expedition. He joined Unesco in 1973. Dr Morcos' main work has been on the exchange of sea water between the Mediterranean and Red Seas through the Suez Canal—data of considerable interest to biogeographers. Hobbies: the history of oceanography (especially 19th-century documents on the seas of the Middle East) and underwater archaeology, in which he has written the only book in Arabic,* Sunken Civilizations, the Story of Submarine Archaeological Discoveries, *and on which he lectured extensively in Egypt. The author lives and works in Paris.*

Alexandria, the city

From the time of its foundation by Alexander the Great in 332 B.C., and for almost one thousand years after, the city of Alexandria experienced uninterrupted growth and prosperity. Within a few decades, it had become the major commercial centre of the ancient Mediterranean world, and for more than three hundred years it remained antiquity's richest and most populous city.

Alexandria was one of the main centres of Hellenistic civilization. It became celebrated for the unique blend of Egyptian, Greek and Roman culture that it achieved and for the leading role it later played in the development of Christian thought—from Coptic times until after the Arabs entered the city in A.D. 640.

The eventful history of ancient Alexandria has left a variety of archaeological remains, including those standing on the surface, buried under the surface or displayed in Alexandria's Graeco-Roman Museum and in other museums of the world. Immensely valuable as these are, they represent only a small fraction of the heritage of a city that occupied such a prominent place in the ancient world. There, monuments and buildings were in many instances products of the creative mixing of major civilizations characterizing the city during its long history.

Yet anyone who has kept abreast of the rapidly evolving field of submarine archaeology and its achievements, both in the Mediterranean and elsewhere, anyone who is familiar with the circumstances in which the city of Alexandria developed, anyone who is aware of the geography, topography and geological history of the region will realize that the story of the archaeological exploration of the Alexandrian metropolis is not over. New chapters remain to be written on the basis, too, of the archaeological discoveries being made nearby under the sea.

To date, exploration for sunken antiquities off Alexandria has been the work of individual enthusiasts, the most successful of whom was probably the Egyptian amateur diver, Kamel Abu Al-Saadat, who died in June 1984 while diving in Abu Qir Bay and to whom this chapter is dedicated. One of the earliest instances of successful submarine exploration was the work of Gaston Jondet, carried out between 1910 and 1915 in the part of the ancient harbour of Alexandria lying to the west of Pharos Island. It was with these considerations in mind that I have chosen Alexandria as an example of what can be achieved in the field of submarine archaeology.

A bit of background

In the year 332 B.C. the eye of the Macedonian leader, Alexander the Great, was caught by an Egyptian fishing village along the Mediterranean coast known as Raqoda or Rakotis—just opposite of which lay the small island of Pharos. Alexander ordered his engineers to draw up plans for a city that would include both the village and the island within its boundaries (Figure 1).

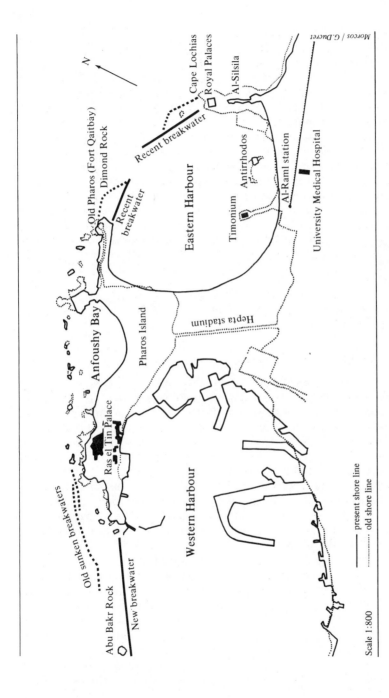

Figure 1. Positions of harbour installations of ancient Alexandria superimposed on chart of the present-day city. (The long, dotted lines represent the original configurations.)

Alexander's engineers linked the island to the mainland with a narrow causeway that they named Heptastadium because it was seven stadia long (about 1,200 metres). This causeway divided the waters of Alexandria into two parts, the Eastern and Western Harbours. The Eastern Harbour was the main port, and the city's palaces, gardens and government buildings were built around it; it handled the more important naval and commercial vessels. These were guided into port by the celebrated Alexandria lighthouse which stood on what is now the site of the Fort of Qaitbay. Over the years, as a result of a number of factors, the Eastern Harbour gradually declined as a seaport while the importance of the Western Harbour grew; it is the latter that has become the major port of contemporary Alexandria.

As the Heptastadium became progressively silted up, it lost its original configuration, and Pharos Island became part of the mainland city of Alexandria. The latter now extends 22 kilometres to the east, stretching nearly as far as neighboring Abu Qir. This community was to be the scene of the naval engagement known as the Battle of the Nile (1798), in which the British admiral, Lord Nelson, destroyed Napoleon's fleet during the French expedition to Egypt.

As the coastal region where Alexandria is situated has subsided since ancient times, many features of the old port, the Eastern Harbour and the town of Abu Qir have been lost to the sea.

The ancient harbour of Pharos

In 1910 while Gaston Jondet, engineer of the port of Alexandria, was studying the possibility of expanding and improving the Western Harbour, he discovered to the northwest of Ras El-Tin the piers of an entire seaport completely submerged at depths that did not exceed (in most places) eight and a half metres. This find aroused considerable interest at the time because of the massive size of the submerged masonry and the fact that this was one of the first discoveries in submarine archaeology.[1]

Jondet pursued his research until 1915. He regarded his find as evidence of the existence of an old ocean port predating the city that had been built by Alexander to the east of this site. This was also evidence that the earth's crust had, at that point, subsided and that the former seaport had been submerged beneath the sea.

Jondet also found that the old harbour of Pharos had been formed by connecting the western tip of Pharos Island at Ras El-Tin to Abu-Bakr Rock, to the west of the island, via a double line of breakwaters extending in an east-west direction parallel to the coast. These consisted of an outer and an inner breakwater, each about 2,500 metres long, placed 200 metres apart. The structures were unusually massive, measuring 10 metres in height and 60 metres across. Each breakwater consisted of two walls 12 metres thick, the space between them being filled with enormous quantities of rock. The walls were built with huge limestone blocks: the upper ones now appear eroded by the effects of sea and wind, while the broad, spacious submerged harbour now lies under water and protrudes over a sea bed that slopes gradually to 10 m depth in the vicinity of the outer breakwater. Jondet showed an outer harbour, lying between two barriers, and an inner Western Harbour containing a small inlet which he called the Private Harbour. A fourth port, for merchant shipping, lay between Al-Anfushi and Ras El-Tin.

The excitement caused by this discovery was reflected in both scientific journals and public reaction; this was the first time in the present century that such gigantic submerged structures had been found. The discovery was also highly unexpected, particularly since there were no indications of the existence of this harbour anywhere in antiquity's surviving written record.

Opinions differed regarding the origin of the rediscovered structures. Some specialists surmised that the breakwaters had been built to protect the coastline in the region of Alexandria during the Graeco-Roman period, while others (including Jondet) speculated that the ruins were those of a seaport dating from Pharaonic times. A third school of thought attributed the construction to the joint work of the Egyptians around 2000 B.C. (XII Dynasty, Senusret II) and the Cretans as a result of the peaceful trade that characterized the Minoan civilization of Crete which preceded that of ancient Greece; these researchers were suggesting, in fact, that the Alexandria underwater structures are the remains of the oldest man-made harbour.[2]

The Eastern Harbour

Perhaps the most picturesque description of the Eastern Harbour as it was during the period of Alexandria's prosperity under the Ptolemy kings is one left by Strabo, a geographer of the first century B.C. According to Strabo, 'On entering the great port, the island and lighthouse of Pharos lie to the right while on the left are seen a cluster of rocks and Cape Lochias, on whose summit a palace stands. As the ship approaches the shore, the palaces behind Cape Lochias astonish one because of the number of dwellings they contain, the variety of constructions, and the extent of their gardens. . . .'[3]

The most important feature of the Eastern Harbour was that it was protected from both the east and the west. To the east stood Cape Lochias, most of which has been lost to the sea and of which only the Al-Silsila promontory remains. To the west lay the Heptastadium and the eastern tip of Pharos Island, upon which the lighthouse of Alexandria once stood and where Fort Qaitbay now stands. A sea wall extending from Cape Lochias protected the harbour's entrance from the north wind and ocean currents.

In the middle of the harbour and towards the southwest, facing the area between the present-day Al-Raml Station and the Al-Silsila promontory, lay Antirrhodos Island—probably named after the island of Rhodes. On this isle and on Cape Lochias along the eastern shore of the harbour stood the royal palaces, all of which presumably have been submerged beneath the sea as a result of geological subsidence. At the southeastern corner of the harbour, where Cape Lochias met the shore, there was a small inner marina reserved for the use of the royal household, known as the Royal Port. At a point near today's Al-Raml Station, a tongue of land jutted into the middle of the Eastern Harbour. At the edge of this promontory, Mark Antony built the Timonium as a place of meditation and seclusion from the world.

Submarine exploration off Alexandria: some opportunities

We have noted that, ever since its foundation, Alexandria has been affected by the two natural processes of silting-up and geological subsidence. These processes have brought extensive changes to the local topography, resulting in the loss of many of the city's ancient monuments. Excavations have also shown that the city's present ground-level is several metres higher than it once was. This has been brought about by the accumulation of material from successive periods. Archaeologists must dig to a depth of six or seven metres in order to reach the remains of the Roman period; they must go even more deeply to reach those of the Ptolemaic era. This usually takes them below the water level, a situation making their work very difficult.

Archaeologists and geologists have estimated total geological subsidence at Alexandria, to date, at 2-4 metres. The most striking evidence of this process is what happened to Cape Lochias at the eastern end of the Eastern Harbour and Antirrhodos Island at its centre, as well as the submergence of the many structures once standing on the shores of this harbour and its islands. For example, at the time that Gorringe transported one of the two obelisks of the Caesareum Temple in Alexandria ('Cleopatra's Needle') to be erected in New York's Central Park in 1879, he reported that there were several columns standing under the waters of the Eastern Harbour. He noted that they were visible on a clear day, that they constituted an obstacle preventing ships from reaching the point ashore where the obelisks stood, whereas we know that in Roman times vessels could sail right to the shore. E. Breccia, a former curator of the Graeco-Roman museum, maintained that the general outline of Antirrhodos Island and the monuments could be seen on a clear day under the waters of the Eastern Harbour.

According to some ancient writers, including Strabo, the entrance to this port was very narrow. Flavius Josèphe also refers to this feature.[4] 'Even in peacetime,' he wrote, 'it is difficult for ships to enter the harbour of Alexandria, as the entrance channel is so narrow and the rocks that lie just beneath the surface of the water make it impossible to steer a straight course. On the left-hand side, the harbour is protected by artificial moles [breakwaters]. To the right lies the island [of] Pharos, on which a mighty tower stands; from its top, a fire blazes to guide incoming mariners. . .All around this island stand enormous, man-made ramparts. As the waves dash. . .and break in furious tumult, the water in the passage becomes turbulent, making it dangerous for vessels to negotiate the narrow entranceway.'

The problem of sediment deposition

There is no doubt that the configuration of the Eastern Harbour is now markedly different from what it was. All that remains of Cape Lochias today is the Al-Silsila promontory and, while the entrance to the port was very narrow in the past, in the course of time it became so wide that a major breakwater had to be built in order to protect the piers.

All these considerations support my own belief that the most interesting remains of ancient Alexandria still lie submerged beneath the sea. It is not difficult to guess, as a consequence, how much scope there is for submarine research in the waters of the Eastern Harbour. There is no doubt that the structures and other antiquities discovered so far in the waters of the Eastern Harbour are only a small sample of the wealth of archaeological remains that lie hidden under these waters.

It is unlikely, however, that many antiquities will be found on the surface of the sea bed (as might be supposed). The silting process and geological subsidence already cited are factors that must be taken into consideration. Accordingly, we should expect to find that most antiquities are buried under a layer of sand, mud and pebbles left by centuries of sediment deposition.

Attempts to explore the Eastern Harbour with the naked eye or by means of bottom scanning techniques, therefore, will probably yield only scanty results. Any success at all will depend mainly on chance and good fortune. Such operations have been successful on occasion, but only where the sedimentary layer was thin or had been removed by the action of currents and the movement of water—with the result that objects lying close to the surface of the sea bed were uncovered.

Archaeological finds in the Eastern Harbour

In October 1963, a group of naval frogmen succeeded in raising a massive statue weighing approximately 25 tons from a depth of 8 metres behind Fort Qaitbay outside the Eastern Harbour. After examining it and removing a growth of seaweed with which it was covered, Dr Henry Riad (then curator of the Graeco-Roman Museum), decided that the object was a statue of the goddess Isis dating from the third century B.C. The news media were quick to announce the find abroad, and the great size and antiquity of the discovery stirred interest round the world.

In a letter received in reply to a question I had put, Dr Riad stated, 'I remember that in 1961 Mr Kamel Abu Al-Saadat, a local amateur diver, came to the Museum and showed me some fragments of antique pottery which he had found under the water. He also indicated two sites where there were similar pieces, lying amid great piles of blocks of stone of various shapes.'

'The first of these two sites was in the vicinity of Al-Silsila, a long strip of land constituting the eastern flank of the Eastern Harbour. This area had once been known as Cape Lochias (the royal palaces stood there in Ptolemaic times) and proved to contain oblong sarcophagi, statues and building fragments.'

'The second site was in the vicinity of Fort Qaitbay at the Eastern Harbour's western edge. (The fort occupies the place where the lighthouse once stood). Different types of statuary, and fragments of crowns and entire buildings lay submerged at this site outside the harbour.'

In 1962, frogmen from the Egyptian navy had begun to bring up such antiquities. The best time of year for such operations is September-October, when the sea is calm and the water clear. A red granite statue of a man in standing position was successfully raised from the Al-Silsila area. The statue's head and parts of its legs were missing, and the length of the surviving portion was 120 cm. The figure wore a kilt-like garment, also draped over the left arm, which is bare; the left hand clutched the folds of the vestment, while the right arm hung down at the side. (The right hand held an unidentifiable object.) The fact that the back of the statue was flat indicated that it had been placed against the wall of a building. The artefact dated from Roman times and was still in good condition—even though it had been immersed in salt water for more than 1,500 years.

A statue more than two millennia old

The frogmen also retrieved a large statue, broken into two parts, from the sea bed at the fort, and also made of red granite. It lacked parts of its legs but was seven metres long and weighed roughly 25 tons. The monument represented a woman standing with her left foot forward and her arms at her sides. The eye sockets were empty, but unquestionably they once held appropriately coloured stones (as was usual in ancient Egyptian art), and traces of the sacred snake were visible on her forehead. The statue had a supporting column at its back. All this shows that the sculptor had been heavily influenced by the canons of ancient Egyptian art.

The lady wore, however, a Greek gown with a knot tied between her breasts; in Graeco-Roman times, this knot was associated with, specifically, the goddess Isis. This deity was also depicted with plaited hair hanging in symmetrical strands down both sides of her head. It is thus very likely that this statue does indeed represent Isis and that it dates from the third century B.C.

It may be of interest to see how the divers managed to raise such a statue. On 16 October 1963, Sami Dessouki, Alexandria correspondent of *Al-Ahram* (a large Cairo daily), wrote the following account:

The young naval officer heading the team of frogmen described the difficulties he and his men had encountered under eight metres of water, where visibility was limited to one metre because of the sewers that empty into the area near Fort Qaitbay. He told how they had worked for a whole month strapping the two halves of the statue with 4.3-inch steel cable. As a first step, trenches had had to be dug in the sea bed beneath the statue's two sections in order to pass the wire ropes. [The team] had lifted one side of the bottom section still under water, then, in order to pass the cables through, two frogmen had risked their lives by standing under the statue. They would surely have been killed if the cable had broken In fact, the cable broke once under the weight of the top section (10 tons).

And the journalist continued,

The statue of Isis is not the only archaeological artefact under the sea. The young officer said that the site also included what appeared to be a ruined temple, scattered granite columns each about a metre in diameter, a great wall with a sphinx four metres long and three wide, carved in relief, four headless statues, and one of a recumbent sphinx –each approximately two metres long.

Another naval diving officer told me that, under the water off the Al-Silsila promontory, he had seen four large sarcophagi similar to ones he had seen in the Museum. Thus has underwater archaeology opened up exciting possibilities for the specialized study of the city of Alexandria.

The lighthouse of Alexandria

It is hardly possible to speak of old Alexandria without referring to its beacon, one of the seven wonders of the world, built in the third century before Christ under Ptolemy II. A three-floor structure rising to a height of probably 120 metres, it cast a bright light that some historians say could be seen for 30 nautical miles. (It has been suggested that the lighthouse may have been fitted with some sort of magnifying device or a reflecting mirror). It is interesting to note that the word for beacon in some European languages is derived from 'Pharos' Island, and that lighthouse technology is known as pharology.

Probably the most authoritative description of Pharos was made by Strabo: '. . .the extremity of the isle is a rock, washed all around by the sea, and has upon it a tower that is admirably constructed of white marble. . . .'[5] Hague and Christie have written that 'it is unlikely that a building so magnificent and so successful in its purpose would have been the prototype. Its importance, grandeur and in fact its inclusion among the seven wonders of the Ancient World have secured for it a position of pre-eminence in the annals of civil engineering.'[6]

The Pharos was represented by engravings on coins and medals, mostly from the mints of Alexandria but also from other places in the Roman Empire. Engraving was severely limited by the size of the coins, of course. Many later drawings are too highly imaginative, and still others are either incompetently done or too small in scale to show detail. The thirteenth century mosaic to be found in the Chapel of St Zeno in St Mark's Basilica in Venice purports to show Pharos as a backcloth to an episode in the life of St Mark; it is likely, however, that the artist was more familiar with the numerous lighthouses along his own native coast than that he should ever have set eyes on the famous Pharos which, by that time survived only in truncated form.[6]

A record height of construction

It is believed that the earlier Pharos Island was larger than what remains of it nowadays. Even though the island was sinking, its facilities remained fully operational until after the Arab conquest, as it was around the year 700 that the beacon's lantern fell from the top of the lighthouse.

Al-Mas'udi has left a reliable, eyewitness account of the lighthouse as he saw it, in 944. He estimated its height to be 230 cubits (a cubit was roughly 50 cm), a figure confirmed by Abdullatif al-Baghdadi, a meticulous scholar who visited Egypt at the time of Salah al-Din (Saladin).

In 955 a violent earthquake brought down the top nine metres of the tower; then, in 1302, during the reign of the sultan al-Nasir Muhammad ibn Qalawun, another major tremor damaged Alexandria's fortifications, including ramparts and lighthouse. After Ibn Battuta paid a visit to Alexandria in 1350, he wrote, 'When. . .I visited the lighthouse, I found it in such a ruined state that it was no longer possible to. . .approach its portal.'

When the sultan al-Ashraf Qaitbay went to Alexandria in 1477, he ordered that a new tower be built on the site of the old; this structure still stands today. One of the best contemporary studies on the architecture of the fabled lighthouse is the large volume by Hermann Thiersch who attempted to determine in 1909 the tower's original construction on the basis of an examination of the many documented descriptions of the Pharos as well as of monuments and minarets which, according to him, had been influenced by the Pharos architecture. Thiersch's picture agrees with the accounts of Arab chroniclers.[7]

Probably the most reliable and detailed description was that written by Abu-al-Hajjaj (also known as Ibn-el-Sheikh)[8], an Arab from Malaga; he visited Alexandria in 1165 as a competent architect and meticulous chronicler. He gave detailed measurements of the three-level lighthouse—the first of which was square, the second octagonal, while the third was cylindrical and topped by a small domed mosque (7 m high, 14 m in circumference). The last was built, after the Arab conquest, at the centre of the roof where the lantern had once stood. It would seem that the tower itself was by then 122 m high from its base and 135.5 m high if the foundation was included. Hague and Christie have called it the tallest roofed structure built until the steel-framed American skyscrapers appeared in the twentieth century.

Questions of material and form

Honor Frost, the British diver and archaeologist, believes that the large statue of Isis and other figures discovered by Kamel Abu Al-Saadat (the diver) in 1961 are almost certainly remains of the famous Pharos; and, although they do not help in making a reconstruction of the tower, they give a vivid impression of its quality and grandeur. In addition, Al-Saadat found the remains of buildings that may be the lost palaces of Alexander and Ptolemy kings—where it is supposed that Alexander's final resting place, a glass sarcophagus, lies. Thus Al-Saadat presented archaeologists with potential locations of two of the most sensational sites in Mediterranean culture. This is spectacular evidence of the growing importance of submarine archaeology in revealing undiscovered gaps in our knowledge of an ancient community such as Alexandria.

In 1968 the Egyptian Government invited (via Unesco) Ms Frost from London together with the geologist, Vladimir Nesteroff from the University of Paris, to examine the Pharos site with a view to its eventual excavation. During six dives in autumn 1968, Ms Frost and Al-Saadat surveyed the shallow water off the Fort. The British explorer wrote:

> . . .the salient finds in an area of some 180 m^2 were planned by direct measurements as well as theodolite readings. Our photographic coverage was doomed in the choppy, cloudy. . .sea, while the shallowness (3-4 m) made it impossible to 'frame' masonry several metres long. . . .[9]

Frost gave a list of seventeen different items found on the site, with a brief description of each. She extrapolated that such evidence would be multiplied a hundredfold after a complete survey. In 1979, an Alexandrian photographer, Bruno Vailati, and Kamel Abu Al-Saadat led an underwater camera crew with the Italian photographer, Paolo Curto, in covering the same site.

Strabo had said that the Pharos beacon was constructed of white marble, yet most of the larger architectural elements found are made of Assuan (Aswan) granite. Frost has noted the 'white marble' is a translation of the Greek *leucos lithos*—brilliant or white stone. Surrounded by the sea, marble would wear badly even when, as Toussoun has shown, it was placed on the sheltered or harbour side of the building. The apparent contradiction may be explained by the polychrome nature of Ptolemaic architecture, whereby the stone was covered with plaster before being painted.

As to the lighthouse's shape, Curto hypothesizes that the structure stood 200 m high (much taller than the original estimate of 120 m), since some of the submerged ruins lie far offshore. Vailati's divers found columns, architraves (the horizontal members atop and joining two columns) and plinths (supporting bases of columns) estimated to weigh 30-50 tons. Curto assumes that the Pharos was probably of a different shape than that usually believed.[10]

One problem that has baffled archaeologists and other scholars is that of the construction of such a high building upon a narrow base—a task that may well have been more difficult than building one of the great pyramids not far away. The fact that one side of the Pharos seems to have collapsed into the sea and that it still lies *in situ,* unaffected by overbuilding, implies that the foundations of the original structure must have been at the water's edge. Frost notes that this form of architectural brinkmanship, dictated by local geography, is characteristic of other ancient harbour buildings found along the Levantine coast, notably the walls of Arados and the offshore harbour at Sidon.

The years are passing and nothing further has been done to explore this important site, a wonderful opportunity for submarine archaeologists. Perhaps the security conditions prevailing between 1967 and 1974 hampered serious attempts to gain access to the Pharos site, causing the project to be forgotten. As recently as May 1984, Honor Frost noted that 'it has been the deepest disappointment for me to have been unable to help survey' the area. 'It would have been easy,' she continued, 'to raise international support for such a project, from both the archaeological world and the press.*

Roundabout Abu Qir

The town of Abu Qir lies east of Alexandria, situated on an ancient site once occupied by three towns. The best known of these was Canopus, which stood at the mouth of the Canopic, or westernmost, distributary of the Nile delta. This was one of the seven distributaries that flowed before the ninth century and of which five have since dried.

Canopus and the surrounding communities constituted a major commercial and religious centre at the time of the arrival of Alexander the Great. From here he sailed across Lake Mareotis to the town of Rakotis (the site of Alexandria). A strong rivalry developed between Alexandria and Canopus that ultimately led the rich merchants as well as the markets to move to the new city. Subsequently, Alexandria's prosperity flowed back in the direction of Canopus and Minotis, nearby. The area prospered, linked to the metropolis by a navigation canal which carried throngs of visitors bound for its temples or in search of entertainment (innocent and otherwise).

It was not long before a number of temples arose, including the Temple of Aphrodite at Abu Qir Point and the Temple of Osiris or Serapis—the remains of which are believed to be among the ruins of Canopus. The first named of these temples became celebrated for its reputed power to protect seamen from the perils of the sea, while the second acquired fame for its curative powers and thus attracted many pilgrims.

Abu Qir's submarine remains

The Temple of Aphrodite (or Zephyron) was built on the tip of Abu Qir Point, a zone subject to rough seas because of this coastal region's variable climate. A chronicler of the time tells us that the religious structure itself was so lofty that it reached almost to the heavens.

---

* Personal communication.

Breccia gives an account of the excavations which he undertook in the Abu Qir area, referring in several passages to the archaeological remains extending from the shore and disappearing under the waves. He describes a number of pools or baths and alludes to the submerged remnants of a massive building that he believes to have had a connection with one of the temples.

Despite this long history of exploration and the wealth of temples, churches and other edifices that have come to light, it appears that the antiquities uncovered to date represent only a small portion of the total. Many of Abu Qir's monuments suffered heavily during the various religious conflicts of the past and, among those surviving, there has been extensive plunder. Some have been used (as in Rome) as quarries, providing cut stone for new construction. Considering what has happened to the shoreline of Alexandria, however, it is clear that the whole story of Abu Qir has not yet been told—that the waters are bound to contain innumerable relics in a much better state of preservation than would have been possible if they had been left exposed on the surface, within the reach of irresponsible people.

In 1859, a Suez Canal Company engineer named Larousse discovered that the ancient Canopic distributary extended for a distance of eight kilometres into the bay. This feature, now appearing on hydrographic charts, probably explains the reference in the writings of Pliny to an island at the mouth of the Canopic distributary. Omar Toussoun has argued that this may be a reference to Nelson Island, which lies 11 km out to sea.

Between 1930 and 1933, partly as a result of observations made by a British Royal Air Force officer flying from the base at Abu Qir, statuary remains were brought up from the floor of the bay. These included a larger-than-life head of Alexander the Great, made of white marble (now Item 23848, to be found in Room 12 of the Museum).[11]

The Canopic distributary was considered to be the major channel for trade in and out of Egypt. What was certainly the oldest port in the country, and perhaps one of the oldest natural harbours in the world, prospered at the mouth of the distributary. Archaeologists disagree as to the exact locations of Canopus, Minotis and Meraclium, but they agree that these three towns lay in the environs of Abu Qir Bay.

The resting place of Napoleon's fleet

In my *Submerged Civilizations*, I devoted two chapters to sunken vessels in, respectively, the Mediterranean and other seas of the world. Most of the hulks discovered in the Mediterranean date from ancient civilizations. Elsewhere in the world, with a few exceptions (such as Viking ships*), most discoveries have been more recent than the fifteenth century—when maritime travel and piracy burgeoned. Some of the naval and merchant vessels that have been

*Viking ships were built around A.D. 1000; some of those found at Roskilde Fjord, Denmark, for example, date from the early eleventh century.

discovered are of great material and historical value. The Swedish naval ship
Vasa that sank in Stockholm Harbour on 3 August 1628 is one example,
while another is the British *Mary Rose,* sunk before the eyes of Henry VIII in
Portsmouth Harbour (19 July 1545). The latter's wooden shell was raised in
October 1982, after its cannon and 17,000 miscellaneous objects had been
salvaged.

It may seem strange to include Napoleon Bonaparte's armada, which has
lain in the waters of Abu Qir Bay since 1798, in a discussion of true antiquities. It is not odd, however, if we take into account the efforts being devoted
to the task of searching the waters of North America, Europe and Australia
for antiquities dating only since the age of the great geographical explorations. The sea battle of Abu Qir (or the Battle of the Nile), although well
documented, is no less important for Egyptian tourism than Al-Alamain (El-
Alamein) lying to the west of Alexandria.

Under Abu Qir Bay's calm waters reposes part of the fleet that carried
Napoleon to Egypt. In one of history's most famous engagements, the British
naval units commanded by Admiral Nelson sank most of Napoleon's flotilla.
The battle was a crucial turning point in the struggle between Britain and
France during the eighteenth century. The wealth of information extant
provides underwater archaeologists with a great opportunity. Anyone wishing
to locate the vessels sunk during combat would do well to begin by comparing eye-witness reports with the accounts contained in the logs of surviving
ships, and by consulting the charts and writings of the experts of the period.
There follows a simplified account of the engagement.

A systematic assault gets under way

In summer 1798 a task fleet of thirteen ships of the line, under command of
Admiral Nelson, was scouring the Mediterranean, seeking the French units
reportedly under sail from Toulon for an unknown destination. On 28 July,
while Nelson was off the coast of Greece, he learned that the French navy
had reached Egypt carrying Napoleon and an invasion force. Nelson headed
for Alexandria, where he arrived on 1 August only to find a few small naval
craft in the harbour. He decided to pursue his search to the east.

At 2:45 PM of the same day and fifteen miles to the east, Nelson's lead
ship signalled the presence of a flotilla lying at anchor off Abu Qir. Abu Qir
Bay, just east of Abu Qir Point, is bounded by a semicircular coastline. Its
floor 'shelves' very gently towards deep waters lying at a great distance from
the shore. To the north, from Abu Qir Point to what came to be known as
Nelson Island, extend some rocky shoals.

After the Napoleonic expedition had disembarked at Alexandria (earlier,
on 1 July) a French task force under Admiral Brueys* sailed to Abu Qir Bay
where the French flag officer believed his fleet to be safe because of natural

---

*François Paul Brueys d'Aigaïlliers (1753-98), vice-admiral.

protection. His ships dropped anchor in line, close to shallow waters full of dangerous rocks, and the fleet set up batteries of cannon on the island. Believing that he was thus completely protected on one flank and that he would have to defend only his seaward exposure, the French commander failed to consider the possibility that his units might be encircled (Figure 2). Although his staff officers were divided as to whether their force should go into action with their sails set or lying at anchor, after some hesitation Brueys opted for the second possibility.

Of the seventeen French vessels, nine carried 74 guns each, three were equipped with 80 guns each, while the mighty flagship *L'Orient* had 120 cannon. Since Brueys expected that he would be attacked only from his seaward flank, he placed his larger ships at the centre of the line in order to protect both its wings. The British fleet of thirteen vessels, each carrying 74 guns, sailed towards Abu Qir in line astern, the first ten ships being followed at a considerable distance by the remaining three.

By 3 PM Brueys was informed of the presence of the British squadrons; he began his preparations on the assumption that the enemy would not arrive until nightfall and would not risk attacking in the dark in waters presenting an array of obstacles and navigational problems. Nelson, on the other hand, estimated his time of arrival at sunset; he possessed no reliable charts of the Bay except a number of French maps of the region. These were in the hands of the captain of the lead ship and played an important part in the subsequent course of events.

Despite all the circumstances that could have justified delaying the attack until morning—and contrary to the expectations of both his men and the enemy, Nelson decided on an immediate assault. The plan he devised was to bring all his ships to bear on a few French vessels at a time, rather than attack the entire force all at once.

The lead British ship, *HMS Goliath,* reached the island as the sun was setting on Abu Qir Bay. She reduced speed and sailed up to the end ship of the French line, then passed between the latter and the gun positions on the island—maintaining a course that kept her from shallow waters. *Goliath* made her way through with such precision that the French thought there must be Egyptian pilots aboard. And just as Nelson had suspected, the French ships were not prepared to fire on the landward side. Four more Royal Navy units followed suit, each emptying her guns into the same French ship, which sank as the fifth British vessel sailed to the attack.

From the ruins of battle, a certain cultural heritage

The British warships then took up positions between the French navy and the shore. As the sixth ship (*HMS Vanguard,* Nelson's flagship) sailed up in her turn, she headed for the other flank and was followed there by the remaining four units. In this way the French fleet was caught in a cross-fire. The rest of the engagement was fought in total darkness broken only by flashes of light

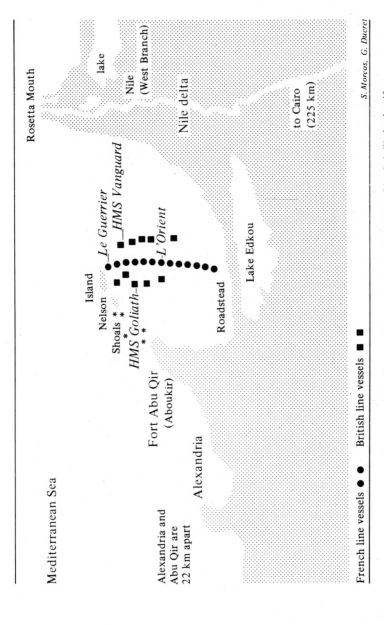

S. Moros, G. Ducret

Figure 2. Details of the positions of the French and British fleets during the battle at Abu Qir. It is likely that if Admiral Brueys' vessels had set sail, instead of fighting while at anchor, French losses would not have been so severe. Note that only 13 French and 10 British ships are shown here. (Reconstructed after J.A. Foex in *Etudes et Sports Sous Marins*, Nov.-Dec. 1984, and R. Grenfell, *Nelson the Sailor*, London, Faber, 1949.)

French line vessels ●●   British line vessels ■■

from the 500 or so guns that were in action across the Bay—providing an awe-inspiring spectacle for people watching from the beach.

The British now concentrated their fire on the seventh vessel, Brueys' flagship, *L'Orient*. As soon as two warships sent by Nelson to reconnoitre Alexandria harbour arrived in the Bay, they went into action. *L'Orient* was soon set afire and, at 9:45 PM the ship blew up with a terrific detonation and began to settle in the water. Within half an hour, the first and second French ships had ceased firing; two hours after that, the first five ships had ceased all resistance. By midnight, the northern half of the French flotilla was completely destroyed.

With the arrival of dawn, *L'Orient* had gone to the bottom. Two other ships had run aground while still three others were listing. Four vessels emerged safely from the encounter. The British captured six ships and sank *Le Guerrier*, too damaged to be towed.

So it seems likely that at least four vessels lie beneath Abu Qir Bay's calm waters, including *L'Orient*. The foregoing account suggests that the French flagship sank at the southernmost point of the French navy's line. As this is the point where that line was closest to shore, the remains should not be difficult to locate, inspect and salvage. Some have maintained that something of the shipwrecks can be seen, and that their mastheads are visible under water on windless days with a calm sea. Because *L'Orient* is thought to have carried the pay of fleet and army personnel transported to Egypt, location and recovery of the ship would be an attractive operation. Substantial archaeological and other technical benefits would accrue from such an effort—which would also stimulate the tourist industry.

The individual efforts of Kamel Abu Al-Saadat succeeded in identifying in 1965 three sites of wrecks, in collaboration with local fishermen. Later, he led excavation missions of four divers. Sami Dessouki's reporting in *Al Ahram* of 19 August 1977 noted that a silver coin bearing the legend 'Ferdinand V, King of Naples,' as well as the date 1798, has been recovered.

Generations later: 'Operation Bonaparte'

In 1978 the French lawyer and diver, Jacques Dumas, became fascinated with the prospect of exploring and rescuing wreckage from the French fleet at Abu Qir. He studied old documents, visited museums and conferred with specialists—thus reconstructing an accurate account of what had transpired on 1 August 1798 and of what might be found beneath the water. After enlisting the support of the Musée de la Marine in Paris and the French Embassy in Cairo, he received Egyptian permission in 1983 to dive and salvage.

In June 1983 the French minesweeper *Vinh Long*, working in close collaboration with the Egyptian navy, found the site of the disaster of *L'Orient*, 8 km from shore in Abu Qir Bay. The wreckage is strewn over a length of 150 m, with guns and anchors scattered about. In the autumn

Dumas, working with Al-Saadat and other Egyptian divers, detected the sites of the 74-gun *Le Guerrier* and the 4-gun frigate *L'Artémise.*.

These discoveries prompted the Egyptian authorities to begin building a museum at Abu Qir to house the finds. French support came from Institut Français de Recherche pour l'Exploitation de la Mer (IFREMER) and the public utility Electricité de France, who gave two laboratories for the treatment of metallic and non-metallic remnants. The Regional Council of Provence-Alpes-Côte d'Azur also provided support, since *L'Orient* had been built in Toulon (whence Napoleon's fleet had sailed to Egypt) about two centuries earlier.

L'Orient had been a floating castle, the largest of its kind at the time. A three-deck vessel, the ship had 120 guns and carried 850 sailors and marines. One should expect the explorations to provide not only items of navigational or military interest, but also items from daily life. Bonaparte had occupied Malta on his way to Egypt, so that precious treasure may have been aboard these vessels. (Bonaparte was two weeks shy of his 29th birthday when his ships went to the bottom of Abu Qir Bay).

By June 1984, a concerted attempt to search for the French fleet began aboard the small French ship, *Bon Pasteur,* with a 12-man team of French and Egyptian divers concentrating on *L'Orient*. This encouraging start was marred by the sudden death of Kamel Abu Al-Saadat, who succumbed to a heart attack. A gallant and committed diver and archaeologist had been lost.

Present and future rediscovery of old technology

Dumas brought up a large silver spoon from one dive, but so far no one has recovered hoards of the gold, silver and jewels fabled to have been seized by Napoleon from the Knights of Malta. A treasure of everyday objects has been found, however—buttons, buckles, wine bottles and kitchen spoons, all preserved in the silt deposited by the Nile nearby. A miniature world of the daily things needed in life (coins, cups, pots, pistols and swords) are slowly adding to a reconstruction of living in the late 18th century.

The silver coins found faithfully identify the era. Some bear Louis XV's profile, some that of Louis XVI, and others carry the 'Union et force' slogan of the French Revolution and the dates *An II, An IV* or *An V*. Buckets of seared bronze plating and splintered wood are eerie reminders of the fiery explosions that sent *L'Orient* to her grave. *L'Artémise*'s binnacle has been recovered, as have been dividers, T-squares and even pencils that can still be used.

The salvage task had two main efforts. A suction pump operates an 'air lift' to draw the muddy water covering the wreckage through a hose. Dumas and a colleague spent much of their time at a sieve on board ship, examining and recording the artefacts recovered. Divers from the Egyptian fleet and the Société Française d'Archéologie Sous—Marine (SOFRAS) manhandle larger items, while Gilbert Fournier undertakes the delicate task of directing the vacuum cleaner-like hose along the sea floor.

A temporary exhibit of the objects removed from below the waters opened in June 1984 inside the restored Fort Qaitbay in the Eastern Harbour. On an average day, the *Bon Pasteur*'s crew hauls up a flour mill (for example), nails, spikes, bullets and bits of lead type in both Arabic and Latin characters—all mixed with stones, coral and sludge sucked from the sea bed. Napoleon had acquired his Arabic-language press from the Vatican, 'borrowed' by his chief scientist, Gaspard Monge (the mathematician and inventor of descriptive geometry). Bonaparte's declarations to the Egyptian population were disseminated via this Arabic press, the first to be used in Egypt. It would be interesting to reconstitute French views on the creation of indigenous mass media during a military occupation.

As to the origins of *L'Orient,* the first authentication of the hulk's identity was made when its rudder was cleared of sand and debris. It had remained in perfect condition (it was copper-clad), with the inscription *Le Dauphin Royal,* its first name when it was launched during the reign of Louis XVI. Plans are now under way to salvage the huge, 12-metre high rudder and a 6-metre long anchor as well as several other parts of the vessel.*

Now, almost 170 years after Napoleon's ill-fated Nile campaign, we shall be learning more about its mechanics thanks to submarine exploration. The French leader had left Toulon with 33,000 men and 300 horses aboard 350 vessels of all kinds. Besides arms, ammunition and the other accoutrements of war, the armada had a scientific commission of 167 scholars under Monge's supervision.[13] The explorations begun in our time will surely continue to reveal technological and other cultural progress of centuries past.

---

* We regret to announce that when this chapter went to press, Attorney Jacques Dumas died suddenly on 22 March 1985, while lecturing in Agadir, Morocco. The above mentioned plans may be subject to delays or alterations as a result of this tragic death.

Notes

1. G. Jondet, Les ports antiques de Pharos, *Bull. Soc. Archéologique d'Alexandrie,* No. 14, 1912; Les ports submergés dans l'ancienne île de Pharos, *Mémoires présentés à l'Institut Egyptien,* Vol. IX, Cairo, 1916; (with Malaval) *Le port d'Alexandrie,* Cairo, Port and Lighthouse Administration, 1912.
2. F. El-Fakharany, The Old Harbours of Alexandria (public lecture series), University of Alexandria, 1962-1963 season (in Arabic).
3. Strabo, *Geography,* Vol. XVII.
4. Fl. Josèphe, *La guerre des Juifs avec les Romains,* Vols. III, IV.
5. Strabo, op. cit.
6. D.B. Hague, R. Christie, *Lighthouses: Their Architecture, History and Archaeology,* Cardiff, Gomer Press, 1975.
7. H. Thiersch, *Der Pharos, Antike Islam und Occident—Ein Beitrag zu Architekturgeschichte,* Leipzig and Berlin, von Teubner, 1909.
8. For a shortened version in English, see M. Asin, M. Lopez Otero, The Pharos of Alexandria, *Proc. Brit. Acad.,* Vol. 19, 1933.
9. H. Frost, The Pharos Site, Alexandria, Egypt, *Nautical Archeology,* Vol. 4, 1977.

10. P. Curto, The Seventh Wonder of the World, the Pharos Lighthouse, *Oceans,* Vol. 14, No. 4, 1981.
11. O. Toussoun, Les ruines sous marines de la Baie d'Aboukir, *Bull. Soc. Archéologique d'Alexandrie,* No. 29, 1934.
12. A good account is given by Jacques Dumas, Sous la mer, les canons de Bonaparte, *Géo* (French ed.), No. 74, April 1985.
13. About a year after the Battle of the Nile, which did not drive the French from Egypt, France's expeditionary force fought the Turks at the Battle of Abu Qir (25 July 1799). On 8 March 1801, Great Britain sent Lieutenant-General Sir Ralph Abercromby to dispossess the French of Egypt; he landed his troops at Abu Qir under severe opposition. Abercromby fell near Alexandria, on 21 March, hit by a spent ball, and died on 28 March. The French forces left later in the same year.

See also

E. Breccia, *Alexandriea ad Aegyptum,* Municipalité d'Alexandrie, Alexandria, and Instituto Italiano, Bergamo, 1914.
E.M. Forster, *Alexandria: A History and A Guide,* Alexandria, Whitehead Morris Ltd., 1938.
S.A. Morcos, *Sunken Civilizations, the Story of Submarine Archaeological Discoveries,* Cairo, Dar-el-Maaref, 1965 (in Arabic).
————, Submarine Archaeological Discovereis in the Mediterranean Sea, *Arch. and Histor. Studies* (Alexandria Archaeological Society), No. 1, 1968 (in Arabic).
Published recently and not cited in this chapter: S. A. Schwartz, *The Alexandria Project,* New York, Delacorte Press - Eleanor Freide, 1983.

International co-operation in oceanic research and between non-governmental and intergovernmental organizations is essentially a twentieth-century phenomenon, in spite of the fact that the first non-governmental conference on oceanography was held in 1853. Initially the co-operation was linked with hydrography, navigation, marine meteorology and (to a lesser extent) fishery resources. The development of this co-operation and the interactions between non- and intergovernmental bodies are traced over the past century.

Chapter 20

Co-operation among non-governmental organizations in fostering oceanic research

F. W. G. Baker

'Mike' Baker is Executive Secretary of the International Council of Scientific Unions (ICSU). In 1957 he became indirectly involved in non-governmental, co-operative oceanic research through the secretariat of the International Geophysical Year; he maintained this interest through contacts with ICSU's Scientific Committee on Oceanic Research and the Special Committee for the International Biological Programme. Dr Baker co-edited the volume of the Annals of the IGY *on oceanography (No. 46) and contributed a chapter on carbon dioxide. His address is: ICSU, 51 boulevard de Montmorency, 75016 Paris (France).*

Background

'Include the sea and make the plan universal, and we will go for it.' This reply by the United States of America to the British Government's proposal for a uniform system of meteorological observations was evoked by M. F. Maury at the Maritime Conference held in Brussels in 1853. This was not only the first of the oceanographic conferences but also the start of international non-governmental co-operation in fostering oceanic research.[1] The meeting arose from the communication of a project prepared by Captain Henry James, F.R.S., to establish 'a uniform system of recording meteorological observations' that seems to have been developed following the Meteorological Conference held at Cambridge in 1845.

The Council of the Royal Society, which has been asked for advice, suggested that

every ship that is under the Admiralty should be furnished with instruments properly constructed and compared, and. . .instructions for making and recording observations as far as their means will allow, should be sent to every ship that sails, with a request that the results of them be transmitted to the Hydrographer's Office of the Admiralty, where an adequate staff of officers or others should be provided for their prompt examination, and the publication of the improved charts and sailing directions to which they would lead,[2]

The first national non-governmental attempts to stimulate co-operation in oceanic research had been made almost 200 years earlier within the framework of the Royal Society. In 1662 Lawrence Rooke prepared *Directions for Seamen Bound for Far Voyages,* in which he asked them to record magnetic dip and variation, weather, comets and other phenomena, the times of tides, directions of tidal streams, to measure variations in salinity in different latitudes and at different depths, etc. Four years later the Royal Society adopted proposals from Sir Robert Moray that all ships should make observations and record their results in two journals: one for the Royal Society and one for Trinity House. *Philosophical Transactions,* No. 24, is devoted to a revised version of Rooke's *Directions.* Thus began not only the idea of standard observations but also the idea of data centres with the data deposited in at least two centres to void the effects of natural catastrophes.

Although the 1853 conference was a success, many years were to elapse before the next oceanographic conference. In the interim there were two major developments: the first was the creation of an international organization of meteorologists, which continued to see ships as suitable platforms for making metorological observations; and, second, the first major voyages of scientific study of the oceans by research vessels such as the *Challenger,* which surveyed the world's oceans, from December 1872 to May 1876, as well as the *Novara, Valditia, Tuscarora, Bache* and the *Blake.* Although the results were published and were available internationally, there was still no international body to bring together the oceanographers. However, non-governmental discussions about oceanography were taking place within the

International Association of Academies (IAA) at the end of the nineteenth century; but the first international organization in oceanography, the International Council for the Exploration of the Sea (ICES) was created in 1902 as a result of an emergency in the West Swedish herring fishing industry. Seventeen years later the IAA, which had become the Association of Interallied Academies as a consequence of the First World War, suggested that an International Association of Physical Oceanographers be formed in 1919 at the first assembly of the International Research Council—the immediate predecessor of the International Council of Scientific Unions.

The International Association of Physical Oceanography (IAPO)

IAPO, a section of the International Union of Geodesy and Geophysics, which was created in 1919 in Brussels at the first general assembly of the International Research Council, held its first general assembly in Paris in January 1921 under the Presidency of H.S.H. Prince Albert I of Monaco. The scientific scope of the association was summarized as follows: morphology of the ocean floor; morphology of the sea surface; movements of water masses; physical and chemical studies of sea-water.

Because there was no formal association for the biological oceanographers, the biologists were invited to participate in the 1922 assembly in a subsection on biological oceanography and the name of the unit was changed to the Section of Oceanography. The biologists' response was so lacking in enthusiasm for this collaboration that in 1930 the name was changed back to IAPO, and its scientific scope was limited to scientific studies of the sea using mathematics, physics and chemistry.

The association did, however, continue its attempts to develop cooperation between other fields of oceanography by inviting the presidents of the following three intergovernmental organizations to be vice-presidents of the Association: (a) the International Council for the Exploration of the Sea (ICES), which was founded in 1902; (b) the International Hydrographic Bureau (IHB), founded in 1919; and (c) the International Commission for the Scientific Exploration of the Mediterranean (ICSEM), founded in 1919. This initiated close co-operation between intergovernmental and non-governmental organizations, which has been of steadily growing importance in fostering oceanic research ever since.

Suggestions were made for research committees for the Atlantic and Pacific Oceans but, because oceanographic research had been actively carried out in the Atlantic by ICES, the International Ice Observation and Ice Patrol Service in the North Atlantic and the North American Fishery Association, IAPO's role was limited. It was, however, responsible for the stimulus that led to the creation of the Consejo Oceanográfico Ibero-Americano in 1929, and it played a lesser role in the creation of the International Committee on the Oceanography of the Pacific in 1923. In addition to stimulating the establishment of regional oceanographic organizations and its co-operation with

Glossary of abbreviations

ACMRR	Advisory Committee on Marine Resources Research
ACOMR	Advisory Committee on Oceanic Meteorological Research
BIOMASS	Biological Investigations of Marine Antarctic Systems and Stocks
CCCO	Joint Committee on Climate Changes and the Oceans
CMG	Commission on Marine Geology
CSAGI	Special Committee for the IGY
ECOR	Engineering Commission on Oceanic Resources
FAO	Food and Agriculture Organization
GEBCO	General Bathymetric Chart of the Oceans
IAA	International Association of Academies
IABO	International Association of Biological Oceanography
IACOMS	International Advisory Committee on Marine Sciences
IAPO	International Association of Physical Oceanography
IAPSO	International Association for the Physical Sciences of the Ocean
IBP	International Biological Programme
ICES	International Council for the Exploration of the Sea
ICSEM	International Commission for the Scientific Exploration of the Mediterranean
ICSU	International Council of Scientific Unions
IGY	International Geophysical Year
IHB	International Hydrographic Bureau
IIOE	International Indian Ocean Expedition
IOC	Intergovernmental Oceanographic Commission
IPY	International Polar Years
JCO	Joint Commission on Oceanography
SCAR	Scientific Committee on Antarctic Research
SCOR	Scientific Committee on Oceanic Research
WMO	World Meteorological Organization

them, the Association played a vital role in the standardization of instruments, terminology and techniques, the issue of samples of standard sea-water, the interchange of research material and the collection of results of observations and research. The Association also co-operated closely with the IHB in (a) studies of mean sea-level and its variations, and in the creation of a network of recording stations, and (b) continuation of the General Bathymetric Chart of the Oceans (GEBCO). It was also instrumental in developing a network of stations and in standardizing observations of tidal elevations and currents and their analysis. It initiated, jointly with the International Association of Meteorology, studies of interactions between the sea and the atmosphere.

The Association continues to play an active role in these studies as well as in activities both within the framework of the geophysical union and in co-operation with other international bodies. Members of the association have been active in the developments that led to the creation by ICSU in 1945 of the Joint Commission on Oceanography with the participation of the geophysicists and the biologists; the JCO was dissolved in 1954 after stimulating ICSU to: play an important role in the renewed production of the General Bathymetric Chart of the Oceans, and in establishing an ad hoc group which suggested the creation of the Scientific Committee on Oceanic Research; to develop a programme of oceanography during the International Geophysical Year; and to create the International Advisory Committee on Marine Sciences and the Intergovernmental Oceanographic Commission, which are referred to below.

The Polar Years and the International Geophysical Year

Although the main scientific fields under study during the first two International Polar Years (IPY), 1882-83 and 1932-33, were meteorology, aurora and geomagnetism, the first two years had oceanographic and hydrographic programmes and six nations participating in the first IPY and eight in the second. The French IPY I expedition at Orange Bay, Cape Horn, recorded the effects of the Krakatoa explosion on its self-recording tide gauge—the first such self-recording instrument to be used in an international programme. The Third IPY, which became the International Geophysical Year (IGY), had a much wider and more co-ordinated programme.

At the first meeting of the Special Committee for the IGY (CSAGI) in 1953, proposals for a programme of oceanic research were so few that the report indicates: 'It does not appear that oceanographic observations are likely to be actively pursued unless additional countries are prepared to participate.' Two of the proposals are, however, worth citing: first, within all sea areas systematic research should be carried out with simultaneous observations of the partial pressure of CO_2 in air and sea in order to establish the amount and direction of CO_2 flux; secondly, certain aspects of the geochemistry of the hydrosphere and the atmosphere should be studied. There was much more enthusiasm at the second and following meetings, but it was not until January 1957 that the 'final' programme was developed at a meeting of the IGY Working Group on Oceanography held at Göteborg under the chairmanship of C. O. D. Iselin.

This programme was outlined under the following headings: Long wave recording—including the generation, propagation and maintenance of surface oscillations with periods between two minutes and two hours; Sea-level recording; Deep water circulation; Polar-front surveys—developed in close association with ICES and the International Commission for the North-West Atlantic Fisheries; Multiple ship measurements—using three or more vessels to make simultaneous measurements of water movements and density distribution; Circulation through straits; Carbon dioxide measurements at sea; Radioactivity measurements; Biological investigations;[3] Bathythermograph

observations from weather ships; Bathyscaphe observations; Study of waves
and swell.[4] An outline of the results of the programme is given in Volume 46
of the *Annals of the IGY*.

There have been mixed comments on the IGY programme in oceanography. G. F. Humphrey wrote:

The International Geophysical Year is the prime example of how the countries of the
world joined together in a co-ordinated effort to learn more about the physical nature of
our planet [but] at the time of the planning of the IGY oceanographers were not yet
ready to organize large co-ordinated projects.

Oceanographic congresses

The first major Oceanographic Congress took place, 31 August-11 September
1959, at the United Nations building in New York. The meeting was sponsored by the American Association for the Advancement of Science (AAAS),
SCOR and Unesco. Roger Revelle, who was closely linked with all three
bodies, was president. The congress' themes were Boundaries of the Sea,
Cycles of Organic and Inorganic Substances in the Sea, the Deep Sea, History
of the Oceans, and Populations of the Sea. Following the success of the first,
a second congress was organized in Moscow, with the IOC and SCOR playing
the key convening roles.

The third congress was held in Tokyo (1970) and the fourth in Edinburgh
(1976). To give an idea of the number of organizations involved in Tokyo,
the international committee for the assembly included representatives of
SCOR, IAPSO, IABO, CMG of ICSU, IOC, Unesco, the ACMRR of FAO, the
ACOMR of WMO, ICES and ECOR. The most recent Joint Oceanographic
Assembly (as these meetings are now called) was held in Halifax, Nova Scotia,
in August 1982; there the major organizational role was played by SCOR,
with strong support from Unesco and the IOC. Although the Assemblies have
become the forum at which new research results and trends are discussed
among the global oceanographic community, it is interesting to note that the
subjects chosen for the first meeting were still featured in the last.

The Convention on the Continental Shelf

The interest in the oceans stimulated by the IGY and the growing concern of
oceanographers that possible political pressures might restrict freedom to
conduct scientific research in the soil of the continental shelf and in the
waters above, led ICSU to make objections to some of the draft articles in the
proposed Convention on the Continental Shelf. These were presented through
Unesco to the UN's International Law Commission. The revised Convention
adopted at the International Conference on the Law of Sea, Geneva, Feb-
ruary - April 1958, met to a considerable extent ICSU's objections.

The efforts to maintain this freedom of oceanic research resulted in the

8th General Assembly of ICSU, Washington, D.C., October 1958, which adopted the following resolution:

The 8th General Assembly of ICSU

resolves that the National Members of ICSU be requested to ask their Governments, when ratifying the Convention on the Continental Shelf, to signify that at the same time they grant general permission to any scientific research vessel to conduct investigations of the bottom and sub-soil of the continental shelf, providing the programme is specifically approved by the International Council of Scientific Unions, which is the international organization acting for national scientific academies; whereby ICSU will guarantee that the investigations are leading to results that will be published openly for the benefit of science,

and whereby the coastal state should be notified sufficient time in advance so that it may if it desires designate a representative to take part in the work,

noting thereby that this proposal is designed to facilitate the operation of Article 5 of the Convention by assisting Governments to identify bona fide scientific research and to avoid diplomatic delays which would jeopardize many types of scientific investigation.

A number of ICSU's National Members communicated this recommendation to their Governments and ICSU transmitted it to Unesco for submission to the UN. None of the signatories signified their willingness to grant this general permission. With hindsight one is tempted to suggest that this would have been totally impossible because it would have meant that ICSU would have been forced to put a scientist on each research vessel to ensure that the results would be published openly. No mean task!

The ICSU-SCOR push for open research

This lack of response did not, however, diminish either ICSU's interest or activities in the problems of scientific oceanographic research in relation to the Law of the Sea. The General Assembly adopted in 1974 the following resolution:

The 15th General Assembly of ICSU

recalling the resolution adopted in 1972 by its 14th General Assembly in support of open oceanic research, intended for the general benefit of mankind and characterized by full and timely availability of research plans and results that are to be used to the best advantage.

recognizes that the conduct of oceanic research is being affected by new restrictions and issues that have arisen in recent years both as regards the extension of zones under national jurisdiction and the open ocean, urges National Members of ICSU: i) to advise their governments of the importance of effective international cooperation in long-term programmes of oceanic research, and ii) to encourage the participation of their scientists in the formulation of national policies. Such policies would improve cooperation and facilitate: a) the further growth of marine sciences, and b) an increase in the competence in this field of the developing countries interested, and would emphasize the need for all countries to assist in these aims, thereby forming a basis for an international agreement at the United Nations Conference on the Law of the Sea,

recommends that the UN and specialized agencies greatly strengthen their assistance to coastal states: a) for the interpretation and application of the results of ocean research, and b) in the training of marine scientists,

requests the Secretary General of ICSU to bring these recommendations to the attention of the organizations concerned of the UN system,

invites SCOR to provide a short statement on the scientific aspects of the possible further development of oceanography under the new ocean regime.

ICSU and SCOR continue in their efforts in support of open oceanic research intended for the general benefit of humankind.

SCOR and the International Indian Ocean Expedition

If the oceanographers were not ready at the start of the IGY, they had made a lot of headway even before it started when, in 1954, in the framework of the Joint Commission on Oceanography a suggestion was put to ICSU that it create a more permanent body. This group was initially called the Special Committee on Oceanic Research (SCOR), then later the Scientific Committee. The original aim of SCOR was to further the co-ordination of scientific activity in all branches of oceanic research and the framing of scientific programmes of worldwide significance. The first meeting of SCOR was held in 1957 at Woods Hole, Massachusetts, United States.

The committee decided that the main objective should be to develop and co-ordinate an international, deep-ocean programme. It drew attention to three long-range problems considered to be critical for the future of humanity: (a) the use of the deep sea as a receptacle for the waste products of our civilization; (b) the oceans as an important source of protein for mankind; and (c) the role of the oceans in climatic change.[5] The Indian Ocean was chosen as the scene for this international expedition because: (a) the Indian Ocean at that time was one of the least investigated and is one of the controlling factors in the development and dissipation of the monsoons, and (b) over a quarter of the population of the world lives in countries bordering the Indian Ocean—a population in which a protein deficient diet is common, and which could benefit from a study which would provide a basis for the wise expansion of fisheries.[6] In 1960 the first meeting of the Indian Ocean Expedition working group was held, drawing up a basic minimal plan. The programme evolved into a large international enterprise and in late 1960 Unesco agreed to cosponsor the expedition. One important feature of the expedition was that it provided educational opportunities for scientists on the ships of countries other than their own and stimulated an interest in marine science.

The history of the International Indian Ocean Expedition (IIOE)[7] gives information about the growth and demise of the International Advisory Committee on Marine Sciences (IACOMS) and the creation of the Intergovernmental Oceanographic Commission (IOC), so I will not repeat it here. Suffice it to say that the IOC and SCOR agreed in April 1962 that 'as from now on the formal responsibility and authority for the co-ordination of the IIOE will be given to the Secretary of IOC.' This began a co-operation between IOC, an intergovernmental organization and SCOR, a non-governmental organization that continues today. These bodies have formed a joint Committee on Climate Changes and the Oceans (CCCO) which is working on one of the problems suggested at the first meeting of the IGY Committee in 1953.

Productivity of the oceans

The question of the productivity of the oceans raised by L. Zenkevitch was pushed back into the international arena when ICSU began its International Biological Programme (IBP) in 1964. Marine productivity was one of the five special project areas and led to a wide international programme the results of which are currently being published, such as the volume on *Marine Production Mechanisms* edited by M. J. Dunbar. Although the IBP terminated in 1974, studies of marine productivity continued. One of the most recent programmes on marine productivity to be launched, by SCOR with the Scientific Committee on Antarctic Research (SCAR) and the International Association of Biological Oceanography (IABO) with the Advisory Committee on Marine Resources Research of FAO and the consultative parties to the Antarctic Treaty, is that on Biological Investigations of Marine Antarctic Systems and Stocks (BIOMASS). The First BIOMASS Experiment has already taken place: the second is in preparation. Information about the investigations is published regularly in BIOMASS *Newsletter*.

Although my article is about organizations, it should not be overlooked that organizations are formed by people. In the early stages of oceanic research these tended to be solitary individuals, but, as oceanographic research developed, the numbers grew and the need for co-operation developed. In one of the major growth periods, in the 1950s and 1960s, it is interesting to see how few oceanographers were involved in the executive bodies of the major organizations. The list of members of the three principal organizations, IACOMS, IAPO and SCOR, gives a total of thirty-two members; but seven oceanographers, G. Bohnecke, A. Bruun, G. E. R. Deacon, M. Eyries, H. Mosley, R. R. Revelle and L. Zenkevitch, all figured in two of the steering committees of these three bodies.* This certainly helped to facilitate exchange of information and avoided unnecessary duplication.

In more recent years there has been a tendency for some kind of cross-fertilization between IABO, IAPSO, IOC and SCOR—with Anton Bruun, George Humphrey, Henri Lacombe and Roger Revelle having been chairmen of two of the organizations. In addition the first two secretaries of the IOC, Warren Wooster and Konstantin Fedorov, both became presidents of SCOR after leaving the IOC. It is interesting to note that two of the magic seven mentioned above, George Deacon and Roger Revelle, who were still (in 1984) *active* in IOC and SCOR. Salt water is a good preservative.   ■

---

* George Deacon died on 16 November 1984.

Notes

1. Although the participants were invited by the United States Government, all the members of the conference stated that they had no authority whatever to pledge their country in any way to the proceedings. *Maritime Conference Held at Brussels for Devising a Uniform System of Meteorological Observations at Sea,* p. 12, Brussels, Imprimerie Hayez, 1853.
2. Minutes of the Council of the Royal Society, 22 April 1852.
3. These biological investigations were suggested by IUBS and by L. Zenkevitch of the USSR, who put forward a proposal for the productivity of the sea; a theme that was taken up later and adopted by ICSU's Special Committee for the International Biological Programme. Zenkevitch later, in 1961, submitted to SCOR a proposal for a proposal for an Experimental Tropical Oceanic Station (ETOS) in the Indian Ocean but this never became reality.
4. *Annals of the IGY,* Vol. IIB, pp. 592-8, Oxford, Pergamon Press, 1959.
5. It is interesting to note that the original SCOR report putting forward these three problems had a curious typing error: commenting on problem 3 'It is probable that the waters of the ocean play a magic role in changing climates'. 'Magic' should, of course, have read 'major'.
6. See the article by S.A. Qasim in this book.—Ed.
7. D. Behrman, *Assault on the Largest Unknown,* Paris, Unesco, 1981.

To delve more deeply

BEHRMAN, D. *Assault on the Largest Unknown—The International Indian Ocean Expedition.* Paris, Unesco, 1981.
DEACON, M. *Scientists and the Sea 1650-1900—A Study of Marine Science.* New York, Academic Press, 1972.
HERDMAN, W.A. *Founders of Oceanography and their Work.* London, E. Arnold, 1923.
MAURY, M.F. *Physical Oceanography of the Sea.* New York, Hooper, 1855.
Procès verbaux. Paris, Association d'Océanographie Physique, n.d. *SCOR Proceedings,* Vol. 1, No. 1, 1965.

A distinguished personality of the Intergovernmental Oeanographic Commission describes how a concerted effort by universities and related institutions can help equip a country to engage in oceanic research programmes suited to its national needs. The author offers the experience of his own country, Mexico, as a model possible to be applied elsewhere. This chapter was presented as Contribution 316 of the Instituto de Ciencias del Mar y Limnologia, at the Mexican Autonomous National University (UNAM).

Chapter 21

The role of universities in building national capability in the marine sciences

Agustin Ayala-Castañares

The author, who is immediate past chairman of the Intergovernmental Oceanographic Commission, was trained in biology, micropalaeontology and marine geology. He is a member of numerous Mexican and International professional bodies. His specific interests concerning oceanography have been in specialized education, long-term science policy and planning, and the conception and building of national infrastructures in the field. Professor Ayala-Castañares' address is: Director, Instituto de Ciencias del Mar y Limnologia, UNAM, Apartado Postal 70-157, Mexico 04510 D.F.

Introduction

In today's world, a country must develop its own scientific and technological capabilities in order to progress economically and otherwise. The process of establishing science and technology requires a 'critical mass' of highly qualified personnel and the governmental will to invest significant funds in education and research over a long period.

Developing nations must base their socio-economic progress on scientific data, concerning the natural characteristics and resources of their territory, in order to establish the foundations of their industrial transformation.

Such a process is particularly valid in the case of the ocean. The sea must be considered integrally, including its origin and history, its physical, chemical, geological and biological processes, and its resources, as well as the social and economic consequences of its exploitation and administration. For these reasons, oceanography in its modern concept is not a science by itself, but a combination of scientific and humanistic disciplines: the marine sciences. These allow us to consider the sea as a whole—with an inter-disciplinary approach—and, in consequence, to make proper decisions concerning the use and management of our marine resources.

Marine scientific research is costly: it requires highly qualified people and very specialized vessels, equipment and installations. In all these aspects, the gap that exists between the marine scientific capabilities of industrialized and developing countries is daunting.

The universities are called upon to play an important role in building the scientific infrastructure required in developing nations by training the necessary number of scientists and technicians. Furthermore, university research programmes in science and technology undoubtedly represent an essential ingredient of development.

Scientific research at the Mexican Autonomous National University (UNAM)

UNAM, as a national institution, plays an important role in the country's life. It has been the major source of professional people required for the technological and administrative servicing of the nation; it has been able to allocate significant funds for research, aimed at finding solutions to national problems and extending cultural benefits. Research at UNAM was formally established in 1929 (when its autonomous legal status within the government was approved) and at which time three previously established scientific research institutes became part of the university. Currently twenty-one insti-tutes and centres are devoted to science and many others to the humanities.

UNAM now plays an important role in the national system of science and technology, mainly because of generous funding by the Federal Government.

Until recently our scientists dedicated themselves primarily to basic problems, but there is now a trend to venture into the applied sciences. Traditionally UNAM's research institutes concentrated on specific fields (biology, chemistry, geology, physics). During the last fifteen years, a trend to establish interdisciplinary units has developed, as exemplified by the Institute of Applied Mathematics and System Research, the Material Research Institute and the Institute of Marine Sciences and Limnology.[1]

Recognition by the government of the value of science and technology increased greatly in 1970, with the creation of the Consejo Nacional de Ciencia y Tecnología (CONACyT)—National Science and Technology Council.

Institute of Marine Science and Limnology (ICML)

This institute is a good example of the evolution of scientific research at UNAM. A small start in marine research was made in 1939 by the Institute of Biology; during the second half of the 1950s additional and more ambitious activities commenced in the Institutes of Geology and Geophysics. The number of scientists in the three institutes was initially small, the physical and financial resources minimal, and no connection existed between the institutes. In 1967, the Department of Marine Sciences and Limnology was established in the Institute of Biology, linking existing elements with part of the marine geology group.

In 1973, the Centre of Marine Sciences and Limnology (CCML) was created, integrating the manpower, installations, equipment and budgetary resources of the Institutes of Biology, Geology and Geophysics (see Fig. 1). Strong institutional support has been given since then, in an attempt to provide proper elements for consolidating the marine sciences.

A great deal of the initial efforts were dedicated to manpower training, both in foreign institutions and locally (with the valuable assistance of experts provided by a large United Nations Development Programme/ Unesco/CONACyT activity, as well as the Multinational Programme of Marine Sciences of the Organization of American States. Most of the research was concentrated in coastal or nearshore studies, in several regions of the country, particularly coastal lagoons. All offshore work was carried out using vessels borrowed from the Mexican navy or on board foreign vessels. In 1976, a graduate programme in marine science was established.

In 1979, an academic evaluation of the CCML was carried out, and as a consequence it was decided to (a) obtain a research vessel to strengthen capability at sea and (b) upgrade the centre within the university structure

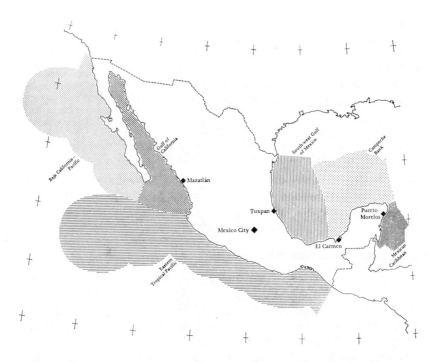

FIG. 1. Generalized map of Mexico and its Exclusive Economic Zone, showing its oceanographic regions and the location of facilities of the Instituto de Ciencias del Mar y Limnología (ICMC).

to the level of an institute. Subsequently (1981) the CCML became the ICML by decision of the University Council. One of its major tasks was to realize the great potential represented by the 10,000 kilometres of coastline, 500,000 square kilometres of continental shelf and more than 3 million square kilometres of Exclusive Economic Zone, declared by Mexico in 1976, when the National Constitution was modified.

Objectives of the institute

The objectives of the institute are to:
Contribute to interdisciplinary knowledge of the seas and continental waters, as well as their resources.
Contribute to the scientific and technological development of the country through the development of the marine sciences and limnology, and to participate in the solution of problems of national significance.
Contribute to the preparation of highly qualified human resources in marine sciences and limnology.

Establish the necessary relations with different academic units within
    UNAM, as well as related institutions, through co-operation and advice.
Stimulate the establishment of institutions and stations in marine sciences
    and limnology in different zones of Mexico.
Contribute to the spread and diversification of the marine sciences and
    limnology.

Areas of research

These include physical oceanography, chemical oceanography and pollution,
geological and geophysical oceanography, biological oceanography and
fisheries biology, and limnology.

Programmes

There are currently more than eighty research projects being carried out in
coastal or marine areas at different locations in the Pacific Ocean, the Gulf of
Mexico and the Caribbean Sea, and in Mexican waters.

Human resources

The total number of employees is 245, of whom 118 are academic staff
consisting of 62 full-time researchers and 56 full-time academic technicians.
There are also 127 administrative and support personnel. In general, the
community is young and significant numbers of graduate students are
finishing their doctorates abroad.

Publications

The institute publishes two journals: *Anales del Instituto de Ciencias del Mar
y Limnología* (annually, since 1974) and *Publicaciones Especiales del Instituto
de Ciencias del Mar y Limnología* (occasionally since 1978).

Teaching programme

The Graduate Programme in Marine Sciences (with specialities in physical
oceanography, chemical oceanography, geological oceanography, biological
oceanography and fisheries) was adopted in its present form in 1976. Special
emphasis is given to the active participation of the students in research under
close tutoring. The average number of students is about twelve per year,
selected from about a hundred applicants. The present student population
of the project is eighty-one, with fifty-eight at Master's level and twenty-
three at Doctor's level. Students come from UNAM, other Mexican universities and from many Latin American countries. Up to now fifty-six
Master's and six Doctor's degrees have been granted.

ICML installations are in the main campus of UNAM in Mexico City: 'Mazatlán' Station at Mazatlán, Sinaloa; 'El Carmen' Station, at Ciudad del Carmen, Campeche; 'Puerto Morelos' at Puerto Morelos, Quintana R oo.

Research vessels

Besides a considerable number of small craft, the institute operates two medium-size vessels specially designed and built for multiple research and training activities in the seas off Mexico.

The first of these vessels became operational in 1981; it has been used for work off the west coast of the country. In order to cover the need for a similar research facility along the east coast, a second vessel was built in 1982. This became operational in early 1983. These modern ships are operated by crews of fifteen and have a capacity for scientific groups of up to twenty persons. They contain ample work space in the form of laboratories, mechanical and electronic workshops, a recording room, a lecture room, and the like. Given their advanced basic equipment and general laboratory and deck capacities, these ships are well adapted to perform the multiple tasks of modern marine research. The operation and expenses of both vessels are shared by UNAM, CONACyT and PEMEX through an interinstitutional agency.

International co-operation

A great deal of the development of the institute is the result of the catalytic effect of international co-operation, both multilateral and bilateral.

Multilaterally, the already mentioned UNDP/Unesco/CONACyT programme has given substantial support. This project, which was begun on a small scale in 1963, grew between 1976 and 1980, and is currently providing expertise, particularly in offshore work and vessel operation. The Multinational Programme of Marine Sciences of the Organization of American States was also useful in this respect. Participation in Unesco's Intergovernmental Oceanographic Commission has provided many useful contacts. In the field of the bilateral co-operation, several projects have been carried out with different institutions: in the United States, the Scripps Institution of Oceanography, Stanford University, Texas A&M University, Louisiana State University, Oregon State University; in the United Kingdom, the University of Liverpool; in France, the University of Bordeaux, the University of Marseille; in Costa Rica, the University of Costa Rica; and numerous others. This co-operation is very useful, and negotiations are in process with Australia, Israel, the Federal Republic of Germany, Colombia, Panama, Brazil, Ecuador, the Soviet Union, and others.

Closing remarks

The ICML has come to reach an important stage as a solid institution. It is hoped that it will be a centre of excellence.

The ICML is not the only Mexican academic institution helping to create an infrastructure in marine science. Others are also making important contributions. The academic institutions are the basis, indeed, for the future of marine sciences in Mexico—as exemplified by the currently intensive use of the new research vessels. This situation is the result of institutional policies as well as governmental decisions.

In the governmental sector, Mexico is beginning to assign high priority to the ocean, as typified by the fisheries bureau being raised to secretariat level, and the significant increase in offshore oil developments, as well as the political decision to sign and support the United Nations Convention of the Law of the Sea.

The way in which UNAM has contributed to the building of a national infrastructure in marine science is unique among developing countries, and may serve as a useful example for other nations to emulate.   ■

Note

1. Limnology is the scientific study of bodies of fresh water, especially ponds and lakes.

To delve more deeply

AYALA-CASTAÑARES, A. *The Enhancement of Marine Science Capabilities, Future Directions*. Bologna, The Johns Hopkins University Marine Sciences Workshop, 1973.
——. Marine Science and Technology. In: K. H. Standke and M. Amandakrishnan (eds.), *Science, Technology and Society, Needs, Challenges and Limitations. Proc. International Colloquium on Science, Technology and Society, Vienna*. Oxford, Pergamon Policy Studies on International Development, 1979.
——. Las ciencias del mar y el desarrollo de México. *Ciencia y desarrollo*, No. 43, March/April 1982, pp. 14–27.
——. *Creating Favourable Conditions for the International Cooperation for the Transfer of the Marine Science and Technology in the Context of the New Ocean Regime*. UNU/IOC/Unesco, Workshop on International Cooperation in the Development of Marine Science and the Transfer of Marine Technology in the Context of the New Ocean Regime. (In press.)
AYALA-CASTAÑARES, A. et al. Estructura y evolución de la investigación científica en la UNAM. *Ciencia y desarrollo*, No. 34, September/October 1980, pp. 33–48.
FLORES, E. Science and Technology in Mexico: Toward Self-Determination. *Science*, Vol. 216, 1983.

INSTITUTO DE CIENCIAS DEL MAR Y LIMNOLOGÍA. *Buques oceanográficos 'El Puma',* *'Justo Sierra'.* Mexico City, Universidad Nacional Autónoma de México, 1982.
PROGRAMA NACIONAL INDICATIVO PARA EL APROVECHAMIENTO DE LOS RECURSOS MARINOS. *Ciencia y tecnología para el aprovechamiento de los recursos marinos (Situación actual, problemática y políticas indicativas).* Mexico City, Consejo Nacional Ciencia y Tecnología, 1982.
SOBERON ACEVEDO, G. La investigación como función universitaria esencial. *Ciencia y desarrollo,* No. 34, September/October 1980.
UNIVERSIDAD NACIONAL AUTÓNOMA DE MÉXICO. La universidad en marcha. *La universidad en el mundo,* No. 21, 1980. (Special number.)

*Discussion of information flow and knowledge transfer is most opportune
in the wake of the Law of the Sea Convention and the demands being made
under the banner of the new international economic order. It is urgent
that the results of oceanic research benefit all nations, via a programme
involving all countries. Development of the concepts and rules applicable to
marine science has not kept pace with technical change making possible
better sensing, observation and measurement; nor has this development kept
pace with the revolution in information technology.*

Chapter 22

The flow of information
and transfer of knowledge

*Geoffrey Leighton Kesteven studied zoology at Sydney University,
Australia. Since graduation, in 1936, he has engaged in fisheries research
and administration with the Australian Government, with the Food and
Agriculture Organization of the United Nations, and with other
organizations and institutions. He obtained a D.Sc. degree in 1948. He can
be reached at the following address: 12 O'Brien's Road, Hurstville 2220,
New South Wales (Australia).*

Although a discussion of information flow and knowledge transfer may be opportune at the moment, it is done at some risk of being trivialized in the details of communication engineering as well as in the politics of sharing know-how. This risk may be reduced somewhat if we agree that beyond information and knowledge our goal is understanding.

Marine science shares common ground with meteorology and astronomy in that most of its experiments are conducted by nature herself, who tantalizes us with a thimble-rigging game; in this game, we must guess under which thimble we shall find not a pea but the identity of the factors nature has changed to generate the phenomena commanding our attention. Marine scientists are thus committed to a regime of observation and measurement, the start of the flow of information, which in all respects is massive.

Because of the great geographic range of oceanic continuities (the Gulf Stream, for example) and the significance of discontinuities (such as the thermocline), the spatial dimensions of the observational regime are considerable. Similarly there is a large range over time, from extremely slow processes (e.g. the northward transport of Antarctic waters) to extremely rapid ones (such as the turbulence of coastal zones). All these signify a great extension in time and space of an impressive array of biotic and abiotic variables.

Some oceanic phenomena are of global dimensions. For instance, El Niño, the phenomena of cold currents once looked upon as more or less the private property of Ecuador and Peru, is now seen to be part of a broad complex comprising events in Malaysia and the Atlantic Ocean, as well as clear across the Pacific Ocean; these phenomena are associated, in general, with atmospheric phenomena. Because of such characteristics, it is often difficult in marine science to specify bounds to a system and to determine how far beyond the boundaries one need go to assess the strength of external forces bearing upon the system and influencing the behaviour of variables within it.

Considering the dimensions of these systems, the continuities within them and the interactions among them, it is clear that the observational regime needs to be large and complex. The regime needs, simultaneously, intense local components and work of broader scope at other levels. But operation and collection of arrays of data are only the beginning. Each datum is written into a table or coded electronically so that it is the subject of a sentence having the form: 'This datum, X, is the value observed of a variable, Y, at position P at time T on day D.' The number of such sentences in any marine research project is great and obviously must be reduced, otherwise the mass of data would be incomprehensible.

Data characteristics and the grammar of science

Putting aside the case in which X is the same at various positions over a space of time, the general practice is to examine the statistical characteristics of the value of X: either alone (in which case P, T and D are 'classifiers'), or in interaction with values of P (in examination of variations of spatial variation of Y with time), or with values of T and D (in examination of variations of Y with time), or with all three. The result in each case is the formulation of a new sentence, and the sentences become progressively more complex and of broader connotation as more sentences are compressed into one. For example, an array of simple sentences can be replaced by a summarizing sentence to the effect that 'this datum, X, is the mean of the values observed of the variable, at Y, at places, P, at times, T, on days, D'. Then, using the grammar and syntax of scientific thought, these sentences are combined into propositions with regard to single variables and sets of variables.

The grouping of these sentences into paragraphs reporting, or conceptualizing, the relations between variables in complex sets (and between sets of sets) and appraising the significance of such relations is obedient to the rules of logic, whether this is accomplished intuitively or by imagination, or by a laborious manipulation of possible arrangements.

The purpose of this elementary revisit to the grammar of science is to emphasize that the 'flow of information' is not a matter merely of transmitting and receiving a stream of figures (even coded data), however sophisticated might be the equipment employed for this purpose (as, for example, the observation by satellites). The subject needs to be discussed at several levels, from the primary sentences of raw data, through various levels or reduction and analysis, to substantive descriptive propositions, which may themselves be of predictive power or components of predictive models.

Examination of the subject through these several levels must be made along three paths: the semiotic, the epistemological and the logistic. These paths intersect at several points. In moving along them, attention must be paid to problems which hinder participation by the less developed countries (LDC) in ocean research.

The meanings of symbols

We find it convenient to understand the word 'symbol' to signify 'a sign to which by common consent a particular meaning has been assigned', so that one who writes or draws a particular symbol may assume that anyone seeing the symbol will understand an assigned meaning. Conversely, a person seeing a given symbol may assume that the intention of its author is that the assigned meaning should be understood.

The flow of information is generally understood to refer to the transmission and reception of messages, whether vocally, by a signalling apparatus, electronically, or by the passing of inscribed material. While this description may serve the purposes of communication engineering, it passes over the fact that the message (*pace* McLuhan) is only the medium and is not the meaning: the message is not information as such. There are, therefore, two aspects of the flow of information to be examined. On the one hand, there are the mechanics, as it were, of the transmission and reception of messages which we shall soon examine along a logistical path; on the other, there is the fundamental task of formulating a message so that, upon being read, it communicates accurately to the reader the information possessed by its originator. We are concerned here, that is, with the rules of formulating messages and not with matters such as the 'corruption' or degradation of messages during transmission.

We can agree at the outset to spend no time on matters such as the effects that characteristics of a medium may have on a reader's understanding of a message transmitted through it; we assume that we are concerned solely with the formulation of unambiguous messages to communicate facts ascertained by agreed procedures, as well as propositions constructed logically with regard to the contents of such messages. By this I mean material that does not need 'body talk' in order to complete the meaning, and whose meaning is not modified by visual or aural 'colour' of the transmitting medium.

The meaning of specific terms

The rules to which we refer relate to (a) the standardization of terms, so that these will serve effectively as symbols within unambiguous messages and propositions, and (b) the construction of sentences, in so far as this ensures that the symbols used retain the meanings given by the rules of standardization. Standardization is required within each language and between languages. Much progress has been made in standardizing symbols used in primary messages relating to physical and chemical variables, but less has been done for biological oceanography. The imbalance is still greater, and widening, at higher levels, more particularly in social sciences.

The problems can be exemplified in fisheries research. Although measurement of 'fishing power' and fishing effort is of crucial importance to studies of the dynamics of exploited fish populations—and has been a major preoccupation of fishery biologists for many years, no standard definition of a unit of fishing power has yet been adopted. Fishing effort (the number of units of fishing power exerted) is recorded in a wide range of units, from the cast of a net to a whole year of operating a boat.

At another level, there is little agreement on how to refer to components of a fishery system, whether by 'unit stock', 'fishing unit', and 'unit fishery', or on terms by which to refer to properties of these units, such as 'standing stock' (of resources) and 'fishing capacity' (of a fishing unit). At still another level, at virtually the summit of fishery biology, there is confusion between 'maximum sustainable yield', 'maximum sustainable economic yield', 'total allowable catch' and similar terms. The situation in ecology is little better, exemplified by the favoured term 'ecosystem', meaning really little more than 'that which is studied by an ecologist'.

Rules for the construction of sentences specify, at primary level, the contextual information—units in which measurements are made, their place, time and date—that must be included in order to give full meaning, as well as the order in which these terms should appear. At ascending levels, rules specify the information that must be included with regard to methods of reduction and compression; they deal, at all levels, with grammar, syntax and logic, or the qualities of clear, direct and unambiguous communication.

Acquiring knowledge, gaining wisdom

A simple fact reported in a primary sentence ('Water temperature was observed to be 15.5 °C at a given place at a certain hour of a particular day') in itself tells us little. The message can be made meaningful only by reading it in conjunction with other informative sentences, or with propositions, telling us whether the observed value is one that (under the circumstances) was to be expected or not—but then, from knowledge of the processes determining water temperature at the observational point, the datum could indicate something about the behaviour of these processes at the time given.

Clearly, this ascent of the slopes of understanding is protracted, often difficult. Successful assault of a peak of knowledge does not assure us of a clear view and understanding of the entire terrain. So knowledge can be seen to stand at different levels—not merely levels of quantity of known facts but also levels of ability to perceive relations and to comprehend. In the language of computer operations, this is a matter of ability to file facts selectively so that they can be easily 'accessed' and of whether we have developed and can use routines by which to review our stock of facts, appraise them, select from them those which are relevant, and test the relations between them.

A transfer of knowledge, then, is not a matter simply of handing over packages of what is known—in textbooks and journals. Just as effective reception of a message requires a capacity to comprehend the information it conveys, so at the receiving end of a transfer of knowledge there must be a capacity to place received knowledge in a larger context and to perceive the significance of the enlargement of meaning that thereby results. Nor can this capacity be created by supply of a software package; it must be developed by each worker as he or she conducts research. That is to say, the capacity is to be gained in the course of winning new knowledge as, conversely, new knowledge is to be won only by those acquiring and developing this capacity.

Impartiality and new paradigms

The efficiency of the exercise of a capacity to assign 'received' knowledge and especially to perceive the significance of relations thus disclosed, is much influenced by the paradigmatic stance of the person exercising the capacity. Thus, to quote from Stephen Toulmin,

Rutherford bowed out from his authoritative position with good grace—admitting candidly that his own training left him unequipped to master the new abstractions [of quantum mechanics], and hopelessly prejudiced in favour of a material model of atoms and fundamental particles, as 'little hard billiard-balls—preferably red or black'.[1]

This aspect of the matter has the effect for those concerned with the transfer of knowledge that their procedures need to be not only impartial with regard to the efforts of workers engaged with different paradigms, but also capable of drawing attention to the implications of new paradigms and of assisting those to whom knowledge is being transferred to maintain flexibility in their appraisal of significance of relations.

The point I have sought to make in the preceding paragraphs is that the body of knowledge grows by a process of which the acquisition of knowledge and achievement of understanding are parts. Much learning leads often to only little understanding, and understanding often lies idle; no new knowledge comes save through understanding reached from a base of knowledge.

The relevance of the foregoing to the future of ocean research may be clearer when it is considered in relation with the importance of the LDCs having a larger and more effective part in ocean research. I presume that I have been asked to deal with knowledge transfer in this special issue of *impact of science on society*, with a view to exploring the entire subject of ocean research but more particularly to examine the problems of transfer of knowledge to developing nations in terms of the importance of the future of ocean research and how this may be accomplished.

System identification, hierarchization of variables

Our argument proceeds as follows:
Ocean research must be simultaneously global and local.

Research in waters of LDCs is required for the world programme.

Research in the waters of LDCs should be conducted with substantial participation by specialists from these countries (if not wholly so).

The conduct of research by scientists from LDCs should not be the execution of observational procedures designed elsewhere; instead, original research capable of contributing new knowledge and of having a part in the development of new research regimes is required.

Therefore procedures of knowledge transfer should be designed to assist LDC specialists to acquire capabilities rather than have them imposed.

Although our discussion deals with general principles, something must be said about an aspect of the procedures of ocean research which is rather particular to it and thus has much effect on the planning of its strategies, including system identification and delimitation, and the hierarchization of its variables. (These remarks concern solely marine biology.) One characteristic of a mature science is that its practitioners are able to identify (and thus delimit) systems—the objects of their study, which are autonomous to a high degree, their behaviour can be predicted from knowledge of them, from observation of their state at the time of prediction, and from measurement of external factors bearing upon them.

Efficient exercise of this ability entails a narrowing of the task of system monitoring by the identification of (a) one or more systemic properties whose behaviour is most reliably indicative of the system's state and (b) the external factors most critically influencing the system. It is unnecessary to elaborate on the economies of observational work and gains in precision that are possible when a science has achieved command at this level, or on the advances that then become possible.

What may be noted is the distance yet to be covered in marine biology. This can be illustrated, using fisheries biology as an example of applied ecology. Workers in this field, confronted with the fact that the unit stocks of single species (of which they have made 'assessments') do not qualify as autonomous systems, are now occupied with models of multiple-species stocks.

Assistance to less developed countries

The many programmes of Unesco and the Intergovernmental Oceanographic Commission must pay much attention to the advances made in science by the improvement of information flow and knowledge transfer, as discussed above. The LDCs can benefit from this progress, but special attention must be paid to the disabilities under which the LDCs labour. This benefits the LDCs directly, but it is also an important contribution to science in general.

One seemingly simple step is to develop the system of oceanographic data centres (ODC) so as to bring them within the reach of LDC institutions and, at the same time, to expand the biological component of the ODC programme. This would be a project concerned with hardware and software, yet it need not imply great proliferation of mainframe equipment. On the contrary, the activity should be based on mini- or even microcomputers at the user level; equipment should be organized 'upwards' through national, regional and worldwide centres. These should include storage, processing and other services appropriate to user capacities and requirements, whether at country or regional level. With respect to software, a first task would be the standardization of terms, followed by the formulation of rules with respect to the use of terms and formulation of sentences.

Another matter is that of translation. An active programme of assistance would not attempt to promote comprehensive translation of all textbooks and journals relating to marine science; it would seek to bring some rationality to scientific literature by arranging for translation of severely edited and reduced material. The Food and Agriculture Organization's programme of Species Synopses and Species Identification Sheets admirably exemplifies the method that could be employed.

The most substantial part of the programme would be that which sought to promote the capacity to acquire knowledge and achieve understanding. To go into detail with respect to this part of the programme is beyond the scope of this article but, briefly, its main drive would be to bring about the participation of LDC scientists in activities at the working-face of original research, in place of pursuing imitative research. The Intergovernmental Oceanographic Commission's programme for the preparation of Marine Science Country Profiles could have considerable effect in this connection. ■

Note

1. S. Toulmin, *Human Understanding* (Vol. 1), Oxford, Clarendon Press, 1972.

The fundamental role of science and technology in the economic and social development of all countries is a basic assumption underlying negotiations within the United Nations for the reshaping of economic relations worldwide. These negotiations have been motivated mainly by the perception of inequalities in the distribution of wealth and power—one of the most conspicuous features of this perception being the differences in the scientific and technological capacities of different countries. The Law of the Sea seeks to redress some of the imbalance.

Chapter 23

Science, technology
and the new Convention
on the Law of the Sea

Maria Eduarda Gonçalves

*Maria Eduarda Barroso Gonçalves holds several degrees (including the doctorate) in law, her specialty being international public law.
Dr Gonçalves has also studied at The Hague Academy of International Law and the Commission of the European Communities. She is currently a staff member of the National Board of Scientific and Technological Research in her native Portugal, where she is in charge of research on the international legal aspects of science and technology, and assistant professor at the New University of Lisbon. She participated, as a member of the Delegation of Portugal, in several sessions of the United Nations Conference on the Law of the Sea. Her address is: Junta Nacional de Investigação Cientifica e Tecnológica, Av. D. Carlos I, 126, 1200 Lisboa (Portugal).*

Introduction

The Charter of Economic Rights and Duties of States[1] is a major juridical instrument, global in scope, emerging from negotiations within the United Nations system. The charter acknowledges the rights of states to benefit, without reference to their levels of development, from progress in science and technology; it provides explicitly for a special obligation of states to facilitate access by developing countries to scientific and technical advance, in ways corresponding to their economies and needs. Developed nations are thus called upon to co-operate with developing countries to help them to strengthen their scientific and technological infrastructures, including research and development. The charter takes into account the interests of developing countries in the formulation of guidance concerning technology transfer.

It is partly because of this background that the United Nations convened the Conference on the Law of the Sea, whose work was recently completed with the adoption, in April 1982, of the Convention on the Law of the Sea.[2] In its preamble, the convention defines as its aim the establishment of a legal order for the seas and oceans, one that would ensure the equitable and efficient use of their resources, and the 'study, protection and preservation of the living resources thereof'. It is explicitly assumed that the attainment of this goal will contribute to a just international economic order, for coastal as well as landlocked countries.

In this context, the conference considered that access to marine science and technology constitutes an essential condition for the effective exercise of the rights and the fulfilment of the duties assigned to states for the exploration, exploitation, management and conservation of marine resources, and, ultimately, for enabling them to take advantage of new opportunities of economic and social development. The conference was also conscious of the fact that participation in scientific research activities and technological development, as well as access to technology, should not be seen as an endeavour separate from the development of national scientific and technological capabilities. All these aspects are interdependent. From the point of view of the developing countries, it is not sufficient to acquire knowledge produced abroad; they need to obtain an adequate degree of autonomous capability, too. This should enable them to make their own choices, adapt imported technologies to local conditions, and participate as partners in the universal process of research and development related to the oceans.

These assumptions led to the introduction in the convention of specific provisions on (a) marine scientific research and (b) the transfer of marine technology; the provisions take into account the special situation of developing countries.[3] Besides the codification of the existing International Law of the Sea, the conference had also as objective its progressive development. It is worth appreciating the innovative aspects of the convention, as these consist of new opportunities opened to states in terms of marine science and technology. Such opportunities find their expression directly in the rights, and indirectly in the duties, assigned to states by the convention—rights and duties determined by the relative situation of states in terms of development and geographical position in relation to the sea.

Marine scientific research: some 'rights and duties'

The basic feature of the new regime of marine scientific research in the Exclusive Economic Zone and on the continental shelf is the jurisdiction which is conferred upon coastal states with regard to such activity.[4] This jurisdiction implies the rights of coastal states to authorize, regulate and conduct marine scientific research.[5] It follows that nations have gained the power to control (at least indirectly) activity taking place in these areas which may lead to the acquisition of knowledge concerning marine resources and ecosystems. This includes the power to require, as a necessary condition for their giving consent to a research project, that foreign entities admit nationals of the coastal state involved to direct participation in the project, grant access to the research results, and assist in their interpretation.[6] The jurisdiction of a coastal state with regard to marine scientific research could be understood as instrumental in the sovereign rights of the coastal states for the exploration, exploitation, conservation and management of marine resources. The jurisdiction, in other words, is the legal condition for the obtention of the scientific basis necessary for such activities.

Coastal states are formally assigned rights, therefore, in marine scientific research. The question is how countries, in particular developing ones, can acquire the necessary means—in terms of human, financial and material resources—to exercise such rights effectively. This seems to be as valid a question with respect to the right to conduct research as that of authorizing research to take place in the Exclusive Economic Zone or on the continental shelf.

Indeed, the pursuit of the goals of coastal countries, especially developing ones, under the new regime implies that conditions be created for the conduct of true marine research. Under the present circumstances, and at least for the immediate term, the industrialized states are most likely to contribute the largest share of the effort to acquire new knowledge. Thus the conciliation of the diverse interests involved in marine scientific research is to a large extent dependent upon co-operation among the states concerned. And this is implicitly recognized by the convention.

International co-operation in marine scientific research, as regulated by the convention,[7] could be constituted as follows: (a) co-operation to facilitate marine scientific research in the interest of the research itself or, seen from another angle, of the state that undertakes research (and indirectly, of the international community as a whole);[8] (b) co-operation specifically aimed at expanding the opportunities for coastal states to participate in marine research and have access to its results, as well as to promote their capabilities for research through training and education.[9]

Limits to the duty to co-operate

In the first case, the obligation to co-operate is incumbent upon all states. Such duty is interpreted without prejudice to the 'rights and powers' explicitly assigned to coastal states; the goal is to induce these states to exercise their rights in a reasonable manner. In the second case, the explicit reference to developing states, as the preferential beneficiaries of the actions foreseen, implies that the duties are to be considered as primarily incumbent upon industrialized states.

Among these duties, there are those to be complied with by states intending to undertake research in areas under national jurisdiction. According to the convention, research entities shall ensure the right of the coastal state (if it so desires) to participate or to be represented in the project and to provide the coastal state at its request with: (a) preliminary reports as soon as practicable; (b) final results and conclusions after the completion of the research; (c) data and samples derived from the marine research project; and (d) assessment and assistance in interpretation.[10]

It should be pointed out, however, that the opportunities thus opened to coastal states are limited. In fact, it does not imperatively follow that the coastal states may intervene in the formulation of the research project in order to introduce adjustments which would virtually make the scheme more consistent with the needs and interests of these states. Technical assistance to be furnished in connection with foreign research projects conducted in the Exclusive Economic Zone or on the continental shelf would, in any case, be part of the arrangement to facilitate access. It could be added that the possibilities of coastal states' representatives participating in research conducted by foreign entities will depend on their endogenous capacity in science.

As was noted above, the convention also stipulates a duty of states to 'actively promote the dissemination of scientific data and information and the transfer of knowledge resulting from marine scientific research especially to developing countries' in addition to 'strengthening . . . the autonomous capability of the countries for marine scientific research through programmes for training and education adequate to their technical and scientific personnel'. This appears to be too vague a formulation, however, to leave any certainty as to its effectiveness.

One may conclude from this brief overview of the regime applicable to marine scientific research—with its special focus on the promotion of scientific and technological capabilities of coastal developing countries—that the mechanisms provided for in the convention will tend to operate complementarily to other national and international efforts.

Technology transfer: rights and duties of states

The Convention on the Law of the Sea provides for the creation of legal and institutional conditions for a balanced distribution of marine science and technology at the world level. It should be pointed out that the convention covers not only the transfer, but also the development, of technology. In accordance with the goals of the negotiations on new international economic relations, what is envisaged is not only that developing countries acquire the technologies needed to carry out maritime activities in improved conditions, but also that they obtain an autonomous scientific and technological capability. Effective development and transfer of marine technology imply that a set of conditions be met both from the point of view of the nature of the mechanism for transferring technology and that of the capacities of recipient countries. The convention reflects an awareness of both these aspects.

With respect to agreements on transfer of technology, the treaty prescribes *inter alia* the duty of states 'to endeavour to foster favourable economic and legal conditions for the transfer of marine technology for the benefit of all parties concerned on an equitable basis' and 'to endeavour to promote favourable conditions for the conclusion of agreements, contracts and other similar arrangements under equitable and reasonable conditions'. The convention thus limits itself, through very general and virtually non-binding formulations, to encourage certain courses of action by states. Under present circumstances, industrialized states agree to adhere to general and flexible international engagements in order to initiate action concerning the trade of technologies. This reduces the potential effectiveness of the provisions for the transfer of marine technology. As a consequence, it will presumably remain (a) the degree of expertise and access to information, and (b) the effective means of states to intervene in the authorization and control of foreign investment and in the transfer of technology that will determine the arrangements made between suppliers and recipients of marine technology.

Given the serious difficulties faced by developing countries (resulting from the characteristics of the market for technologies, together with the low level of their scientific and technological capabilities), non-commercial mechanisms and procedures are needed to enlarge the scope of scientific and technological knowledge available for the benefit of such countries. There is, as well, a need to create foundations to help develop required infrastructure.

The functions of international co-operation

International co-operation, directly between states or through international organizations, is also sought. According to the convention (and similarly to what was suggested for international co-operation in marine scientific research), this co-operation can be seen from two points of view: (a) co-operation to promote the transfer of marine science and technology in general,[11] and (b) co-operation to promote the development and transfer of marine science and technology specifically directed to the needs and interests of developing countries, thus implying the creation of the necessary human, material and institutional infrastructures.[12] The latter is the principal focus. The objectives and measures envisaged for this international co-operation clearly point to satisfaction of the interests of developing countries. Such objectives and measures are not unknown in co-operation: their innovative aspects are the special emphasis which is put on the urgency of expansion and reinforcement of programmes and on the initiation of activities to help respond to the new needs faced by developing countries. This implies special responsibilities on the part of developed, technologically advanced states.

As far as international organizations are concerned, they appear—together with direct bilateral and multilateral relations among states—as further instruments of co-operation. In fact, international organizations, particularly the specialized agencies of the United Nations, have had mainly a role of promotion and co-ordination of co-operation in marine scientific research and in the development of transfer of marine technology. This has been done by providing for training and education, equipment and so on, and assisting those most in need of development of their potential in marine science and technology. The importance of references to the functions of 'competent international organizations' in the convention is symptomatic of the increased role these are asked to play in the implementation of the accord's objectives. The resolution on 'Development of National Marine Sciences, Technology and Ocean Service Infrastructures', adopted by the conference, should also be taken into account, particularly as it recommends that 'international organizations within their respective fields of competence [assist] developing countries in the field of marine science, technology and ocean services'.[13]

A final point emphasized by the convention, when dealing with co-operation among international organizations, reflects the multi- and interdisciplinary character of ocean affairs.[14] Economy and efficiency also imply rationalization, through increasingly effective co-ordination at the global and regional levels, of the activities of the various international bodies responsible for marine science and technology.

Growing demands from developing countries will be presented to international organizations, in particular to those belonging to the United Nations system. The situation arising from the extension of the limits of national jurisdiction by coastal states is already stimulating international co-operation in this field, because of the perception by these states of the difficulties they face in terms of the newly created maritime zones. Reinforcement of the efficiency of these institutions is urgently needed to respond to new imperatives.

The sea bed: rights and duties of the authority

The regime applicable under the new Law of the Sea to the development and transfer of marine science and technology, as well as to the promotion of scientific and technological capabilities required for exploring and exploiting the mineral resources of the sea-bed area beyond the limits of national jurisdiction, presents some special features. These follow from the legal status of the area—defined in the Convention on the Law of the Sea as the 'common heritage of mankind',[15] and particularly from the functions assigned to the International Sea-bed Authority. The last is a new international entity created for administering the area.[16]

The nature of the regime for the area led to special rights and responsibilities being assigned to the International Sea-bed Authority. These rights and responsibilities should be viewed in connection with the nature of this organization as administrator and supervisor, as well as with its functions to be undertaken through the Enterprise.[17] These responsibilities require that the authority take measures to acquire technology and scientific knowledge relating to activities in the area.

The authority is therefore entitled to carry out marine research concerning the area and its resources, and to enter into contracts for this purpose; it is to promote and encourage the conduct of scientific research in the area, then co-ordinate and disseminate the results of such research and analysis when available.[18] The authority should, in addition, encourage prospecting of the area.[19]

Duties of states and private entities to co-operate

Activities in the area will be 'organized, carried out and controlled' by the authority. 'Activities in the area' means, for the purposes of the convention, 'all activities of exploration for, and exploitation of, the resources of the area'.[20] This excludes marine research not intended to be the basis for exploitation of the resources of the area. Nevertheless, it seems legitimate to conclude (from the legal nature of the area and from the regime applicable to it) that at least moral obligations are incumbent upon researchers. This applies to co-operation and information exchange to further the opportunities for the authority's personnel and that of developing countries to participate in such research and to have access to its results.

In addition, a general obligation to promote international co-operation in scientific research in the area is imposed on parties participating in international programmes[21] and to ensure that programmes are developed through the authority and other international organizations, effectively disseminating the results of research and analysis.[22]

Access to technologies for exploring and exploiting the sea bed raises problems of a special nature. These technologies are of recent date coming from a limited number of enterprises in a few highly industrialized countries; some technologies are not yet commercialized. In the absence of compulsory mechanisms applicable to the relations between the authority and national operators, the authority would not dispose of the required technology; it would not be able to initiate exploitation of the mineral resources of the area. The feasibility of the regime as applied to the area could then be in question. Special provisions have been inserted in the convention, however; these provide for undertakings by national operators, to make available to the Enterprise the technology used when carrying out activities in the area under contract with the authority.

Promoting scientific and technological capabilities

It is obvious that access to scientific and technical knowledge for exploring and exploiting the sea bed is not a sufficient condition for the authority to engage in research and exploitation of sea-bed resources, nor in the administration and control of related activities, in a rational and effective manner. An essential additional requirement is the availability of manpower with the appropriate scientific, technical and managerial capabilities; these will enable the authority not only to understand available scientific data, technologies and other know-how, but to develop technologies further.

The administrative and operational responsibilities incumbent upon the authority call for special focus on education and training in deep-sea mineral development. This should be seen in connection with the duty of the authority to pay due attention to the importance of recruiting its staff on as wide a geographical basis as possible.[23]

The convention establishes a general obligation of states to promote programmes for the transfer of technology to the Enterprise and to developing countries with regard to activities in the area, as well as measures directed towards the advancement of the technology of the Enterprise and the domestic technology of developing states. This is to be done by providing opportunities to personnel from the Enterprise and from developing states for training in marine science and technology and for their full participation in activities in the area.[24] Specific engagements are incumbent upon operators in the area to be defined in more precise terms in their plans of work, particularly concerning trainees and transfer of data.[25]

Opportunities for creating national scientific and technological capabilities, especially in developing countries, relating to activities in the area may emerge indirectly from the convention. In view of the international system based on the concept of the 'common heritage of mankind', the pertinent provisions are meant to create the scientific and technological conditions for (a) the authority to have the appropriate capabilities and (b) the Enterprise to have the required technologies, thus ensuring effective implementation of the objectives of the regime for the area.[26]

Conclusion

When compared with the international legal situation existing before its adoption, the Convention on the Law of the Sea offers some new opportunities to states, particularly to coastal and to developing nations, to participate in scientific research as well as to have access to marine technology. Both these issues have merited specific treatment in the convention.

Opportunities for participation in marine research, which translate into rights of states—and, therefore, benefit from a significant degree of juridical imperativeness—will presumably have a secondary impact on the development of the scientific and technological capabilities of developing countries. Opportunities for access to marine technologies, though extensively formulated, appear to be juridically of less far-reaching scope because of the vagueness of the wording of the respective provisions. The provisions consist more of wishful claims by states than of subjective rights, properly speaking.

With respect to the regime for the sea-bed area, the potential benefits arising from the convention for developing countries in the domain of marine science and technology will depend on practical intervention by the International Sea-bed Authority.

In all cases, the effective pursuit of the objectives of the convention in ocean science and technology will depend on the will of states, as reflected in adequate functioning of the national and international institutions having responsibilities in research and technological development and their transfer in respect to the sea.

This naturally implies adjustments in the structures, means and activities of the organizations concerned, particularly of the United Nations system. In this context, the initiative undertaken by the Intergovernmental Oceanographic Commission of Unesco in adopting a Comprehensive Plan for a Major Assistance Programme to Enhance Marine Science Capabilities of Developing Countries deserves special mention, as it offers an important potential for the future.   ■

Notes

1. Charter of the Economic Rights and Duties of States, adopted by
   Resolution 3281(XXIX) of the General Assembly of the United Nations,
   12 December 1974; see also *The Vienna Programme of Action on Science and
   Technology for Development*, New York, United Nations, 1979.
2. Convention on the Law of the Sea, adopted 20 April 1982 at the Eleventh
   Session of the United Nations Conference on the Law of the Sea, Drafting
   Committee, 7 June 1982 (Working Paper, 1). (The convention was signed
   10 December 1982.)
3. Convention, op. cit., Part XIII (Marine Scientific Research), Part XIV
   (Development and Transfer of Marine Technology) and Part XI (The Area),
   Articles 143, 144; Annex III, Articles 5, 14 and 15, *inter alia.*
4. Ibid., Article 56.
5. Ibid., Article 246.
6. Ibid., Articles 248, 249.
7. Ibid., Section 2 of Part XIII.
8. Ibid., Articles 243, 255.
9. Ibid., Article 244, second paragraph.
10. Ibid., Article 249, first paragraph.
11. Ibid., Article 266, first paragraph, Article 270.
12. Ibid., Article 266, second paragraph; see also Articles 268, 269.
13. Resolution on Development of National Marine Science, Technology and
    Ocean Service Infrastructures, submitted by Peru on behalf of the Group of 77
    (Doc. A/CONF.62/L.127, 19 April 1982).
14. Convention, op. cit., Article 278.
15. Ibid., Article 136.
16. Ibid., Article 137, second paragraph; Article 153, first paragraph.
17. Ibid., Article 158, second paragraph; Article 170; Annex IV, Statute of the
    Enterprise. The Enterprise is the operational arm of the authority,
    responsible for direct exploration and exploitation of the resources of the area.
18. Ibid., Article 143, second paragraph.
19. Ibid., Annex III, Article 2, paragraph 1(a).
20. Ibid., Article 1, Use of Terms.
21. Ibid., Article 143, third paragraph.
22. Ibid., Article 143, paragraph 3(a), (b) and (c).
23. Ibid., Article 167, second paragraph.
24. Ibid., Article 144, second paragraph.
25. Ibid., Annex III, Article 14, 15.
26. Ibid., Part XI, Article 150.

PART 4.

MANAGEMENT
AND
GOVERNANCE

The words of the Portuguese poet who said 'today is the first day of the rest of your life' come to mind with the adoption and signature by 120 countries of the United Nations' Convention on the Law of the Sea. This historical event is the culmination of a long process of negotiation among states, and—indeed—the beginning of a new era of relations between man and the sea. Its test will be the effectiveness of the realization of the Convention's principles and rules, of the dreams and aspirations that inspired the new Law of the Sea.

Chapter 24

Institutional arrangements for the new ocean regime

Mário Ruivo

Secretary of the Intergovernmental Oceanographic Commission, the author is a marine scientist by training, specializing in the management of living resources. Dr Ruivo was director of FAO's Division of Fisheries until 1974. Since, he has been senior advisor to the Minister for Science and Technology, director-general for Marine Resources and Environmental Research in Lisbon, and head of the Portuguese Delegation to the United Nations Conference on the Law of the Sea. Address: Unesco/IOC, 7 place de Fontenoy, 75700 Paris (France).

The new ocean regime is a development with far-reaching consequences for the conduct of oceanic affairs at the national and international levels. It is also a symbolic event, in that it reflects the important changes that have been shaping the world community since the Second World War (characterized mainly by the process of decolonization and self-determination) as well as by the drive for a more just and equitable economic order, built on a foundation of peaceful and friendly relations between countries, regardless of their political and economic systems.

The long negotiations at the Third United Nations Conference on the Law of the Sea (UNCLOS, 1973-82), preceded by the preparatory work of the United Nations Committee on the Peaceful Uses of the Sea Bed and the Ocean Floor Beyond the Limits of National Jurisdiction (1968-73), took place at a time when, in parallel, other historical negotiations were being conducted within the United Nations. These included those leading to the adoption of the Declaration and the Programme of Action on the Establishment of a New International Economic Order, adopted by the General Assembly at its sixth special session; The Treaty on Peaceful Uses of Outer Space (1966); the United Nations Conferences on the Human Environment (1972); and on Science and Technology for Development (1979). UNCLOS was visibly influenced by some of these negotiations, and some of the concepts developed therein, adjusted to the specific conditions and requirements of international co-operation in the field of oceanic affairs, were later reflected in the text of the United Nations Convention on the Law of the Sea.

This process, by which some concepts originally formulated in general terms were made more specific and action-oriented, was completed by provisions defining the mechanism for implementation. An example of this is the recognition of the sovereignty of coastal states over the living resources in their Exclusive Economic Zones and the new rights granted to them for the protection of the marine environment. The philosophy of a new international economic order inspired the regime for the 'area' (the sea bed beyond the limits of national jurisdiction) and contributed to the adoption of the concept of the common heritage of mankind with respect to the minerals located there, as well as an international regime for the management and exploration of the area.

Organizing and apportioning research; technology transfer

Among the many features of the new ocean charter that constitute innovations with respect to the 1958 Convention on the Law of the Sea, mention should be made of the regime for marine scientific research and the transfer of marine technology. This innovation derived from the growing perception by states of the value of science and technology, first to economic and social development and secondly as sources of power and an element of the independence of states in the modern world. Science is a fundamental component of a technological civilization. Scientific activities generate the information required for many kinds of economic activity. Without scientific data, modern states, government administrations and industrial management could not perform their functions effectively or achieve their goals.

The relevance of science and technology to development found its global recognition in the United Nations Conference on Science and Technology for Development which led to the Vienna Programme of Action (1979). In this programme, high priority was given to self-reliance through the strengthening of the national scientific and technological capabilities of developing countries, an objective which calls for the improvement of the international institutions in that field (that is, the United Nations system).

Some of the issues raised at the United Nations Conference on Science and Technology for Development were also addressed by the Convention on the Law of the Sea, which defines a set of principles and rules on science and technology in ocean affairs (see Parts XIII and XIV). The Convention includes a comprehensive regime for the conduct of scientific marine research and the transfer of knowledge and technology. Some principles and rules establish the duty of states to co-operate directly with each other or through the appropriate international, regional or subregional organizations.

The rights accorded to coastal states—and the corresponding duties of states engaged in research—for granting or refusing their consent to research to be conducted in their Exclusive Economic Zones, create favourable conditions for the participation of local scientists in research activities, training, access to acquired knowledge, and so on. It should be noted in this context that, besides references in the convention to 'regional centres',[1] to be established or reinforced for the promotion of marine scientific research and technology, and related training and transfer of knowledge, mention is also made of national centres to be created at the local level for the same purpose. The functions assigned to these institutions—and the need for further elaboration of the concept of regional centres—reflect the concern for participation by, and the aspiration for real partnership among, developing countries in scientific and technological international marine programmes and activities in the future.

Institutional and operational evolution

At the conference, the preparation by the United Nations Secretariat of a Directory of International Organizations Competent in Ocean Affairs, as requested by the Portuguese delegation, provided a useful panorama of the existing arrangements; it contributed to an increase in the perception of the richness of the institutions available, as well as of the intricacy of their relationships. As a consequence, the importance of their optimal utilization, through appropriate co-ordination, was highlighted. Steps have already been taken or are under way, by a number of organizations to adapt themselves to the new demands, and to be able to fulfil the functions assigned to 'competent international organizations' in the convention. This is demonstrated by the adjustments taking place in geographical areas, and the jurisdiction and function of regional fishery bodies established directly by governments or in the framework of the Food and Agriculture Organization (FAO). The same could be said of the recent change in the name of IMCO (the Intergovernmental Maritime Consultative Organization), which reflects the evolution of the organization and the growth of its role with respect to, for example, protection of the marine environment from pollution caused by ships or dumping, which led to the establishment of the IMO Marine Environment Protection Committee.

Following this trend, Unesco, through the Intergovernmental Oceanographic Commission (IOC), initiated a study of the implications of the convention, in which its articles were taken as indicators of problems requiring solution through international co-operation in the fields of ocean science and services, and the related training and education. The aim is to strengthen its programmes and undertake the structural adjustments needed, including, eventually, improvements in the statutes of the commission.

As immediate steps, the IOC has established a new category of regional subsidiary bodies—the regional subcommissions—the first of which has just been created for the Caribbean and Adjacent Regions (IOCARIBE) as a successor to an association of the same name. It is expected that such subcommsisions will progressively replace the existing regional programme groups responsible for the promotion and co-ordination of marine scientific research and related activities, including the building up of regional centres and the establishment of networks.

Comparable institutional adjustments can also be observed at the national level. In fact, traditional administrations, established usually on a strictly sectoral basis (for example, fisheries, the merchant navy) are confronted with difficulties in coping with the new problems resulting from the multiple use of the oceans and their interactions. To deal with these difficulties, two trends can be identified: one is towards the establishment of a sectoral administration dealing with ocean affairs as a whole; the establishment of Ministries of the Sea (as in France), falls in this category. The other trend is based on the use of the existing sectoral administrations operating under a body having the power to formulate national policy in ocean affairs and to co-ordinate its implementation. This is the case, for example, of the Department for Ocean Development established in India.

The universality of science and marine co-operation

In the more specialized field of marine scientific research a general trend, which has been encouraged by the IOC, is the strengthening or the establishment of a marine science co-ordinating body, namely a National Oceanographic Commission, formed by representatives of universities, departments of state and other institutions deeply involved in marine scientific research. Such a body is intended to mobilize the available relevant national resources and to act as a focal point for international co-operation.

The pre-eminence found in Unesco's second Medium-Term Plan (1984-89) of activities of the Organization and, in particular of the IOC and the Division of Marine Sciences of Unesco *vis-à-vis* the oceans and their resources, and the strengthening of the Commission as a major instrument for the realization of the programme—especially in connection with a Comprehensive Plan for a Major Assistance Programme to Enhance the Marine Science Capabilities of Developing Countries—should be regarded in the context of the broad institutional process under way.

As institutional evolution proceeds, and the International Seabed Authority and the Commission on the Limits of the Continental Shelf are established after the entry into force of the Convention, new results can be anticipated. One could foresee the need, by the end of this century, to review the arrangements under which the United Nations operates in the field of international ocean affairs (as was proposed in 1978 at the Third United Nations Conference on the Law of the Sea).

The promise of the future

A recent study on 'cross-organizational programme analysis' prepared by the United Nations in the field of marine affairs offers a good panorama of the way the system is operating; it shows that, in spite of some minor areas of inevitable (and even desirable) overlapping, no substantive duplication of effort exists. It is to be foreseen, however, that in this process the risk of conflicts of competence may emerge, calling for full use of available mechanisms of co-ordination; one of these is the Inter-secretariat Committee on Scientific Programmes Relating to Oceanography (ICSPRO) and the active intervention of interested states. One should also keep in mind that, besides the United Nations, there are non-governmental organizations operating in the field of ocean affairs (including marine science), as does the International Council of Scientific Unions. These should continue to play an important role in their interaction with the intergovernmental bodies of the United Nations.

It is therefore essential that such institutions also undertake a parallel effort of adjustment to the new conditions, in order to become (as much as possible) universal and representative of the world scientific community. For historical reasons, many such organizations still consist predominantly of representatives or scientists of developed countries. In fact, scientists—as I had the occasion to emphasize in a letter published in *Impact of Science on Society* (No. 3, 1979)— must also learn to adapt to the new rules of the ocean game. The new ocean regime has a potential for expansion of scientific and engineering research in the seas through the active participation of individuals and institutions of all countries, both industrialized and developing, in a spirit of true partnership.

I am pleased to be able to include this paper in the present volume because the book offers an overview of the many major issues involved. In the variety of their analyses and opinions, my fellow authors provide an encouraging, convergent view that mankind has, de facto, entered a historical phase intended to ensure peaceful uses of the oceans.   ∎

Note

1. The workshop held in September 1982 under the auspices of the United
   Nations University, the Intergovernmental Oceanographic Commission and
   Unesco (Division of Marine Sciences) attempted to interpret this concept and
   to develop guidelines for a strategy for the establishment of regional centres and
   networks.

*Until a little more than twenty years ago, there was no suitable
intergovernmental mechanism to co-ordinate and rationalize scientific
research in regard to the world ocean. The Intergovernmental Oceanographic
Commission was created in 1960 to accomplish this, and one of the
commission's leading scientists here documents its achievements and
future plans.*

Chapter 25

Partnership in
intergovernmental co-operation:
for a better understanding
of the oceans

Hans Ulrich Roll

*Hans Ulrich Roll was born in Danzig (now Gdansk); he was trained in
mathematics, physics and biology at the Universities of Göttingen and
Danzig, then gained a doctorate of science at Berlin (meteorology,
oceanography). After serving ten years as a naval meteorologist,
Dr Roll spent another seventeen as a principal civil meteorologist of
his country (mainly at Hamburg). He is past president of the German
Hydrographic Institute, honorary professor (since 1962) of marine
meteorology at Hamburg University and since 1973 chairman of the German
Committee for Marine Research and Technology. From 1965 to 1982, he
represented the Federal Republic of Germany on the executive council
of IOC and served four years as its first vice-chairman. He was taught
in Brazil and the United States and is the author of numerous scientific
publications, among them the textbook,* The Marine Atmosphere. *His
address is: Rögenfeld 34, 2000 Hamburg 67 (Federal Republic of Germany).*

How to define partnership in oceanography?

The term 'partnership' has increasingly been used in describing international or intergovernmental co-operation in marine scientific research, so it seems appropriate to analyse and study this usage more in detail.

Partnership is normally considered as a contract between persons associated in business: a partner is a person carrying on business in association with another person or persons. This rather general definition of partnership can be made more specific by saying that the co-operation should take place under equal terms. This means that two or more partners engaged in a common undertaking provide equal intellectual as well as material contributions, enjoy equal rights and benefits and, of course, run the same risks.

There is a further, perhaps more sophisticated, form of partnership in which the partners hold different shares in the common business and, consequently, are distinguished by different rights, risks and benefits. For each partner, however, a certain balance exists between his contribution and risks, on the one hand, and his rights and benefits, on the other. Otherwise, a partnership would not develop.

The question now arises as to whether or not these forms of partnership have been realized in intergovernmental co-operation in the field of marine science. Before we can answer this question, we must investigate why and how intergovernmental co-operation was introduced in oceanography. This can be best illustrated by summarizing the history of oceanographic sciences.

Ocean research originated on a national basis

The voyages of the great discoverers up to the eighteenth century greatly contributed to our knowledge of the oceans, at least concerning conditions at the sea's surface. But marine research in its proper sense (concerned with the real nature of the oceans down to the sea floor) began only about 130 years ago. One of the centres where scientific interest in the ocean developed and grew was the University of Edinburgh, where Robert Jameson was professor of natural history. Among his students were Charles Darwin and Edward Forbes. The latter went dredging in the Mediterranean Sea and arrived at the conclusion that, below a depth of about 425 metres, life could not exist in the sea. He called this the Azoic Zone, but his findings were soon challenged. Charles Wyville Thomson, also from Edinburgh University, carried out some dredging in the deep Norwegian fjords as well as in the waters between Scotland and the Faeroes, and was able to find evidence of abundant life in Forbes's Azoic Zone. Later, west of the Ile d'Ouessant, he succeeded in discovering rich animal life at a depth of 3,700 metres. This new insight, which clearly contradicted the findings of Forbes, most certainly called for further examination and general confirmation.

Wyville Thomson, who in 1870 became professor of natural history at Edinburgh, approached the Royal Society of London for assistance in requesting governmental support for a long scientific cruise across the oceans, whereby final clarification should be obtained about life in the deep sea. The British Government responded favourably to this request and furnished the steam corvette H.M.S. *Challenger*, which set out in December 1872 for a three-and-a-half-year voyage round the globe, in an attempt to examine

Glossary of abbreviations

CICAR	Co-operative Investigations of the Caribbean and Adjacent Regions
CSK	Co-operative Study of the Kuroshio and Adjacent Regions
GIPME	Global Investigation of Pollution in the Marine Environment
ICES	International Council for the Exploration of the Sea
ICSPRO	Inter-secretariat Committee on Scientific Programmes Relating to Oceanography
ICSU	International Council of Scientific Unions
IDOE	International Decade of Ocean Exploration
IGOSS	Integrated Global Ocean Service System
IHO	International Hydrographic Organization
IOC	Intergovernmental Oceanographic Commission
IOCARIBE	IOC Subcommission for the Caribbean and Adjacent Regions
IODE	International Oceanographic Data Exchange
LEPOR	Long-term and Expanded Programme of Oceanic Exploration and Research
SCOR	Scientific Committee on Oceanic Research
TEMA	Training, Education and Mutual Assistance
UNCSTD	United Nations Conference on Science and Technology for Development
UNDP	United Nations Development Programme
WMO	World Meteorological Organization

the ocean depths. Further scientific problems and other disciplines were added to the original biological objectives, so that a thorough investigation of the nature of the ocean would be made.

This voyage opened the first phase of ocean exploration, characterized by long, wide-ranging cruises by research ships from many seafaring nations, often circumnavigating the globe and collecting scientific data and samples from vast sea areas. These voyages were planned and accomplished in a similar way to the *Challenger* expedition.

There are three particulars which should be pointed out when referring to early oceanographic investigations. First, these undertakings were entirely national: no international or intergovernmental co-operation or partnership was involved. Secondly, the results of such expeditions, however, were published internationally, i.e. they were put at the disposal of all nations. Thirdly, it was necessary that an appropriate oceanographic infrastructure should exist in the nations concerned; this means that sufficient resources in personnel and material facilities were available.

This way of initiating and executing oceanographic expeditions was still valid during the second phase of ocean exploration (1925–45) when the German research vessel, *Meteor*, undertook the first systematic and detailed investigation of a limited oceanic area, the South Atlantic Ocean. The German Atlantic Expedition was followed by similar cruises carried out by ships of other countries in other oceans. These undertakings were still of purely national character; only their results were made available internationally.

How international co-operation came about naturally

After the Second World War, it gradually became obvious that a national, single-ship approach was not sufficient for acquiring a good knowledge of oceanic processes. The oceans are vast and restless, with a great variety of currents not only at their surface but also in the depths. It is thus unlikely that measurements taken by a single ship, that is, taken at different places at different times, were coherent enough to be integrated into a realistic picture of the oceans' circulation and stratification. Oceanographers became aware of the necessity to measure representative oceanic properties simultaneously at many places if they wished to gain a complete understanding of the dynamics of the ocean. This could be ensured only by establishing a network of oceanographic stations occupied and operating simultaneously.

Such requirements could not be fulfilled by a single research vessel; they called for a closely co-ordinated joint observational programme, implemented by a number of ships or other measuring platforms. It was obvious that such an effort could hardly be made by a single country. The pooling of the national facilities of several countries, i.e. international co-operation, became necessary in order to achieve further progress in marine research. This, then, was the starting-point for the third phase of ocean exploration characterized by worldwide and intergovernmental co-operation in joint programmes, which began in earnest after the Second World War.

European co-operation began early

This short historical account of oceanography is a little biased in so far as it leads one to believe that, before the Second World War, marine science operations at sea were of national character only. In order to rectify this, I should add that my presentation refers to general development. There is an important exception, related to a particular field of marine research and to one region. In Europe, at the onset of the technological age and (consequently) the rapid extension of sea fisheries, it was recognized by the turn of the twentieth century, that international co-operation was needed to provide sea fisheries with appropriate scientific data on the abiotic and biotic environmental factors influencing the life of fish. Discussions among the leading experts of northern European countries led to the establishment of an intergovernmental organization, the International Council for the Exploration of the Sea (ICES). In 1902, ICES began by a simple exchange of notes between the governments concerned and, since 1964, the council has been based on a formal convention. In the beginning, eight European countries belonged to and financially supported this council; at the present time, eighteen European and North American states co-operate. ICES's headquarters is in Copenhagen, Denmark.

The main features of ICES are the following:

ICES's main objective is to assist sea fisheries in the North Atlantic Ocean by furnishing appropriate scientific advice.

ICES, through its scientific committees and expert groups, offers opportunities for presentation and discussion of results of marine research, for personal contacts among marine scientists, and for exchange of technical information.

ICES, while limited in mission, represents a coherent body of member states all at the same stage of scientific and economic development.

ICES's structure is such that its decisions are made essentially by active marine scientists, while the officers of the council act as their trustees.

ICES is a flexible organization, adaptive to new scientific, economic and political requirements.

ICES can be considered as a first example of partnership in intergovernmental co-operation in marine research. The chief reasons for its successful functioning for more than eighty years seem to be that its member states are (a) equally developed in the field of marine science and (b) united by a common interest in providing advice and assistance to sea fisheries in the North Atlantic Ocean and its marginal seas. Apparently an acceptable balance exists between contributions and benefits for all ICES member states.

Thus it is not surprising that ICES was the first intergovernmental organization to start co-ordinated and regular research cruises, with the aim of assembling the necessary basic information on temperature and salinity in the ICES area. Multiple ships' programmes were developed in the southern Kattegat before the Second World War. Thereafter, such operations were greatly extended within the framework of the International Geophysical Year, 1957–58. ICES organized the Atlantic Polar Front Survey (1958) in which twenty-two research vessels from eight countries participated. They investigated the variable boundary zones between warm, highly saline branches and extensions of the Gulf Stream in the south, and cold low-salinity waters of polar origin in the north. This mixing of water masses is an important factor in the abundance of fish and in the weather and climate of Europe.

A similar co-operative effort organized by ICES, was the so-called Overflow Expedition in 1960, whose purpose was the study of the outflow of Arctic bottom water over the submarine sill between Iceland and the Faeroe Islands; this is of considerable significance in the renewal of Atlantic bottom water. In 1973, the Overflow Expedition was repeated in an extended zone which included the outflow between Iceland and Greenland; thirteen research vessels from eight member states participated.

Intergovernmental marine co-operation on a global basis

Within ICES, intergovernmental co-operation in exploring the sea is limited to industrialized countries and to the North Atlantic Ocean. During recent decades, stimulated particularly by the negotiations of the United Nations Conference on the Law of the Sea, a general interest in the nature and resources of the oceans has grown markedly all over the world. Further, it has become a recognized demand that the marine environment be preserved, treated with care, and protected against pollution and misuse. This growing awareness has called for a comprehensive, worldwide approach to marine science. A new global system of international co-operation thus had to be developed—machinery that would permit the participation of all nations interested in the ocean, promoting worldwide co-operation and co-ordination in marine science.

The Intergovernmental Oceanographic Commission (IOC) was established in 1960 within the framework of Unesco. IOC began with a membership of 40, mostly industrialized, countries, and now comprises 110 member states which are at varying stages of economic development. Consequently, the task faced by IOC has always been quite different from that of ICES. Apart from the fact that IOC is not a regional body like ICES, but a global organization, it must pay due regard to the composition of its membership, which is characterized by a two-thirds majority of developing countries. These need primarily advice and assistance in (a) building their oceanographic infrastructure and (b) the development of their coastal zones and ocean areas beyond, the exploration, economic exploitation and environmental protection of which will fall under their own sovereignty or jurisdiction once the new international ocean regime is firmly established.

The industrialized countries, on the other hand, are more interested in promoting problem-oriented marine research in order to know more about (a) the nature and resources of the ocean and (b) the functioning and interdependence of its complex physical, chemical and biological systems. How can such somewhat diverse objectives be brought under a common heading? How can real intergovernmental partnership, which appears to be well established within ICES, also be achieved within IOC?

IOC's two decades of work are impressive

During the more than twenty years of its existence, IOC has developed a comprehensive spectrum of activities which can be subdivided into three main areas.

Ocean sciences: promotion and co-ordination of co-operative investigations regarding the nature and resources of different oceanic areas.

Ocean services: organization of oceanographic services to the scientific community, governments and the public, comprising oceanographic data banks and data exchange, a network of oceanographic stations, dissemination of oceanographic products such as analyses and forecasts of oceanic conditions, a tsunami[1] warning system.

Training, education and mutual assistance (TEMA), with the aim of developing real partnership among IOC member states so that they can participate adequately in IOC's activities.

It is not intended to give a full account here of IOC's activities. Regarding the ocean sciences sector, suffice it to say that programmes of co-operative research have been carried out in the Indian Ocean, 1959–65; the tropical Atlantic Ocean, 1963–64; the Kuroshio and adjacent regions of the western Pacific Ocean, 1965–77; and the Caribbean and adjacent regions, 1967–76.

In 1969, IOC conceived the Long-term and Expanded Programme of Oceanic Exploration and Research (LEPOR), designed for implementation within several decades. LEPOR is being updated by paying due regard to new scientific insight, such as that obtained during the International Decade of Ocean Exploration (IDOE) from 1971 to 1980, the 'acceleration' phase of LEPOR. One of the major projects of IDOE/LEPOR is the Global Investigation of Pollution in the Marine Environment (GIPME), which is connected with worldwide marine-pollution monitoring programmes. IOC has participated in several meteorological programmes by providing the oceanographic input and is currently engaged in preparing the oceanographic component of the World Climate Research Programme. Recently, IOC has concentrated its efforts on developing scientific programmes related to living and non-living resources of the ocean.

With respect to ocean services, it should be mentioned that IOC has established a smoothly functioning system for the International Oceanographic Data Exchange (IODE). Together with the World Meteorological Organization (WMO), IOC is developing the Integrated Global Ocean Service System (IGOSS) aiming to collect and disseminate current information on the state of the ocean. In the Pacific Ocean, IOC manages the Tsunami Warning System providing timely warnings of huge, disastrous waves generated by seaquakes. Jointly with the International Hydrographic Organization (IHO), IOC has published the *General Bathymetric Chart of the Ocean*, presenting the best information available on the morphology of the sea floor.

It must be stressed that the activities mentioned above are of a catalytic nature. IOC's function is to initiate, organize and co-ordinate co-operative programmes. The implementation of these programmes depends entirely on the readiness and capability of IOC's member states to commit themselves to adequate participation and concerted action. The scientific results obtained by such joint efforts may be published, the paramount importance of intergovernmental co-operation in true partnership being thus clearly documented.

Bearing in mind the heterogeneous composition of IOC as well as the commission's intention that all member states participate fully in IOC's programmes, IOC must pay great attention to education, training and mutual assistance in marine science. Before entering into this aspect, we should examine another form of international co-operation in marine research, namely the non-governmental one.

The non-governmental approach to partnership

In general, scientists are united by a collegiate sentiment towards each other. This solidarity results from the commonality of their scientific interest and studies. Marine researchers, in particular, are forced to look to others for co-operation, because the topic of their investigations—the ocean—is so vast and the costs of such research are so high that the pooling of expertise and effort is essential. Co-operation at the scientist-to-scientist and institution-to-institution levels has proved to be very effective; it has led to a form of international co-operation wherein different countries are represented by national groups of scientists and not by government delegations. In oceanography, the relevant non-governmental body is the Scientific Committee on Oceanic Research (SCOR), established in 1957 under the umbrella of the International Council of Scientific Unions (ICSU). Affiliated with SCOR, national committees for ocean research have been formed in thirty-five countries, twenty-six (74 per cent) of which are industrialized countries and only nine (26 per cent) developing countries or countries on the threshold of industrialization.

This international body appears to be sufficiently homogeneous to achieve good partnership, the unifying agent being scientific interest and judgement. Naturally, government support is needed when a huge co-operative exercise in ocean research is to be carried out. Thus, the importance of SCOR lies in its function of providing sound scientific advice to IOC on request and offering incentives for progress. The scientific guidance offered by SCOR is, in a word, indispensable to IOC.

Intergovernmental partnership: the present situation

From the beginning of IOC's existence, education, training and mutual assistance have received considerable attention in order to improve the participation of member states in the commission's working programmes. But the fact that much still needs to be done in this respect is documented by the attendance of member states' representatives at the last four sessions of the IOC Assembly from 1975 to 1982. Of the industrialized member states, 88 per cent were represented by marine scientists or had such experts among their delegates; only about 9 per cent of them were exclusively represented by diplomats, while 3 per cent were absent. The corresponding numbers for the developing countries among the IOC member states were quite different. The percentage of these countries represented by marine scientists amounted only to 36 per cent in 1975 and dropped to 26 per cent in 1982, while the percentage of such countries exclusively represented by diplomats rose from 36 per cent in 1975 to 46 per cent in 1982; the percentage of developing countries not able to attend ranged from 17 per cent to 28 per cent.

These figures signify that the developing countries apparently have difficulty in sending marine scientists to sessions of the IOC Assembly, where the main decisions of the organization are taken. Instead of being represented by marine scientists, these countries are represented by diplomats or else they are absent. This is not helpful for establishing the 'scientist-to-scientist' contact, so important to partnership. Also, it could be that ocean research has become so significant that it can no longer be left to marine scientists alone. Oceanography is emerging from the quiet, purely scientific sphere and entering the turbulent zone of public interest; here, politicians and diplomats have more say than marine scientists. As long as such development is general and affects more or less all member states and the composition of their delegations to IOC meetings, it must be considered as unavoidable. But if this trend towards diplomatic representation is characteristic of only the developing countries, it might make the establishment of partnership 'under equal terms' among member states somewhat difficult.

Unfortunately, the inadequate attendance of marine scientists from developing countries at sessions of IOC's Assembly can be taken as indicative of the overall participation of these countries in the commission's scientific programmes.

A comprehensive plan emerges

In order to remedy this situation, IOC has recognized that it is essential, first, to help create in developing countries the awareness that a sound scientific basis and oceanographic infrastructure are needed, and, second, to bring about appropriate actions at national, regional, and global levels if those countries wish to: (a) achieve their national goals in ocean affairs; (b) participate fully and on equal footing in IOC's programmes of ocean sciences and services; and (c) accept the opportunities offered and the responsibilities laid down by the new ocean regime.

Therefore, at the twelfth session of its assembly in November 1982 IOC developed and adopted a Comprehensive Plan for a Major Assistance Programme to Enhance the Marine Science Capabilities of Developing Countries. The following actions have been proposed therein at the different levels.

National. The first important step is to generate interest in marine research within the governments of coastal states and to develop the political will of these states to engage in marine science. Marine research and technology should be given high priority because of their economic importance. This is clearly demonstrated by the new ocean regime, whereby coastal states obtain, in their Exclusive Economic Zones and on their continental shelves, sovereign rights to (a) explore, exploit, conserve and manage natural resources as well as (b) exercise jurisdiction with regard to marine scientific research and the protection and preservation of the marine environment. This implies, among other things, that coastal states must have sufficient capability to judge the applications of foreign countries wishing to perform marine research in their economic zones and on their continental shelves.

All coastal states should establish a high-level national oceanographic co-ordinating body, with sufficient power to play an effective role in the formulation and implementation of a marine science policy. The most important (but also most difficult) measure concerns the building of an adequate national oceanographic infrastructure, composed of trained marine scientists and technicians as well as the necessary research facilities. While foreign advice and assistance will be needed for this purpose, it is essential that endogenous intellectual reserves, natural creativity and capabilities for oceanographic work be activated, and that stability of interest and governmental support be ensured.

Regional. The relative cultural homogeneity of states in a given geographic region and the similarity of their needs in marine science strongly recommend regional co-operation. National capabilities in marine science and ocean services could be developed and strengthened: mutual stimulation and cross-fertilization can be expected through participation in regional oceanographic activities. Pooling of limited national resources and joint operation of costly facilities (research vessels, research centres) could be beneficial to all participating countries. Foreign aid from international agencies can also be mobilized for regional exercises.

It is in the maritime regions that intergovernmental partnership in marine science can be realized most successfully. IOC provides good examples of such development. The Co-operative Investigations of the Caribbean and Adjacent Regions (CICAR), carried out from 1967 to 1976, ended with the wish of the countries of the region to establish an IOC subcommission for the Caribbean and adjacent regions (IOCARIBE), in order to continue and develop further co-operation in marine science. After positive experience was gained during a pilot period of several years, the IOC Assembly decided in 1982 to respond favourably to the request of this region. A similar regional mechanism was set up in the western Pacific Ocean, upon termination of the Co-operative Study of the Kuroshio and Adjacent Regions (CSK) in 1979.

Global. As far as composition, expertise and experience are concerned, IOC is the only intergovernmental organization having the capability to deal with all aspects of marine scientific research and related technical aid on a global scale. Also by its mandate to act as joint specialized mechanism for the United Nations organizations, co-operating under the ICSPRO agreement[2] when discharging certain of their responsibilities in the marine field, IOC has to play a central role in implementing the comprehensive plan. What can be, and has been, done?

Bearing in mind that IOC's role is basically of a catalytic nature, we cannot expect that the commission would be able to provide funds for establishing oceanographic infrastructures in developing countries. But IOC can provide advice and act as mediator between member states desiring assistance and potential donor countries or international mechanisms, e.g. Unesco, the United Nations Development Programme (UNDP) and the United Nations Conference on Science and Technology for Development (UNCSTD). Unesco, in particular, being the leading agency in the field of science and education among the ICSPRO agencies and operating a division of marine sciences, should be used fully in order to meet the pertinent need of developing countries.

ICSPRO's plan

The comprehensive plan includes the following forms of assistance:
Medium-term and long-term fellowships
Training courses
Provision of international consultants and experts
University-to-university exchange programmes of students and
   visiting professors
On-the-job training in foreign laboratories in the region or beyond
Use of research vessels
Provision of literature, equipment and supplies needed for planned
   teaching and research programmes

There are many examples of such assistance organized by IOC or Unesco
and financed by voluntary contributions paid by IOC member states into
the IOC fund-in-trust. These forms of assistance should be enlarged and
strengthened.

Outlook

Let us now return to the beginning, where it was asked whether partnership—co-operation under equal terms—had been achieved in oceanography.
In my view, the answer is a qualified 'yes'. Partnership appears to be established in certain regions where adequate homogeneity with regard to the
economic and scientific development of states exists. Partnership is further
realized in the purely scientific field, among scientists, where scientific goals
and viewpoints prevail and harmonize.

On a global scale, the intergovernmental co-operation in marine science
which began to develop some twenty years ago has made good progress.
It can be stated that co-operative investigations of the oceans, initiated
and co-ordinated by IOC, exert a stimulating effect on member states, in
particular on developing countries. When the International Indian Ocean
Expedition was organized in the late 1950s, it was the social—quite apart
from the scientific—challenge of an ocean surrounded mostly by developing
countries that provided the main motivation to marine investigations. A
major intergovernmental research effort in the Indian Ocean was envisaged
as an encouraging and beneficial influence on the countries of the region.
These expectations have come true; from the twenty-three countries participating in that co-operative investigation, eleven could be considered

as developing countries which since have made remarkable progress in oceanography.

With the continuing application of the principles and concrete proposals for mutual assistance given in the comprehensive plan described earlier, it is to be hoped that partnership in intergovernmental oceanographic co-operation will one day affect the entire globe. ■

Notes

1. Large ocean waves of seismic origin, capable of vast destruction.
2. ICSPRO, the Inter-secretariat Committee on Scientific Programmes Relating to Oceanography, was created in 1969 in order to organize co-operation among organizations of the United Nations family concerned with oceanic programmes.

To delve more deeply

BUECKMANN, A. *On the History of ICES*. ICES Council Meeting, 1973. (General Assembly Report, 6.)

LINKLATER, E. *The Voyage of the Challenger*. London, John Murray, 1972.

ROLL, H. *A Focus for Ocean Research, Intergovernmental Oceanographic Commission, History Functions, Achievements*. Paris, Unesco, 1979. (IOC Technical Series, 20.)

SCIENTIFIC COMMITTEE ON OCEANIC RESEARCH. *SCOR Handbook*. Paris, International Council of Scientific Unions, August 1981.

Most of the new Convention on the Law of the Sea deals with the future international regime related to exploration, then exploitation, of the ocean bed beyond a state's continental shelves. The regime concerns chiefly (at least, in the immediate future) the polymetallic nodules found on the sea bed; it will apply eventually to all the ocean's resources—including any that might be discovered in the future. Economic benefits derived from any of these would be distributed equitably.

Chapter 26

The sea bed, international area*

* Reproduced from the *UN Chronicle*, Vol. XIX, No. 6, June 1982, with permission.

The longest part of the Convention, and the one to which the conference devoted more of its time than to any other, concerns the future regime for exploring and exploiting the bottom of the deep ocean in areas beyond the continental shelf of any state. Of main economic interest in this area at present are polymetallic nodules lying on or just below the sea bed at great depths, composed of manganese, copper, cobalt and nickel—though the convention extends to all resources of the area, including any which may be discovered or may become economically exploitable in future.

The Convention would establish a 'parallel system' for exploring and exploiting the deep-sea bed. Under this system, all activities in the area would be under the control of the International Sea-Bed Authority, which would be authorized to conduct its own mining operations through an organ called the Enterprise. At the same time, the authority would contract with private and

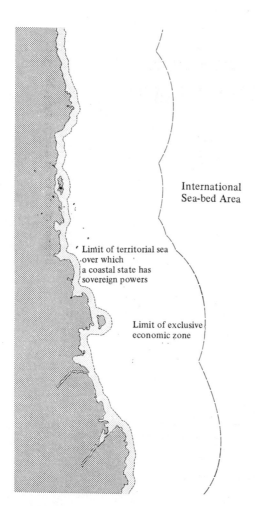

International
Sea-bed Area

Limit of territorial sea
over which
a coastal state has
sovereign powers

Limit of exclusive
economic zone

Of all the minerals known to exist in and under the seas the most valuable at the moment is oil: about one-fifth of total production comes from the continental shelf within 200 miles of the coast of about seventy-five countries and under relatively shallow water. Many such areas have not yet been tapped, and vast reserves are suspected under much deeper water. Techniques for mining under deeper and deeper water have progressed in leaps and bounds in recent years.

Nodules rich in nickel, manganese, copper and cobalt are the most valuable resource now known in the International Sea-bed Area. They lie on the ocean floor under water thousands of feet deep.

state ventures to give them mining rights in the area, so that they could operate in parallel with the authority. The resources of the area would be managed as a 'common heritage of mankind'.

Private consortia

To meet the concerns of a number of Western industrialized countries whose private consortia have already begun exploring the deep sea bed, the Conference adopted Resolution II, setting out a scheme for protecting the investments of such firms by virtually guaranteeing that they will receive sea-bed mining contracts after the convention enters into force, as long as they are sponsored by a state party to the convention. The scheme was expanded during the negotiations, so that it now encompasses state or private enterprises from Japan, India and the Soviet Union as well as five consortia from countries of North America and Western Europe, with room left for future investments by developing countries.

The rules, regulations and procedures that will govern sea-bed activities and other aspects of the system are to be drafted by the Preparatory Commission.

The sea-bed part of the Convention and the annexes associated with it were the subject of 18 pages of formal amendments submitted by the United States, along with six other industrialized Western countries having a particular interest in sea-bed exploitation—Belgium, France, the Federal Republic of Germany, Italy, Japan and the United Kingdom. Four of these countries— all but France and Japan—also proposed amendments on the draft relating to investments.

Who will benefit?

The financial and other economic benefits derived by the Authority from sea-bed mining would be distributed to parties to the convention 'on a non-discriminatory basis'. The interests and needs of peoples who have not attained full independency or self-government would be taken into particular consideration in connection with sea-bed activities.

Sea-bed policy

The first objective of sea-bed policy listed in the convention is 'the development of the resources of the Area'. This phrase was initially suggested by seven Western nations, who wanted it in a separate new article. It has been placed in the general article on sea-bed policy.

Another stated objective of sea-bed policy is 'the promotion of just and stable prices remunerative to producers and fair to consumers for minerals produced both from the resources of the Area and from other sources'. The seven Western countries sought unsuccessfully to delete the reference to other sources.

Production limitation

The Convention states a sea-bed production control policy whose basic aim would be to encourage sea-bed production with the least possible harm to land-based producers of the same minerals. To the extent that economic hardship for such producers could not be avoided, a compensatory scheme would be set up for their benefit.

A production control formula would allow sea-bed miners to produce up to 60 per cent of the projected annual increase in the world's demand for nickel, leaving the existing market as well as the remaining 40 per cent of the projected increase to land-based producers. The Enterprise would be entitled to produce up to 38,000 tons of nickel a year from the total sea-bed production ceiling to be established by the authority. The authority would be empowered to enter into commodity arrangements covering all sea-bed production.

Exploration and exploitation

The convention provides that all sea-bed activities must be carried out either by the Enterprise or by private and state entities 'in association with the authority'.

The whole range of sea-bed activities, as well as other aspects of the system's operation, are to be governed by rules, regulations and procedures to be approved by the assembly of the Authority on the recommendation of the Council of the Authority.

Qualifications of applicants

An annex to the convention spells out in general terms the qualification standards which would be required of applicants for sea-bed contracts. They would have to be a state party to the Convention or an entity sponsored by such a state. They would have to meet financial and technical standards to be defined in advance by the Council of the Authority, and would have to agree to accept the Authority's control over sea-bed activities and comply with the technology transfer requirements set out in the text.

Reservation of sites

The mining system is to work in this way: a state, or a private or public firm under its jurisdiction, would put up to the Authority an area of the sea bed it regarded as having commercial possibilities. The Authority would allot half of this area for exploitation by the would-be miner, under contract with the Authority; it would reserve the rest for use by the Enterprise or by developing countries, basing its choice on data supplied by the applicant on the polymetallic nodules in each of the two sites.

Once a site was assigned, the mining contractor would need two more approvals before he could operate in the international area: a plan of work, authorizing him to develop the mine-site, and a production authorization, permitting him to produce up to a specified amount of minerals from that site each year.

Plans of work, in the form of contracts between the Authority and the miner, would have to be approved by the council of the Authority. Provided that the applicant was financially and technically qualified, approval of a plan of work would normally be almost automatic, as long as the plan was endorsed by the Council's legal and technical commission. To reject the Commission's favourable action would require a consensus decision by the Council. This provision was placed in the text at a previous session to meet the concerns of industrialized countries which wanted a reasonable assurance of access to the area.

The Convention also contains what has been called an 'anti-monopoly' clause, aimed at ensuring that no single nation obtains too large a share of the sea bed as a whole or of any sizeable part of it. The production authorization, specifying the tonnage which each contractor could mine, would be issued by the council's legal and technical commission, within the total limit placed on all sea-bed production to ensure that it did not exceed its market share.

Technology transfer

To ensure the Enterprise's access to technology, the convention would oblige contractors to make available to that body, on commercial terms, the know-how they employed in their mining ventures. If the Enterprise found it could not obtain such technology on the open market, it would have the option of buying it from a contractor, provided he was the owner. In the case of technology owned by a 'third party' and used by the contractor under a licence or similar arrangement, the contractor would be bound to 'acquire, if and when requested to do so by the Enterprise and whenever it is possible to do so without substantial cost to the contractor, a legally binding and enforceable right' to transfer such technology to the Enterprise. The technology transfer options open to the Enterprise would also be available to a developing country which the authority had authorized to exploit a

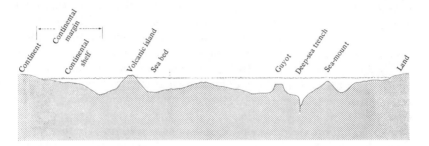

A look at the earth under the sea.

'reserved' area of the sea bed. These obligations would remain in effect for ten years after the Enterprise began commercial production.

The Enterprise

The sea-bed mining arm of the Authority would be a commercial venture under the overall control of the assembly of the Authority and Council of the Authority but with its own statute and a governing board to direct its business operations. The funds required for its initial mine-site—perhaps $1,000 million or more—would have to be borrowed. Half of this sum would be loaned, interest free, by the states that became members of the authority, while the rest would be borrowed on the financial market, with the loans guaranteed by the same states. The scale of assessments for the United Nations regular budget would be used to determine how much each state would lend and guarantee. States would make staggered payments to the Enterprise as it needed the money.

The so-called 'reserved area' of the sea bed would be restricted to the Enterprise or to developing countries.

The assembly

The general policies of the Authority are to be fixed by an assembly consisting of all parties to the Convention. The text describes it as the Authority's 'supreme organ', with power to approve the budget on its submission by the Council.

The Council

The Council of the Authority would consist of thirty-six members elected by the assembly according to a formula set out in the convention. Half of them would come from one of four major interest groups, while the rest would be elected in such a way as to ensure equitable geographical representation in the council as a whole. The four groups are: the largest investors in sea-bed mining (four states), the major consumers or importers of minerals found on the sea bed (four states), major land-based exporters of the same minerals (four states) and 'special interests' among the developing countries (six states).

The 'special interests' category of developing countries would include states with large populations, the landlocked or geographically disadvantaged, major mineral importers, potential producers of the minerals in question, and least developed states. No overall geographical distribution is specified, but at least one country from each region would be elected from among the eighteen members chosen on the basis of geography.

In the Council, the most important questions would be decided by consensus rather than voting. Consensus—defined as the absence of a formal objection—would be required for approval of the rules, regulations and procedures for sea-bed mining, whether by the Enterprise or by state and private entities. Other matters would be resolved by voting majorities of two-thirds or three-quarters, depending on the nature of the issue. Generally speaking, the more important the issue for the functioning of the mining system, the larger the majority that would be needed for the Council to act on it.

The Council: powers and functions

The Council's task would be to execute in specific cases the policies set out in the convention and defined in general terms by the assembly. Among the most important of these would be approval of the rules, regulations and procedures governing the entire sea-bed system.

Added to the text at the end of the session, on the president's proposal, was a provision that, in adopting rules, regulations and procedures, the council must give priority to those covering polymetallic nodules. Rules, regulations and procedures for exploration and exploitation of any other resource would have to be adopted within three years after a member of the Authority requested that this be done.

Council organs

An economic planning commission and a legal and technical commission are to assist the Council. Each would have fifteen members elected by the Council with due regard for geographical balance, or more if the Council decided to expand their membership.

Economic planning commission

This body is to review supply, demand and prices of sea-bed raw materials, make recommendations to the council on adverse effects of sea-bed mining on land-based producers, and propose a compensation system for such producers.

On the proposal of the collegium, the Conference accepted a proposal originally made by the Group of 77, that two of the commission's members be chosen from among 'developing countries whose exports of the categories of minerals to be derived from the [sea-bed] Area have a substantial bearing upon their economies'.

Legal and technical commission

This body would make recommendations to the Council on plans of work (contracts) for sea-bed activities, supervise activities in the area, recommend environmental measures, calculate ceilings for overall sea-bed production and recommend production authorizations for individual contractors.

Funds of the Authority

The Authority would be financed by assessed contributions from its member states, the earnings of the Enterprise, receipts from a tax scheme on sea-bed contractors, and possible loans and voluntary contributions.

On the proposal of the Group of 77, introduced into the text at the suggestion of the president, a compensation fund for land-based mineral producers likely to be most seriously affected by sea-bed production was added to the list of sources of the authority's funds.

Observer status for liberation movements

All states parties to the convention would be members of the Authority. The Conference agreed to an amendment by Iraq that would have the effect of permitting national liberation movements with observer status at the Law of the Sea Conference to have such status with the Authority. This amendment replaced an earlier version suggested by the president to the same effect, as part of his compromise proposals on participation of liberation movements.

Legal immunities

The Authority would be exempt from direct taxes imposed by states on transactions within the scope of its official activities, including its purchases of goods and services. It would also be immune to legal process.

Review conference

At the insistence of the developing countries, which did not want to commit themselves indefinitely to the 'parallel' mining system in its initial version, the text contains a procedure for a fundamental review of the system fifteen years after the start of commercial production. Under this procedure, if agreement on changes was not reached by a review conference within five years after the start of this review, the conference could approve amendments that would take effect for all states parties after ratification or other formal acceptance by three-quarters of them. Thus, it would be easier to modify the Convention under this review procedure than under the normal procedure for amendment, which would require consensus in the Council as well as a two-thirds vote in the assembly of the Authority, followed by acceptance by three-quarters of the states parties.

The three-fourths majority was a change added at the end of the session on the president's proposal. The previous text provided for a majority of two-thirds in both the review conference and the ratification procedure.

The review conference will take its decisions according to the rules used by the Law of the Sea Conference, unless it decided otherwise. There would be no voting until all efforts at consensus were exhausted, and then it would take a two-thirds majority for decisions on substantive decisions (including a simple majority of all delegations participating in the session).

Dispute settlement

The Sea-Bed Disputes Chamber would handle disputes between states parties to the Convention on the interpretation or application of the sea-bed clauses, as well as disputes involving the Authority and contractors. States could also submit their disputes to a special or an ad hoc chamber of the tribunal.   ■

See also

David Cronan, A Wealth of Sea-Floor Minerals, *New Scientist,* 6 June 1985. Economic and political pressures are a new impetus for countries with coastlines to investigate their offshore mineral resources.

Here is presented a summary chronicle concerning the long process of negotiation, during the 1970s and early 1980s, leading to the signing (in late 1982) of the Law of the Sea Convention. Sponsored by the United Nations, this agreement seeks to rationalize claims to the sea's resources and to establish an equilibrium between the potentials in marine research and technology possessed by industrialized and developing nations, whether they be maritime powers or not.

Chapter 27

The Convention on the Law of the Sea and the new international economic order

René-Jean Dupuy

The author is currently professor at the Collège de France, Secretary-General of the Hague Academy of International Law and, since 1973, a full member of the Institute of International Law. In 1968, Professor Dupuy founded the Institute of Peace and Development Law at the University of Nice. He is one of the world's leading legal analysts of sea legislation and space law. Professor Dupuy participated in the Third United Nations Conference on the Law of the Sea, and he published a book on the subject, L'océan partagé *(Paris, Pedone, 1980). He has also published* The Law of the Sea, Current Problems *(The Hague, Sijthoff-Nijhoff, 1974).*

Introduction

The Third United Nations Conference on the Law of the Sea began its practical work in Caracas in June 1974, that is, at the time of the discussions which led to the adoption of the Charter of Economic Rights and Duties of States, and in the year also in which the principles of the new international economic order were proclaimed. It was therefore to be expected that this revision of the Law of the Sea and redefinition of its rules would form part of the general effort which the international community then seemed to be making to establish the international economic system on more equitable foundations.

In developing countries, the philosophy of the new international economic order is based on two concerns: first, their desire to be themselves as a result of their own efforts and, secondly, their desire to be themselves with the support of others.

The first of these concerns leads a nation to rediscover its personality, and to generate itself the efforts that are needed to achieve a self-centred type of development. It serves as a stimulus for action to promote self-reliance.

The second concern involves an appeal to international solidarity. It is based on the idea that a nation cannot really be itself unless it attains a certain level of living. It must have more in order to be more; and, to help it to have more, contributions must be made by the rich countries, which must bow to the requirements of international law in the interests of social justice. This new legal order must include rules which impose on developed countries certain obligations towards peoples suffering from underdevelopment.

One way of describing this dual objective of the new international economic order would be to say that developing countries wish to nationalize everything on their own soil, but to internationalize the activities of everyone elsewhere. The first objective stems from their desire to recover their own resources, whether such resources have been stolen from them through colonization or other forms of foreign exploitation or whether—as is the case with living resources in adjacent waters—they are being plundered by foreign companies. For developing countries, ownership and sovereignty are inseparable: *imperium* without *dominium* is nothing. But a country's recovery of its resources is not merely a matter to be regulated by national law; its right to recover its resources must also be recognized by international law. Thus, after a phase of claim-making and unilateral legislation, recovery of resources is sanctioned by international law. This is precisely the process that has been followed in regard to the Law of the Sea, with the extension of the rights of the coastal state both over its continental shelf and over its Exclusive Economic Zone.

On the other hand, in matters not related to the recovery of national resources but rather to the conduct of third states, and especially industrial countries, the only course open to developing countries is to take action at the international level in the hope of securing, through multilateral negotiations, the adoption of a set of rules which will oblige the rich countries to act in certain ways in their relations with poor countries. This is the underlying consideration in all the work of UNCTAD,[1] whether aimed at: (a) the establishment of a generalized system of preferences; (b) the integrated programme for commodities; (c) a code of conduct for the transfer of technology; or (d) a code of conduct for transnational corporations. The primary objective is to put an end to a regime of economic liberalism which favours the stronger.

In the context of maritime law, this policy of the Third World countries took the form of an attack on the age-old principle of the freedom of the high seas, but the struggle was concentrated largely on the question of the International Sea-bed Area and the ocean floor beyond national jurisdiction. In this case, the developing countries were to benefit from the unexpected introduction of a new principle—the definition of this area as the 'common heritage of mankind'—and the establishment of an international authority responsible for administering the resources of the area in the interests of all mankind but taking into particular consideration the interests of the developing countries.

As the conference completed its work on 30 April 1982, it may be useful now to consider to what extent the convention it adopted embodies the principles of the new international economic order which it sought to introduce into the Law of the Sea.

Nationalization of resources and ocean space

It has often been remarked that the extension of the powers of the coastal state over maritime zones dates back to the Truman Declaration, and that accordingly this trend was started by one of the major industrial powers and not by a developing country. This is undoubtedly true; but it must be remembered also that, immediately after the Truman Declaration, some Latin American countries presented claims which related not only to the sea bed and the subsoil thereunder but also to the column of superjacent waters. This was the position taken by countries with a coastline on the Atlantic, such as Brazil and Argentina. In 1947, those with a Pacific coastline, such as Chile, Peru and Ecuador, were to make an even more unprecedented claim. Since these countries had no continental shelf in the geological sense of the term, they extended their sovereignty up to the 200-mile limit, which enabled them to reach the Humboldt Current whose chilly waters are rich in living species.

It must be noted that the claims of these states were based on the need to meet their development requirements by obtaining access to resources and ocean space that they regarded as belonging to them. What was now being advanced, in place of limited and functional sovereign rights invoked by the President of the United States, was a territorialist concept with a very powerful appeal which was later to prove irresistible. It may be added that the Latin American states with a Pacific coastline also used an argument of an equitable nature: they claimed to be correcting both the inequalities produced by nature and those resulting from defects in the organization of international society.

As we know, the figure of 200 nautical miles became a political, operational and dynamic watchword, which spread to other continents. Many developing countries claimed jurisdiction up to the 200-mile limit, even though they had no current such as the Humboldt Current to reach; but they all justified their claims by invoking the right of peoples to development. The fact that they also invoked considerations relating to the natural management of resources is irrelevant here. A coastal state that issues regulations for the protection of species and the conservation of the environment is, of course, acting in a interest wider than its own national interest; but it is none the less the state which benefits first and foremost from such protection.

'Nationalization' of ocean waters

The extent of the powers attributed to the coastal state under the terms of the convention has led to the conclusion that it was the coastal states that gained most at the Third Conference on the Law of the Sea. Since they have the power to determine the allowable catch in their Exclusive Economic Zones, and also to determine their own capacity to harvest, they would seem to reign supreme over the exploitation of living resources. Also, since the convention gives the soil and the subsoil of the economic zone the legal status of the continental shelf, a number of states which had no continental shelf in accordance with the depth criterion recognized by the Geneva Convention of 1958 now have their continental shelf. For instance, a country which, in accordance with the 200-metre depth criterion, had a continental shelf extending outwards for 50 nautical miles, has now acquired another 150 nautical miles under the concept of the economic zone. It is clear therefore that the extent of the nationalization effected is considerable; but it is questionable whether it is enough to introduce a new international economic order on the seas.

The extent of the nationalization effected is basically the result of two factors. On the one hand, with the introduction of the economic zone, the line where the high seas begin is situated further out from the coast, and, on the other hand, the rights of the coastal state over resources are inevitably bound to develop in practice into rights over the zone.

The economic zone is not part of the territorial sea, and is not part of the high seas either. This double-negative definition reflects the compromise on which the concept of Exclusive Economic Zone is based. The countries that proposed it, be they the moderate Latin Americans (Mexico, Venezuela, Colombia) who proposed it under the name of '*mar patrimonial*' (patrimonial sea), or the Africans who suggested it under the name of 'economic zone' sought in this way to obtain the support of major maritime powers by eliminating from the new concept the territorial element which would have obliged them to comply with the restrictive regime of innocent passage. This explains also why the freedoms of navigation and overflight and of the laying of submarine cables and pipelines are maintained in the zone.

At the same time, all the coastal states refused to have the economic zone regarded as part of the high seas, and there was therefore a difficult debate at the conference on the legal nature of the zone which, as its proponents pointed out, had important practical consequences. If the zone was to be regarded as part of the high seas, that meant that the governing principle was the principle of freedom and that the rights of the coastal state, restrictively listed in the convention, were to be interpreted as exceptions to the rule of freedom; and, as every exception is a restrictive interpretation, this was tantamount to saying that the powers of the coastal state in the zone were to be confined within a regime of an exceptional nature.

On the other hand, if the economic zone is regarded as a concept *sui generis* and if the zone does not in any case form part of the high seas, then the governing principle is the principle of the competence of the coastal state. That competence must be presumed; and the freedoms of navigation and overflight and of the laying of submarine cables and pipelines can be regarded only as residual freedoms of an exceptional nature. It was this second approach that was incorporated in the convention. Article 86 describes the high seas as 'all parts of the sea that are not included in the exclusive economic zone, in the territorial sea or in the internal waters of a state, or in the archipelagic waters of an archipelagic state'. There is evidence here of a concept which might be described as a 'national jurisdiction zone', comprising a zone of full sovereignty—i.e. the internal waters and the territorial sea—and a zone of economic sovereignty—the exclusive economic zone—in which the coastal state has rights over the resources.

The compromise thus established on the legal status of the zone is a fragile one. The logic of the new international economic order induces states to recover ownership not only of resources but also of territories. An owner is never altogether at his ease if the resources he owns are situated in an area that he does not own. This reaction is even more natural in the case of states which are motivated by territorial obsessions.

Territorial rights and discretionary powers

It seems that with the passage of time the subtle compromises reached by the conference will be incapable of preventing the powers of the coastal state from being exercised in fact over the zone as well as over its riches, as happened inevitably in the case of the continental shelf, in spite of the precautions taken in the 1958 convention to recognize only sovereign rights and not sovereignty over the shelf. The territorial attraction was the stronger; it is always irresistible. The following two developments may therefore be anticipated:

A transformation of functional competences into territorial rights. It should be emphasized that the intention here is not to criticize coastal states, but merely to seek out potential factors in this expansion of their rights.

It should be noted first, that some of these factors—those which involve an elementary control naturally induce states to take possession of an area. For instance, under Article 60 the coastal state has the exclusive right to construct and to authorize and regulate the construction, operation and use of artificial islands and installations. The article stipulates that this right shall include jurisdiction with regard to customs, fiscal, health, safety and immigration laws and regulations. This means that the coastal state enjoys a power of a spatial nature, which it exercises with respect both to its own nationals and to companies based in other states.

However, the territorial scope of its powers is even more evident in matters relating to the environment. Anti-pollution measures would seem to be more concerned with the concept of rights over the zone than with rights over resources. The coastal state does not of course enjoy exclusive competence for protecting the marine environment from pollution from vessels; the flag state in this case retains its traditional rights. This is a logical consequence of the maintenance of freedom of navigation in the zone; but the fact remains that from now on, in a zone hitherto regarded as part of the high seas, the competence of the flag state will have to compete with that of the coastal state.

With regard to scientific research, the coastal state exercises strict control and may at its discretion require the cessation of any research undertaken in the zone. As administrator of the economic zone, the coastal state is normally invested with powers of coercion and enforcement (Article 73).

It is in the context of fishing that the extent of nationalization of resources is most striking. Now, as we know, nearly all the high fertility areas fall within economic zones, and the world fishing regime has as a result been radically changed. The predominant position accorded by the convention to the coastal state in the economic zone as an area gives the lie to the theory that the rights of the coastal state relate to resources and not to the zone. The privileged position of the coastal state is obvious when we consider the situation of third fishing states with regard to the economic zone. Articles 63 and 69 do indeed give third parties an opportunity of obtaining access to the exploitation of the living resources of the zone, but such participation is subject to the conclusion of contracts with the coastal state. Conclusion of agreements of this kind will not always be easy: it is hardly possible to force the coastal state to conclude them.

Efforts by landlocked and geographically disadvantaged countries to submit disputes arising from this difficulty to the dispute settlement procedures envisaged in the convention have been unsuccessful, and these countries have had to content themselves with possible recourse to conciliation, the conclusions of which are not binding. But the coastal states wished to remain masters of the definition of their fishing policies in their zones. In fact, the control they exercise within their zones is very close to sovereignty. A further example is to be found in Article 72, which states that rights which a coastal state awards in its economic zone to landlocked states cannot be transferred to third states, either by lease or licence or by the establishment of joint ventures, without the express consent of the coastal state.

It will be noted accordingly that the coastal state also has discretionary powers.

A transformation of exclusive competence into discretionary power. In various cases the convention itself confers such rights on the coastal state, e.g. the right to determine the allowable catch, the right to determine its own capacity to harvest the living resources or the right to fix conditions for the admission of foreign scientific research teams. The provisions of Article 297 are sufficient in themselves to illustrate the scope of the powers of the coastal state. Since the discretionary nature of these powers is recognized, coastal states are not obliged to submit to the compulsory procedures for the settlement of disputes (see in particular paragraphs 2 and 3).

Thus, the regime of the Exclusive Economic Zone does unquestionably sanction the nationalization both of areas and resources, and, from this standpoint, it does satisfy one of the requirements for the establishment of the new international economic order.

Abundant resources: the need for balance

There are, however, comments which need to be made here. First, though it was the developing countries that initiated this recovery of national resources, many industrial countries have realized that they, too, have abundant resources off their coasts, and they, too, have established their Exclusive

Economic Zones. Long before the end of the conference, there was a general tendency among states to enact national laws giving themselves economic zones.

Secondly, this generalization of the advancement of the coastal state has raised doubts as to whether these gains made by the coastal states are consistent with the balance required by the new economic order. The new order does, of course, demand above all a change in relations between developing and industrial countries, but it also calls for co-operation among the developing countries themselves. Does the convention of 30 April 1982 really provide for and contribute to co-operation between them?

The co-operation that is needed between developing or industrial coastal countries and landlocked or geographically disadvantaged developing countries is certainly envisaged and advocated in the provisions of the convention. Yet one element that militates against it is the fact that the rules for co-operation are abstract and general, whereas the states to which they' are addressed are in very specific situations. One striking event which occurred at the conference as a result of this was the split in the Group of 77 in the Second Committee. States organized themselves into groups on the basis of their individual relations with the marine environment and no longer on the basis of the geographical or socio-economic factors which usually determine the formation of groups in the United Nations. There was thus a group of coastal states, and also a group of geographically disadvantaged states; and each of these groups included both developing and industrial countries. The difficulties encountered in the establishment of general rules on the basis of compromises are therefore understandable.

The efforts to determine the extent of the continental shelf provide a particularly notable example of the impossibility of establishing a rule of general application, and of the need to deal individually with particular cases from the very outset of the negotiations. It was clear that the concept of the Exclusive Economic Zone would not absorb the concept of the continental shelf, contrary to the wishes of certain states, particularly some petroleum-producing countries which do not have a shelf extending beyond 200 nautical miles. States that did have one managed to keep it, but it was still necessary to fix the outer limit of the extended shelf. The discussions were made very difficult by the conflicting and often complex situations of different states. Whereas in 1958 a twofold depth criterion had still been adequate, Article 76 of the convention contains no less than eight provisions, each of them applying to different actual situations. It establishes three principal and two subsidiary rules and mentions three special cases, the last of them an extreme case (the convention here is legislating indirectly for a single state, Sri Lanka).

However, regardless of the extent to which the individualism of the coastal state has been recognized, the convention has tried to foster the international co-operation demanded by the new international economic order. As we have seen, the new order calls for international regulation of the conduct of states in their mutual economic relations; and several provisions are aimed at limiting the legal individualism which states wish to exploit to their advantage, and developing their sense of solidarity with third states. In this connection, mention may be made of two sets of rules in which the idea that countries that are more privileged in certain respects—even if they are developing countries—should accept certain obligations towards their disadvantaged neighbours is combined with the desire to protect the objective interests of the international community by rational management of fish stocks. The aim here is to guarantee the preservation of a certain seafood reserve for mankind and to protect the marine environment.

With regard to the exploitation of living resources, there are a number of rules designed to ensure that neighbouring countries, especially those which are landlocked or regarded as geographically disadvantaged, can participate to some extent in the exploitation of the surplus. This was the result of the general climate which prevailed in the negotiations on the Law of the Sea, as in all negotiations carried on within the framework or under the auspices of the United Nations. Every state, and particularly a state that feels that it is deprived in certain respects, appeals to other states to take its specific situation into account and draws comparisons to illustrate the injustices and induce other states to agree to corrective measures. The coastal states themselves adopted this course in order to justify their appropriation of the ocean space they needed for their development. In so doing, they contrasted the exploitation and the poverty they had suffered with the opulence of the maritime powers, which were sovereign and could send their vessels anywhere at will. Not surprisingly, other countries, even more disadvantaged because they were landlocked or had a limited coastline, turned to the coastal state and drew attention to the unfairness of their situation. On this question, too, there followed the kind of dialogue, or exchange of invective, to which the international community resorts, especially in a field in which traditional law—which had never known such a community—left every country free to act for itself.

In fact, equilibrium or disequilibrium?

The convention attempts to limit the individualism of the coastal state in three ways:

It provides that coastal states are to consult the competent international organizations, whether subregional, regional or global. For the purpose of determining the allowable catch in its economic zone, and also for ensuring the protection of the environment, the coastal state must consult institutions whose members will—the assumption is—include interested developing third states.

With regard to the exploitation of fishing surpluses in economic zones, the convention gives a privileged place to developing countries with particular characteristics (Article 70).

Some wished to go further and impose an obligation on certain coastal states in the interests not only of their underprivileged neighbours but of the whole international community, whose destiny depends on development. Nothing ever came of the ambitious proposals submitted in the early 1970s by certain movements or by Malta for the introduction of taxes which the international machinery managing the common heritage was to levy on all maritime activities, including activities undertaken on the high seas; but the idea remained that states which have a continental shelf extending beyond 200 nautical miles must, by way of compensation for this natural privilege, make payments or contributions in kind in respect of the profits derived from the exploitation of the non-living resources of the part of the shelf beyond 200 miles (Article 82). After the first five years of production, payment is to be made annually at a rate increasing gradually over seven years from 1 to 7 per cent of the value or volume of production. This rate is more in keeping with the wishes of the fifty-three other countries, since the previous maximum figure had been only 5 per cent. This contribution is organized as a tax, from which some countries are to be exempted—e.g. developing countries that are net importers of a mineral resource produced from their shelves. Payments are to be made to the authority, 'which shall distribute them to States Parties to this Convention, on the basis of equitable sharing criteria', taking into account development needs and especially the needs of the least developed or landlocked developing countries.

These various provisions are not considered adequate by the landlocked or geographically disadvantaged countries, which believe that the regime of the economic zone increases inequalities among states. However that may be, the requirements for the establishment of a new economic order appear to be reflected more satisfactorily in the system of the International Sea-bed Area beyond national jurisdiction.

The international area of the sea bed
and the new international economic order

In the First Committee, which had the task of defining this regime, the solidarity of the Group of 77 was re-established since the principal objective was undoubtedly to construct one of the sectors of the new order. Also, the conditions for doing so were particularly favourable, since the sea bed had been proclaimed to be the common heritage of mankind and the 1970

Declaration of Principles had stipulated that particular interest be given to the interests of the developing countries.

The concept of mankind is an extremely rich and dynamic one. It has a value transcending space, since it embraces all peoples on the planet, without discrimination. And also, it has a significance transcending time, since mankind does not include only persons living today but bears within itself the seed of future generations. Consequently, those who manage the common heritage today must regard themselves not as owners who are entitled to use or abuse it as they wish, but as stewards who have to see to it that the domain is preserved and administered in the interests of generations to come.

Accordingly, the concept of mankind is not static; it is an inducement for economic and social development. In the opinion of developing countries it is a concept of great value, since it prohibits the appropriation of resources and areas forming part of the common heritage. Thus, the concept of mankind provides for the poor—for those who do not yet have the technology and the financial means to exploit these resources themselves—a guarantee that they will retain their right to take part in this exploitation at a future time. Another consequence is that if the industrial powers exploit these resources, they may do so only in the general interest and not exclusively for their own benefit.

The system based on the concept of the common heritage of mankind is thus in complete contrast with that established in 1959 for the Antarctic and confirmed in 1980, i.e. collective appropriation by a small number of industrial countries, and in 1979 the developing countries decided to request that the moon also should be proclaimed a common heritage, and to advance the same claims in respect of other resources, such as the radio-electric frequency spectrum.

It is therefore not surprising that the charter of the seas and oceans should state that 'All rights in the resources of the Area are vested in mankind as a whole, on whose behalf the Authority shall act' (Article 137, paragraph 2, of the convention). The conduct of member states is governed entirely by rules defining the status and the objectives of the common heritage, to which Article 138 adds the objective of 'promoting international co-operation and mutual understanding'. This vision of harmony is, we suspect, powerless to prevent or even to gloss over confrontations between states, especially between North and South. Even so, it is impossible to deny that the interests of the international community are present at the heart of the system administering the area since, over and above disputes between states, and without denying them, the regime established is designed to secure the participation of all parties in the functioning of an international public authority invested with operational powers on behalf of mankind.

In this case, the idea of public service is given concrete expression, since the convention establishes a machinery of integration. This will not of course in itself put an end to tensions and conflicts, but the community will live within the system with both of them, as the Europeans do in their own community; the only difference will be that, with a membership consisting of a large number of highly dissimilar states, the authority will have to overcome more serious contradictions and remove many obstacles in order to find compromises on which a consensus can be reached. The community exists from the moment the members of the system are unable any longer to ignore one another (and how can they do so in a system which they freely entered?) and are willing to work together to achieve their goals.

How international is the new economic order?

Such indeed is the process involved in the establishment of the new international economic order except that, in the particular case of the authority, the system will have an institutional apparatus more operational than that of UNCTAD. While UNCTAD is above all a forum for negotiation for the purpose of formulating principles, the authority is to be a power centre for the actual management of common resources.

Signs of the new economic order are clearly discernible in Part XI of the convention but, in order to be realistic, it is essential to remember that the conference did not, on 30 April 1982, secure recognition of the system by the majority of the industrialized states which are capable of undertaking the exploitation of nodules in the near future.

The desire to align Part XI with the objectives of the new international economic order is quite clear. During the earlier sessions, the regime governing exploitation was the subject of debates which were often heated. It is sufficient to mention here that the Group of 77 managed to limit the guarantee of access to sites: activities in the area are to be carried out under the control of the authority, which has a genuine margin of discretion in issuing contracts. The conclusion of contracts is subject to the acceptance of certain prior commitments by applicants for production authorizations, who are bound by certain obligations to transfer technology to the enterprise. Guarantee of access may also be made conditional on the establishment of joint ventures with the enterprise or with developing countries. These principles were keenly debated at the ninth session. The Group of 77, on the initiative of the African countries, proposed that the contractor be obliged to transfer not only the technologies of extraction, but also the technologies for processing modules. Potential contractors are strongly opposed to this, since the text also gives the authority discretionary power to terminate or suspend a contract if the contractor fails to fulfil his obligations. The differences of opinion which arose over Article 5 of Annex III are typical of those which occur in a highly developed system of integration, since the negotiators in these cases go into great detail in regard to rules and regulations. The same is true of the provisions concerning the financial terms of contracts (Article 13 of Annex III) and those referring to the financing of the enterprise.

With regard to the latter, the discussions centred on the obligation of the industrial countries to provide resources for the enterprise, in the form of loans and guarantees, to enable it to begin initial exploitation. Those who will have to bear these costs asked for a sufficiently precise ceiling, and a compromise was finally reached.

It was also very difficult to reach agreement on the principles to be applied by the authority in determining a management policy, and particularly in establishing the rate of production. Everyone knows that states which are already land-based producers of the minerals that are likely to be extracted from the international area are strongly opposed to production from the sea bed, which may cause prices to collapse. When the concept of the common heritage was first introduced, the Third World showed some measure of enthusiasm for exploiting the common heritage; a study of the economic data has led it to revise its position. Apart from the least advanced countries which have no mineral resources, the majority of states in the Group of 77 are now motivated largely by anti-production considerations. Some industrial countries, such as Canada, share their determination to accept only a highly controlled production, and a whole series of highly complex provisions has therefore been introduced to prevent such exploitation from affecting national production.

The clash of interests between land-based producers and potential undersea producers is reflected in the adoption of a regime which is based on flexible solutions that ensure a balance—subject to adjustment—between the parties concerned. A glance at Article 151 will reveal the complexity of the provisions designed to link the development of undersea production with that of nickel consumption, and to limit the former to a volume corresponding to a 3 per cent growth of the latter with, in addition, a safeguard clause to protect land-based producers if the growth in consumption is very low.

Derivative legal accords and some conclusions

This community of the sea-bed area, highly institutionalized and invested with regulatory power, will develop a derivative law, which will itself evolve from the continuing disputes but will help to consolidate the system.

The system certainly appears to be a particular entity within the convention. It provides for its own regime of penalties—namely suspension of the right to vote in the case of states in arrears with their contributions, and suspension of exercise of rights and privileges of membership of a member state 'which has grossly and persistently violated the provisions of this Part' (Part XI, Articles 184 and 185). Also, provision is made for a review procedure, separate from the review procedure for the convention and relating only to the provisions governing the system of exploration and exploitation of the resources of the area (Article 155). A conference convened by the assembly of the authority fifteen years from the beginning of the year in which the first commercial production commences will consider whether the provisions of Part XI have actually benefited 'mankind as a whole' and fostered 'healthy development of the world economy'.

Clearly, these formulations, with their ring of harmony, reflect an informal agreement that was reached following a statement by Dr Kissinger who, in order to induce the Group of 77 to accept certain concessions, agreed that provision should be made for the regime of exploitation to be reviewed at the end of twenty-five years. The developing countries were able to secure not only a reduction of this period but also the acceptance of a procedure which departs from the traditional attempt to reach a consensus. If, five years after its commencement, the conference has not reached agreement, it may decide, during the ensuing twelve months, by a two-thirds majority of the member states, to adopt and submit for ratification, accession or acceptance such amendments to the system as it deems necessary and appropriate. This recourse to a vote appears to have been regarded as an incentive to reach a consensus before the five-year period of possible negotiations expired; but it seems to be assumed also that, after the 'system' has been in operation for twenty years, it will have reached a point of no return, and that the community of the area will be sufficiently consolidated to refrain from parting company with its minority. Whether or not this assumption is borne out will depend on circumstances that are at present unforeseeable. It will be noted, however, that the majority originally envisaged was three-quarters and that the Group of 77 managed to have it reduced to two-thirds.

These few illustrations from the convention are evidence of the tremendous difficulties that were encountered in the efforts to introduce the new international economic order into the management of the resources of the sea bed. The new order calls for continuing negotiation, but most of the industrial countries capable of exploiting nodules in the near future were afraid of being subjected to an authority over which, in their view, the Group of 77 would have too much control.

We know how long and difficult were the debates on the structure of the authority, and especially on the decision-making procedure in the council. In August 1980 there seemed at last to be agreement on the provisions of Article 161, but discussion on these compromises was subsequently reopened.

The refusal of the United States, the Federal Republic of Germany and the

United Kingdom to sign the agreement will seriously affect the establishment of the International Sea-bed Area.[2]

The conference provided for a transitional regime in which concerns in keeping with the principles of the new international economic order are reflected in the statute of the Preparatory Commission and in the system for the protection of preparatory investment. In particular, the financial expenditure to be incurred and the transfer-of-technology obligations to be assumed by the pioneer investors will be fairly considerable.

However, the adoption by the major industrial powers of national legislation governing the activities of their companies, together with the refusal of some of them to accede to the convention, deprives the authority of a sizeable financial contribution, since the states that voted for the convention represent 35 per cent of the budget of the United Nations.

It would, of course, be premature to conclude that the attempt to establish a new international economic order in regard to the Law of the Sea has been a complete failure. In the recovery of resources by coastal states and in the internationalization of the sea bed for the benefit of mankind, results which were inconceivable twenty years ago have been achieved or brought a stage nearer. There is still a long road to travel, however, and it is impossible so soon after the Third Conference to say how long it will take to cover the greater part of it.   ■

Notes

1. The United Nations Conference on Trade and Development, a permanent body within the United Nations.
2. The document was signed by 116 states on 10 December 1982 (Japan was to sign shortly afterwards).

An eminent international jurist traces the long historical development of international legislation dealing with the sea and its resources. He then interprets for the reader some of the many possible connotations for the future of the recently signed Convention on the Law of the Sea.

Chapter 28

Emerging concepts of the Law of the Sea: some social and cultural impacts

M. C. W. Pinto

Born in Colombo in 1931, the author studied law in Sri Lanka and at Cambridge University. He has practised before the Supreme Court of Sri Lanka and is a barrister of the Inner Temple (London). Mr Pinto has also been counsellor at law at the International Atomic Energy Agency, the World Bank, and the Ministry of Foreign Affairs of his country. Sri Lankan ambassador to the Federal Republic of Germany and to Austria (1976–80), the author has been intimately involved with the United Nations Law of the Sea since 1968; he was also chairman of the International Law Commission in 1980. Ambassador Pinto is currently secretary-general of the Iran–United States Claims Tribunal at The Hague.

Historical development of the Law of the Sea

Feared and loved, often deified, from time immemorial the sea has been a part of man's consciousness. Over the millennia of man's use and abuse of the oceans and their resources, regulation became inevitable first at the level of the group or community, later at the city, nation and state levels, and finally, in this century, at the global level. Already in India in the fourth century B.C., the Mauriyan kings had established the office of Superintendent of Ships responsible for regulating maritime transport and trade, as well as fishing and mining, and all over the ancient world rules and practices evolved governing the uses of the sea. One of the oldest collections of these rules, the Rhodian Sea Law, though pre-Christian in origin, survives today in a form given to it in the eighth century.

But is was the nations of Western Europe with their large merchant and military fleets which, from the fourteenth century on, laid the foundations of what is today called the Law of the Sea. In 1493, Pope Alexander VI divided the 'undiscovered' world between Spain and Portugal by a line drawn from pole to pole 100 leagues west of the Azores, so that entry into each region by others required their permission. So it was that the Spanish Ambassador at the English Court sharply protested at the intrusion into these waters of Sir Francis Drake and the *Golden Hind* in 1580, provoking from Queen Elizabeth I of England a classic statement of what was to become a basic principle of the Law of the Sea: 'The use of the sea and air', said Elizabeth, 'is common to all: neither can any title to the ocean belong to any people or private man, for as such as neither nature nor regard of the public use permitted any possession thereof.'

As England defied one Catholic monarch, the mighty Dutch East India Company was to defy another. In 1605, acting on the advice of its brilliant young counsel Hugo Grotius, the company moved into the East Indies, shattering the monopoly claimed by Portugal. It was Grotius, regarded by many as the father of modern international law, who later expounded and publicized concepts of free trade and navigation that were later interpreted as supporting a doctrine of 'freedom of the seas'. This doctrine implied that the seas and their resources were open to use and exploitation by all countries without distinction. What was ignored by thus focusing on a notional equality, or lack of legal distinction among all nations, was of course the factual distinction that only a handful of states had the maritime skills and financial capacity to take advantage of this grand concept. This 'freedom' made the high seas the arena for the competitive aspirations of nations with large navies and merchant fleets. It was complemented by their insistence that the breadth of the belts of sea adjacent to coastlines (conceded to be within national jurisdiction and called the territorial sea) should remain narrow, say three miles in extent, thus maximizing the area available for their purposes.

Such was the international law and the international legislative process of the time. As the adventurous, grasping tentacles of empire moved swiftly to encircle the globe, legal doctrine—supporting the aims of the few powerful states of Western Europe—evolved to become part of the cultural bonds of imperialism. For some three centuries after Grotius, in fact until a conference at The Hague in 1930 under the auspices of the League of Nations, there was no collective effort to examine and clarify the principles of what the world had been asked to accept as the international Law of the Sea.

The United Nations Conference of 1958

To the vast majority of peoples for whom this had been a dormant or quiescent period in a cyclic history, the freedom of the seas had little positive significance. To communities subjected to hegemonic influences and occupied with the business of survival, the sea's challenge remained at a primitive level. Only a few countries, only those who had in fact proclaimed and enforced the doctrine of the freedom of the seas, felt a widening awareness of the sea's wealth and possessed the capacity to exploit it.

But this doctrine of freedom in its pristine form could last only as long as populations around the world were relatively small or inadequately organized, and their demands for food and other resources from the sea correspondingly weak. With the phenomenal increase in populations, and the emergence and re-emergence into the mainstream of international life of states that had taken no part in the evolution of the 'freedom of the seas', there came the spread—however slow— of awareness of the sea's resources and, correspondingly, the demand for opportunities to participate in winning and sharing them, and in the making of laws that would establish and safeguard those opportunities.

The Hague Conference of 1930, while it focused expert opinion on issues of the Law of the Sea and pioneered a collective approach to them, could in the circumstances of the time, accomplish little. Participation was confined to forty-eight states, dominated by those of Western Europe. The first politically broad-based effort to develop and codify the Law of the Sea was undertaken by the United Nations in 1958 on the basis of proposals worked out by the International Law Commission in a seven-year study. In the same year eighty-six states joined in drafting four separate international agreements on the Law of the Sea: the Convention on the Territorial Sea and the Contiguous Zone, the Convention on the High Seas, the Convention on Fishing and Conservation of the Living Resources of the High Seas, and the Convention on the Continental Shelf. Important as this contribution was to the development of the law, and of the legislative process,

onspired to undermine the stability of the solutions reached and,
'ent, these agreements did not win widespread adherence.

Several factors undermined the authority of these legislative efforts.
First, the process of decolonization proceeding apace under the aegis of the
United Nations, compelled recognition of increasing numbers of states as
members of the world community. Second, group politics as a means of
mobilizing solidarity and canvassing commonly held positions enhanced
the negotiating strength of the poorer countries, and the Group of 77
(established during the first United Nations Conference on Trade and
Development in 1964, the group now numbers more than 125 countries),
and the non-aligned movement began to play a major role. Third, spectacular
advances in marine science and technology offered new ways of exploiting
the oceans, particularly hydrocarbons at great depth, and the hard minerals
on the ocean floor (polymetallic nodules containing substantial quantities
of manganese, nickel, copper and cobalt, among other useful elements),
as well as new military uses of the sea and sea bed. Fourth, the poorer coastal
states, apprehensive of the accelerated depletion of marine resources through
the application of the new marine technologies in the context of a doctrine of
freedom of the seas which had little meaning for them, responded by
progressively developing the legal content of a freedom of their own: they
claimed countervailing rights of resource and environmental jurisdiction
in offshore areas of hitherto unprecedented extent. Fifth, the major maritime
powers were anxious to open negotiations on three issues of importance that
had remained unresolved at the 1958 Conference: (a) the maximal breadth
of the territorial sea, for which they favoured twelve miles; (b) a regime
to govern the passage of ships through straits used for international navigation; and (c) fishery limits and certain other questions connected with the
regulation of fishing.

An attempt to reach agreement on these topics at the Second United
Nations Conference on the Law of the Sea, attended by eighty-five states
in 1960, ended in failure.[1]

New trends and concepts

Malta's initiative at the United Nations in 1967 to have the sea bed placed
beyond national jurisdiction and the resources of that area declared to be a
common heritage of mankind, and the drive towards another comprehensive
conference on the Law of the Sea which the initiative triggered, was timely
and epoch-making. With the focus on the oceans and their resources brought
about by the United Nations Committee on the Peaceful Uses of the Sea Bed
and the Ocean Floor beyond the Limits of National Jurisdiction (1968–73),
and subsequently by the Third United Nations Conference on the Law of
the Sea (1973–82), man's view of the planet and his relationship to it will

never again be the same. The mass of information on the oceans and their resources, released and disseminated on an unprecedented scale by the conference, enhanced and sharpened as never before the level of awareness and interest of decision-makers and public alike. To the poorer countries, the so-called new members of the world community, the conference conveyed anew and amplified the timeless challenge of the oceans; it also provided a forum in which their claim to the opportunity of taking part in the acceptance of that challenge, and to share equitably in the ocean's wealth could be demonstrated, encouraged in a co-ordinated and efficient manner, and ultimately secured, not as a matter of chance or charity, but as a matter of legal right.

The claims of these countries showed two basic trends: first, developing countries with substantial coastlines sought to expand their national jurisdiction, at least for resource, environmental and security purposes, beyond the confines of a territorial sea which they had been encouraged to think of as necessarily narrow—according to perhaps a pre-ordained, natural scheme of things; and, second, all developing countries sought to participate on the basis of equality with other states in the regulation of activities beyond national jurisdiction in order to ensure: (a) that resources in which they perceived an individual or communal interest would not be exploited ruthlessly for selfish ends, but rather in a rational and equitable manner, without waste, and taking into account the needs of the world community as a whole; (b) that their right to exploit these resources, on an equal footing with all states, would not be subverted by excessive and destructive exploitation by transnational companies and state enterprises which might enjoy a temporary technological advantage at any given time; and (c) that the technology which had been developed within a few countries, would be made available to all on terms they could afford, so as to enable them to make a reality of the rights of participation to which they laid claim.

From the first of these trends developed concepts like the Exclusive Economic Zone, a belt of sea 200 miles wide adjacent to the territorial sea, in which the coastal state has sovereign rights to the exploitation and management of natural resources, as well as rights in respect to conservation of the marine environment and the regulation of scientific research; the 'continental shelf', conferring extensive, but defined rights of resource jurisdiction on the coastal state in regard to the prolongation of its land mass under the sea; and the 'archipelagic state' which protects the marine resource, security and environmental jurisdiction of states composed of numerous islands in the interstitial areas of sea and sea bed. From the second trend developed concepts of joint management and co-operative endeavour with respect to marine resources outside national jurisdiction. In the case of living resources, this meant the promotion of international co-operation through a system (largely fragmented) of collective fishery management;

in the case of non-living resources (mainly the minerals of the deep-sea bed), this meant recognition of the basic principles of a common heritage of mankind, of an area and its resources that were to be managed on a co-operative basis through a new International Sea-Bed Authority with comprehensive powers.

The sea bed beyond national jurisdiction
as a common heritage of mankind

Resolution 2749(XXV) of the General Assembly of the United Nations, adopted without dissent in 1970, declared the area of the sea bed and the ocean floor beyond the limits of national jurisdiction, as well as the resources of this area, to be the common heritage of mankind. The content of the Declaration of Principles, as it came to be known, is quite familiar, but it is useful to recall the understanding of the overwhelming number of states of the world community as to the interpretation of the 'common heritage' concept, and further, the fact that they deemed that the principles contained in the declaration are binding on all states as principles of customary international law.[2]

For the majority of states, the Declaration of Principles proclaimed the community's recognition that the deep-sea bed and its resources had a reasonably clear though evolving legal status: they were—both the area and its resources—the common heritage of mankind. This meant, *inter alia*, that the minerals of the area could not be freely mined. They were not, so to speak, for the taking. The common heritage of mankind was the common property of mankind. The commonness of the 'common heritage' is a commonness of ownership and benefit. In their original location, the resources of the area belong in undivided and indivisible shares, to all countries—to all mankind, in fact, whether organized as states or not. 'Touch the nodules at the bottom of the sea', they seemed to say, 'and you touch my property. Take them away, and you take away my property.'

It followed, therefore, that before anyone could mine the nodules and win them from the bottom of the sea, mankind must consent to the system chosen for doing so. Because the community as a whole had agreed in 1970 that this was the absolute, fundamental rule, that this was in fact the law, the states became engaged for some ten years (under the aegis of the United Nations) in trying to decide what method or system of resource-taking should have mankind's approval. Thus, the declaration of 1970 records the clear understanding of states that 'All activities regarding the exploration and exploitation of the resources of the area . . . shall be governed by the international regime to be established; [that] the international regime . . . shall be established by an international treaty'; that this treaty shall be 'of a universal character' and that it should be agreed upon by the community

'continental shelf', conferring extensive, but defined rights of resource jurisdiction on the coastal state in regard to the prolongation of its land mass
under the sea; and the 'archipelagic state' which protects the marine resource,
security and environmental jurisdiction of states composed of numerous
islands in the interstitial areas of sea and sea bed. From the second trend
developed concepts of joint management and co-operative endeavour with
respect to marine resources outside national jurisdiction. In the case of
living resources, this meant the promotion of international co-operation
through a system (largely fragmented) of collective fishery management;
in the case of non-living resources (mainly the minerals of the deep-sea bed),
this meant recognition of the basic principles of a common heritage of
mankind, of an area and its resources that were to be managed on a co-
operative basis through a new International Sea-Bed Authority with comprehensive powers.

The sea bed beyond national jurisdiction
as a common heritage of mankind

Resolution 2749(XXV) of the General Assembly of the United Nations,
adopted without dissent in 1970, declared the area of the sea bed and the
ocean floor beyond the limits of national jurisdiction, as well as the resources
of this area, to be the common heritage of mankind. The content of the
Declaration of Principles, as it came to be known, is quite familiar, but it is
useful to recall the understanding of the overwhelming number of states
of the world community as to the interpretation of the 'common heritage'
concept, and further, the fact that they deemed that the principles contained
in the declaration are binding on all states as principles of customary
international law.[2]

For the majority of states, the Declaration of Principles proclaimed
the community's recognition that the deep-sea bed and its resources had a
reasonably clear though evolving legal status: they were—both the area and
its resources—the common heritage of mankind. This meant, *inter alia*,
that the minerals of the area could not be freely mined. They were not,
so to speak, for the taking. The common heritage of mankind was the
common property of mankind. The commonness of the 'common heritage'
is a commonness of ownership and benefit. In their original location, the
resources of the area belong in undivided and indivisible shares, to all
countries—to all mankind, in fact, whether organized as states or not.
'Touch the nodules at the bottom of the sea', they seemed to say, 'and you
touch my property. Take them away, and you take away my property.'

It followed, therefore, that before anyone could mine the nodules and win
them from the bottom of the sea, mankind must consent to the system
chosen for doing so. Because the community as a whole had agreed in 1970

reflected in the declaration and under Part XI of the United Nations Convention on the Law of the Sea.

The new international economic order (NIEO)

The common heritage concept was already three years old and its elaboration far advanced when, on 1 May 1974, the General Assembly at its Sixth Special Session adopted Resolution 3201, the Declaration on the Establishment of a New International Economic Order, and Resolution 3202, containing a Programme of Action for its implementation. Later the same year, the Assembly adopted Resolution 3281, the Charter of the Economic Rights and Duties of States. At its Seventh Special Session in 1975, the General Assembly adopted Resolution 3362, 'Development and International Economic Co-operation', setting forth in concrete terms certain measures 'as the bases and the framework for the work of the competent bodies and organizations of the United Nations System' for the achievement of the NIEO.

In the principles of the NIEO set forth in the original resolution, adopted in the aftermath of the use of the 'oil weapon', developing countries included the features they held to be essential: recognition of the interdependence of developing and developed countries, and the obligation of co-operation between them; assistance to, and preferential treatment of, the developing countries; permanent sovereignty over natural resources; the conservation of natural resources, and environmental controls; the sovereign equality of states, and the democratization of decision-making in economic matters, implying universal participation and one-state one-vote; just and equitable relationships between prices paid to developing countries for their exports, and the prices paid by them for their imports; and promotion of the transfer of technology to the developing countries.

The declaration became at once the inspiration and universal guide of all developing countries for multilateral negotiations on global economic issues. Although it contained no mention of the 'common heritage', many of the declaration's principles reflected concepts that had already been canvassed at the Conference on the Law of the Sea, especially those connected with the international regime for the deep-sea bed. The focus given by the declaration now clarified and gave fresh impetus to the development of those concepts, ensuring their incorporation in the new Convention on the Law of the Sea.

The new Law of the Sea from a cultural perspective

Thomas Carlyle once wrote: 'The great law of culture: let each become all that he was created capable of being.' There can be little doubt that the modern international Law of the Sea was designed to enhance opportunities for development of human potential in all countries, and that the United

Nations Convention on the Law of the Sea offers the world for the first time a comprehensive framework of rights and duties that could, if observed justly and in good faith, bring about improvement of the quality of life of all peoples through the conservation and use of the seas and their resources.

It would not be feasible to attempt here a detailed, let alone comprehensive, survey of the cultural impact of the new Law of the Sea. But we may note some of its significant features and consider their possible influence on the social order of the future. Both the legislative process which brought into being the United Nations Convention on the Law of the Sea and the substantive content of the convention present unique, and possibly portentous, features. The legislative process was characterized by (a) universal participation by all states of the world community on the basis of sovereign equality and (b) a holistic approach based on a perceived interrelationship between the issues. The latter encouraged recourse to a 'package deal', and an emphasis on the search for consensus (conceived as the absence of formal objection) as the best guarantees of the viability of the new law.

The convention incorporates new regulatory concepts and norms aimed at conserving the sea's resources and ensuring the spread of benefits derived from them to all on the basis of need; creating a duty of international co-operation primarily for the benefit of the developing countries; promoting the dissemination of technology and scientific knowledge, prerequisites for the exercise of the new rights and performance of the new duties with respect to the sea and its resources; and the institutional resolution of conflicts in accordance with predetermined rules and agreed procedures.

If there is a single thread running through the convention and the legislative process that gave it birth—a single principle that seems to inspire and pervade the whole of this endeavour, it is that of the interdependence of the states of the world community. It matters not whether this is a principle properly to be considered cultural, moral, social or legal. What is important is that the community of states implicitly acknowledge the principle during elaboration of rules regarding the seas and their resources, a subject which must in one context or another touch the lives of peoples everywhere. The interdependence of all states is as old as statehood itself, yet as new as the new international economic order.[7] Through provisions which give it expression in the form of the rights and duties of states, the convention represents a giant step toward a just and ordered world, and cannot fail to have a salutary impact on the mores and habits of thought of the community as a whole.

A few legal interpretations

Writing in 1625 Hugo Grotius observed:

There is no state so powerful that it may not some time need the help of others outside itself, either for purposes of trade, or even to ward off the forces of many foreign nations united against it. In consequence we see that even the most powerful peoples and sovereigns seek alliances.[8]

A decade before, Francisco Suarez, the Spanish theologian, had written:

Mankind, though divided into numerous nations and states, constitutes a political and moral unity bound up by charity and compassion; wherefore though every republic or monarchy seems to be autonomous and self-sufficing, yet none of them is, but each of them needs the support and brotherhood of others, both in a material and a moral sense. Therefore they also need some common law organizing their conduct in this kind of society.[9]

What are the characteristics of this common law of which Suarez spoke? A preliminary answer was offered in our own century by the Chilean jurist, Alvarez. Judge Alvarez, in his individual opinion in the Anglo-Norwegian Fisheries case said,

The starting point is the fact that, for the traditional individualistic regime on which social life has hitherto been founded, there is being substituted more and more a new regime, a regime of interdependence, and that, consequently, the law of social interdependence is taking the place of the old individualistic law.

The characteristics of this law, so far as international law is concerned, may be stated as follows:

(a) This law governs not merely a community of states, but an organized international society.

(b) It is not exclusively juridical; it has also aspects which are political, economic, social, psychological, etc. It follows that the traditional distinction between legal and political questions, and between the domain of law and the domain of politics is considerably modified at the present time.

(c) It is concerned not only with the delimitation of the rights of states but also with harmonizing them.

(d) It particularly takes into account the general interest.

(e) It also takes into account all possible aspects of every case.

(f) It lays down, besides rights, obligations towards international society; and sometimes states are entitled to exercise certain rights only if they have complied with the correlative duties. (Title V of the Declaration of the Great Principles of Modern International Law approved by three great associations devoted to the study of the law of nations.)

(g) It condemns *abus de droit* [violation of the law].

(h) It adapts itself to the needs of international life and develops side by side with it.[10]

Judge Alvarez was noting the close of an era during which the majority of norms of international law were concerned with the adjustment and limitation of national sovereignties and their claims to jurisdiction; an era when the most general, substantive principle of international law was that a state was legally free to act in any way it pleased (Alvarez's 'individual except to the extent that such act was contrary to one or other of the norms of international law, norms that consisted mainly, if not wh prohibitions.

Despite Alvarez's earnest attempts to secure recognition of a 'law of social interdependence', the court ignored it. Nor did the theme seem to hold much interest for scholars, though there were, indeed, notable exceptions.[11] A reason for this lack of enthusiasm may be that, to a lawyer, the scope and context of a 'law of social interdependence' would seem far from clear. Of what relevance is it to his familiar field of rules and norms derived from custom, treaty or judicial decision? Can it be fitted into Article 38 of the Statute of the International Court of Justice? The answer to these questions would probably be in the negative.

Conclusions as to derived legal content

But the United Nations Convention on the Law of the Sea may well have brought recognition of a law of social interdependence a step closer to realiz-ation in concrete and practical terms. In my opinion it may be possible to distil, from the complex rules and institutions established by the new convention, four principles presented here without elaboration as the derived legal content of interdependence:

First, the principle of sharing and integration, which would require the widest possible spread of the benefits of marine resources,[12] including technology and scientific knowledge,[13] to all countries (whether coastal or landlocked) with a view to their rapid integration into a global marine economy. This is seen as the opposite of an earlier regime based on the rigid exclusiveness of property rights which retarded such integration.

Second, the principle of co-operation in marine activities,[14] which requires all states to collaborate with one another even when a position of relative economic strength or legal security might support a policy of no action. This is seen as the opposite of an earlier regime based on unregulated competition, with the winner taking all, and subject only to rules pro-hibiting action that would cause damage to another state.

Third, the principle of non-reciprocal conduct, which would require benefits from resources to move to those in need of them,[15] without necessarily requiring a material equivalent in exchange. This is seen as the opposite of an earlier regime which insisted upon a narrow mutuality as the basis of contract.

Finally, the principle of decision-making by application of consensus-oriented rules. These rules require postponement of voting until all efforts at securing a consensus have been exhausted.[16] This is seen as a devel-opment of the constitutional principle of majority rule to permit the maximal impact of minority opinion short of the veto.

All four principles find tentative expression in the new Law of the Sea. They (and, indeed, other such derived principles) will need further study and development based on the practice generated by their application under the convention. What sanctions will be applied upon failure to implement, for example, a general obligation of the technologically advanced countries to 'encourage and facilitate the transfer to developing countries . . . of skills and marine technology' (Article 273)? What incentives can be envisaged for inclusion, for example, in national legislation, to achieve this? Who could institute legal action upon breach of this obligation? How would we quantify damages—or, for that matter, secure specific performance? To move to another line of questioning, is this a different kind of law altogether, conceived in generalities and supported by hopes? Is it, indeed, law at all?

We end with these questions, and with a tentative response: Yes, it is law; and yes, it is a law conceived differently from municipal law and, indeed, from some of the international law of the past. It is the rudiments of the international law of interdependence foreseen by Grotius, and by Suarez before him. The development of this law by practical and institutional means under the Convention on the Law of the Sea may enhance to a greater extent than ever before the social and cultural impact of international law. ∎

Notes

1. For a history of these developments, see Ann Hollick, *U.S. Foreign Policy and the Law of the Sea*, Princeton, N.J., Princeton University Press, 1981.
2. As to the source of the binding nature of the provisions of the declaration and the illegality of sea-bed mining beyond national jurisdiction in the absence of a treaty generally agreed upon, see 'The Legal Position of the Group of 77', UN Conference on the Law of the Sea, 23 September 1980 (Doc. A/CONF. 62/106).
3. See the resolution on the question of unilateral legislation on sea-bed mining adopted by the Ministers of Foreign Affairs of the states of the Group of 77, 23 October 1979 (Doc. A/34/611).
4. These were things *quae publico usui destinatae sunt* (which are meant for public use); see *Corpus Juris Civilis*, Vol. I, Book II, paragraphs 1–4, Berlin, Mommsen, 1905.
5. A view attributed to Neratius. *Corpus Juris Civilis*, Vol. XXXI.1.14, Berlin, Mommsen, 1905.
6. For a discussion of the collective ownership and the regulation of individual appropriation of things conceived of as *res publicae*, see Giuseppe Branca, 'Le cose extra patrimonium humani juris', in *Annali Triestini di Diritto Economico e Politico*, Vol. XII, 1941, pp. 909 et seq.
7. The Declaration on the NIEO proclaims the determination of states to work

for the establishment of a new international economic order, based on equity, sovereign equality, interdependence, common interest and co-operation among all states. Recalling recent grave economic crises, the declaration elaborates upon the principle of interdependence as follows: 'All these changes thrust into prominence the reality of interdependence of all the members of the world community. Current events have brought into sharp focus the realization that the interests of the developed countries and those of the developing countries can no longer be isolated from each other, that there is a close relationship between the prosperity of the developed countries and the growth and development of the developing countries, and that the prosperity of the international community as a whole depends upon the prosperity of its constituent parts. International co-operation for development is the shared goal and common duty of all countries.'

8. H. Grotius, *De jure belli ac pacis* (trans. F. W. Kelsey), paragraphs 21–22, New York, Prolegomena, 1964. (Reprint.)

9. F. Suarez, *Tractatus de Legibus ac Deo Legislatore*, Book II, Chapter 19, paragraph 5, 1612.

10. International Court of Justice, *Reports of Judgements*, etc., 1951.

11. See for example, B. Röling, *International Law in an Expanded World*, pp. xii–xiv, Amsterdam, 1960; A. McDougal and B. Burke, *The Public Order of the Oceans*, New Haven, Conn., Yale University Press, 1962.

12. United Nations Convention on the Law of the Sea (UNC) (Doc. A/CONF.62/122(UNC)), Articles 136 (common heritage of mankind); 140 (benefit sharing); 148 (participation of developing countries in activities in the Area); 150(c), (d), (g), (i) (resource policy); 160.2(g) (benefit sharing).

13. UNC, op. cit., Articles 143, 242 (international co-operation for the promotion of scientific research); 144, 266, 269–78 (transfer of technology); 254 (landlocked states and marine scientific research); Annex III, Article 5 (sea-bed mining technology).

14. Ibid., Articles 273, 278; Annex III, Article 5(5).

15. Ibid., Articles 140(2); 202 (technical assistance to developing countries regarding environmental protection); 203 (preferential treatment for developing countries); 254 (landlocked states and marine scientific research); 269–78 (transfer of technology).

16. Rules of Procedure of the Third United Nations Conference on the Law of the Sea (Doc. A/CONF.62/30/Rev.3), Articles 37, 38, 39 and the 'Gentlemen's Agreement'; UNC, op. cit., Articles 155(3), 159(6)–(10), 161(7)–(8).

The life-supporting environmental regimes we call ecosystems require 'management' as much by man as by nature if we are to reconcile the dual imperatives of use and conservation of the planet's resources. The understanding of exactly how ecosystems work—or break down—is as important to the total marine milieu as it is to terrestrial regimes. Here, a marine biologist explains the how and why of wise management of these large natural systems.

Chapter 29

Marine ecosystems: research and management

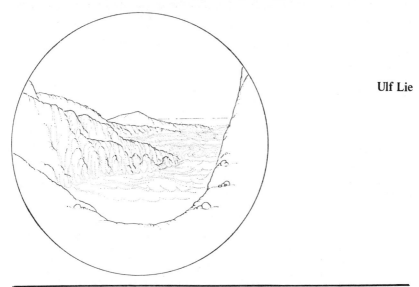

Ulf Lie

Ulf Lie obtained his candidate's and doctoral degrees at the University of Bergen, Norway. He has taught at the Universities of Bergen and Washington (Seattle), served as a programme specialist in Unesco's Division of Marine Sciences, and taught (since 1976) at the Institute of Marine Biology in Bergen. His areas of concentration have been zooplankton and plankton-feeding fishes, quantitative benthic studies, mathematical ecology and ecosystem analysis. For the past six years, Dr Lie has also chaired the Norwegian National Committee of the Scientific Committee on Oceanic Research. His address is: Institutt for Marinbiologi, Biologisk Stasjon, 5065 Blomsterdalen (Norway).

Definitions

Although 'ecosystem' has become a household word for many people during the last decades when there has been a growing concern for the protection of the environment, few can give a clear definition of the term. It seems that common usage agrees well with Eugene Odum's definition: 'An ecosystem is a functional unit of physical and biological organization with characteristic trophic structure and material cycles, some degree of internal homogeneity, and recognizable boundaries.'[1] This implies that ecosystems are composed of both living and non-living components.

The definition refers to 'functional unit', and in ecology it is customary to consider a hierarchy of functional units ranging from the organism (or the individual) to the population which is a group of organisms, the community which is a set of populations, and the ecosystem which is composed of different communities and the interacting components of the environment.[2]

Drawing boundaries

Odum's definition of an ecosystem refers to 'recognizable boundaries', and this points to the delimitation problem which is central at all levels in the hierarchy of ecological functional units. As a rule, there is no problem with the delimitation of an organism, but there are exceptions. How does one define an individual of grass or of marine taxa such as Hydroida and Bryozoa? The boundary problem is more pronounced at higher levels of the hierarchy of ecological units. There has been, and is, considerable controversy in fisheries research and management of fish stocks because of problems related to the delimitation of specific populations, and the term community as used in modern marine ecology has taken on a completely new meaning because of the problems related to distinguishing one community from neighbouring communities. The dilemma is reflected in a relatively recent definition of community by Fager as 'a group of species which are often found living together'.[3]

The delimitation of ecosystems is particularly difficult in studies of offshore systems where there are no natural boundaries. For example, if the object of study is the marine ecosystem under a 100 square kilometre area in the centre of an ocean, the boundaries may be drawn at the pertinent longitudes and latitudes. In order to study relationships among the living and the non-living components of the ecosystem we need to measure the fluxes between these components across the artificially drawn boundaries. In a continually changing and moving environment this is not a trivial problem, and the definition of the boundary conditions cannot be taken lightly. It is conceivable that the fluxes that occur across the boundaries are both stronger and more variable than those generated within the ecosystem, and lack of precision in the delimitation of the boundary conditions may therefore obscure the true behaviour of the system.

Because of this delimitation problem, ecosystem analyses have been particularly successful where nature itself defines the boundary conditions. The ideal aquatic ecosystem is a pond, and the marine ecosystems which come closest to this ideal are coastal systems such as coral reefs, estuaries, coastal lagoons, mangrove systems, bays and fjords. To be sure, the total

ocean also has 'recognizable boundaries', but this ecosystem does not satisfy the criterion of 'internal homogeneity' in Odum's definition. Coastal upwelling systems have been the object of detailed ecosystem analyses, though such systems are not well delimited topographically. However, the upwelling systems are driven by relatively constant and predictable meteorological conditions, they are fairly well delineated from neighbouring water masses, and they occur in well-defined geographical areas.

Coastal ecosystems

The relative ease of delimitation is not the only reason why there has been particular attention paid to coastal ecosystems. The economic importance of these systems has also been a strong impetus for research into coastal ecosystems. Estuaries, bays and fjords provide good natural harbours, and therefore the centres of population in coastal states are often found in such areas. Examples are London, Hamburg, Amsterdam and New York, but similar examples can be found in practically every coastal state. A typical feature of coastal systems is the high organic productivity and the associated rich fisheries and mariculture. Inorganic nutrients which stimulate the growth of plants are transported to the coastal ecosystems by rivers, and in the shallow coastal areas there is also a rich supply of nutrients from the bottom sediments. The resulting high organic production explains why about 50 per cent of the world's fish catch is taken in coastal waters.[4]

The two typical features of coastal ecosystems: their proximity to dense areas of population and their importance for fish production, constitute together an area of potential conflict between users with differing interests. Users are city planners, industry, fishermen, tourists, etc., and the activities of each of these influence those of other users. Such conflicts call for rational management schemes which minimize the harmful effects for individual users and make the best use of the coastal ecosystems for society at large. Good management depends on scientific information, and coastal ecosystem research is therefore central to the problem of coastal zone management.

Although coastal ecosystems are particularly interesting both from scientific and socio-economic points of view, there are also oceanic systems which attract the attention of marine scientists. Examples of such systems are Arctic and especially Antarctic ecosystems, and the so-called oceanic fronts and eddies. These are also systems with high organic production and economic importance, but the oceanic systems differ from the coastal ones with regard to both delimitation and management. Unlike the coastal ecosystems, the oceanic systems are outside areas of national jurisdiction, and management is therefore a complex matter of multinational agreements.

Examples of coastal marine ecosystems

Obviously, since ecosystems are identified on the basis of their topographic, meteorological, oceanographic and biological components, there are virtually an indefinite number of different ecosystems in the oceans. However, it is possible to classify groups of local ecosystems into a small number of typical systems, and some examples of such typical marine coastal ecosystems are given below. Some systems, such as estuaries, coastal lagoons and mangrove systems are characterized by strong interaction with terrestrial systems, and they have therefore been grouped under the heading 'Inshore systems' below.

Coral reefs

The coral reef is a marine ecosystem familiar to many, either by personal experience or through books or films. These ecosystems are widely distributed in the Indo-Pacific region, particularly in the South Pacific Ocean, but some of the major coral reefs in the world are found in the Red Sea and in the Caribbean Sea.

The characteristic organisms of the coral reef ecosystem are the corals themselves, a group of primitive animals which are represented by a number of different species on the coral reefs. The reef-building corals have an exoskeleton of calcium carbonate, which, when the animal dies, forms the foundation for new corals to build on. The result of this process is a continual growth of the coral reefs, and the coral is thus an active agent in land formation.

Three types of coral reefs are recognized: *the fringing reef* which grows as an extension of the land masses, *the barrier reef* which is located at some distance from the shore with a lagoon in between, and the *atoll reef* which is a more or less circular structure of coral reefs with an enclosed lagoon, but with no visible land masses. Charles Darwin speculated that the various reef structures were related, but represented different stages in the development of coral reefs.[5] He proposed the theory that the atolls were formed when the land masses surrounded by barrier reefs subsided, and the subsidence rate did not exceed the growth capacity of the corals. An alternative hypothesis is based on rising sea-level as a result of the melting of the ice after glacial periods, but the result of the two hypotheses is the same. The land masses become submerged, which leads to barrier reefs formed by fringing reefs and finally atolls formed by barrier reefs.

The classical description of the coral reef as an ecosystem was provided by Eugene and Howard Odum.[6] Although more recent research has added detailed information about the structure and function of the coral reef, the basic description has not been altered. The high primary production (i.e. the plant production) of the coral reefs, which is considerable,[7] —to be exact, 2,000–5,000 grams of carbon per square metre per year, usually written 2,000–5,000 g C m^{-2} y^{-1}—is contributed by a variety of different plants: coralline algae, filamentous algae, benthic and planktonic microalgae, but the most interesting and the most important primary producers on the coral reefs are the *zooxanthellae*. These are dinoflagellates which live in symbiosis with the corals inside the animals. The photosynthetic products of the

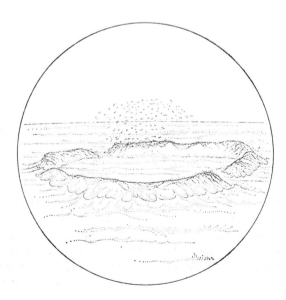

zooxanthellae are taken up directly by the soft tissues of the host corals, and studies have indicated that nearly 90 per cent of the carbon requirements of some corals could be supplied by the symbiotic zooxanthellae.[8] Other species of corals are carnivorous, i.e. they capture zooplankton organisms from the surrounding water masses.

The corals are the major consumers of primary production on the coral reefs, and they form in turn the food for the majority of the other species on the reef. Organic material excreted by the corals serve as nutrients for heterotrophic bacteria, which constitute the food for a variety of coral-reef inhabitants. The coral-reef fishes are particularly interesting and important. Some species consume a large part of the algae growing on the reef, while other species specialize on zooplankton or bottom fauna, including the corals themselves. The quantity of fishes on a coral reef may average about 500 kilograms per hectare,[9] and since the fishes are mostly small and short-lived, the annual fish production may reach 100–300 kilograms per hectare.

Diversity
and destruction

A common feature of the coral reef ecosystem is its diversity of plants and animals. High species diversity is characteristic of mature ecosystems,[10] where each species has its well-defined role to play, and where there is an intense interaction between the species and an efficient utilization of the organic production. Thus, a mature ecosystem does not export its production to surrounding ecosystems, and is not dependent upon import from them. The tight cycling of matter within the coral reef ecosystem and the efficient flow of energy explains how their very high organic production can be maintained in oceanic areas characterized by poverty in nutrients and life.

It has been postulated that high species diversity guarantees ecosystem stability,[11] because the many species provide numerous pathways for energy to flow through the communities. However, since the species in highly diverse, mature ecosystems are very specialized, this is not necessarily the case, and it has been argued on theoretical grounds that simple systems are indeed more stable than complex systems.[12] There are numerous examples from around the world that the high-diversity rain-forest ecosystem has been completely destroyed by lumbering practices, resulting in denudation and soil erosion. Similar destruction of the coral reef ecosystems was witnessed in many areas of the Indo-Pacific region during 1965–75, resulting from an explosive growth in the population of the starfish *Acanthaster planci*, which is a normal predator on corals. Scientists are not in agreement as to the causes for the population explosion of *Acanthaster*, but the conclusion of Endean[13] that removal by man of the natural predators on the starfish, particularly the giant triton, destroyed the ecological balance, is widely supported. The example of the *Acanthaster* destruction of coral reefs has demonstrated clearly that the high species diversity ecosystem is not a guarantee for stability, particularly if perturbations are brought about by man.

The destruction of the coral reefs is not a trivial matter, there is more at stake than the disappearance of one of the world's most spectacular biological systems. The fringe reefs and barrier reefs function as wave breakers for coastlines and tropical islands, and destruction of these reefs could lead to serious shore erosion. It is therefore of considerable importance for human activities that the dynamics of the coral reef ecosystems be studied and understood.

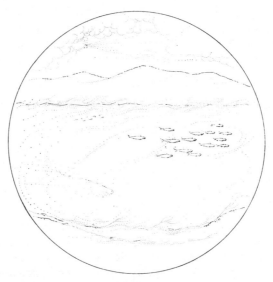

The coastal
upwelling ecosystem

Coastal upwelling is a phenomenon which occurs when the wind blows parallel to the coast, with the coast to the left in the northern hemisphere and to the right in the southern hemisphere, driving the surface waters outwards from the coast. The surface water is then replaced by deeper water flowing upwards near the coast. The phenomenon occurs everywhere, but the upwelling will be of short duration and limited magnitude in areas where the wind direction is variable. On the other hand, in areas where the wind systems are more or less constant over long periods of time, the upwelling phenomenon will result in massive replacement of surface waters, and the physical, chemical and biological properties of the coastal waters will be strongly influenced by the process. Such areas are found particularly on the eastern boundaries of the world oceans, where more or less permanent high-pressure systems set up alongshore winds driving towards the equator. Well-known upwelling systems are located on the western coasts of North and South America and northern and southern Africa. The upwelling system off the Arabian coast and Somalia represents an exception from this general pattern in that it is located on the western boundary of the Indian Ocean. This upwelling system is generated by the south-west monsoon.

Although the meteorological systems which generate the coastal upwelling are large-scale phenomena which occur over distances of hundreds of kilometres and last for many months, the upwelling process itself is on a smaller scale, occuring over distances of 5–50 km and lasting from one to ten days.[14] Therefore, although the major upwelling systems may be considered rather constant, they are locally variable in character. Even the large-scale phenomena are not completely predictable. For some years, at three- to ten-year intervals, warm and nutrient-poor water from near the equator flowed southwards along the South American coast, overlaying the deeper fertile water masses. The wind systems and the upwelling may continue as during normal years, but the upwelled water is now warm, impoverished tropical water. This 'El Niño' phenomenon has had drastic consequences for organic production and fisheries in the area.

The importance of coastal upwelling ecosystems is derived from the transport to the surface of nutrient-rich deeper water, which leads to a very high primary production by the planktonic microalgae. The nutrients are in part introduced to the upwelling system through deep lateral transport from adjacent water masses, but a significant part of the nutrients in coastal upwelling ecosystems is derived from regeneration of nutrients from organic material within the system. Micro-organisms release inorganic nutrients to the water masses or to the sediments by their action upon sinking dead plants and animals, the released nutrients being brought to the surface again by the upwelling process. This decomposition process leads to low oxygen concentrations in the deeper waters in upwelling regions.

Productivity, topography

As the upwelling systems are found at low latitudes where the surface light conditions are good, the stage is set for a rich growth of plants. Thus, the primary production in the Peru upwelling system (about $1,000 \, g \, C \, m^{-2} \, y^{-1}$) is about three times that of high-latitude shelf ecosystems,[15] and about twenty times that in oceanic waters.[16] The high primary production is utilized directly by fish (anchovies), or by crustacean zooplankton which in turn are eaten by fishes such as sardines and mackerels. Thus, the upwelling ecosystem is characterized by very short food chains, only two to three steps, which again results in a high food-chain efficiency, i.e. a high percentage of the energy bound in the plants is channelled to fish harvested by man. Therefore, some of the major fisheries of the world occur in areas of coastal upwelling, and Ryther[17] noted that 50 per cent of the fish catch in the world is taken in upwelling regions which represent only 0.1 per cent of the total ocean area.

The coastal upwelling ecosystems are not clearly defined by topographic features, but by the surrounding water masses. As these water masses are constantly changing in location and in physical/chemical properties, the upwelling ecosystems are defined by dynamic rather than rigid boundaries. Since the upwelling ecosystems are driven by large-scale meteorological phenomena, it was suggested by Margalef[18] that the upwelling can be understood only in the frame of models covering a large space, perhaps all the ocean. Still, the upwelling ecosystems are distinct systems, characterized particularly by high primary production, sustained over long periods of time. The good surface light conditions and the rich supply of inorganic nutrients stimulate plant production, which seems to be limited by water turbulence transporting the plants downwards to depths with unfavourable light conditions. The dynamics of the ecosystem seems thus to be driven by physical rather than by biological factors as proposed by Margalef.[19] Unlike the coral reef ecosystems, the upwelling ecosystems are characterized by import (nutrients from surrounding water masses) and export (fish harvested by man and birds), and the species diversity in the ecosystem is low.

The upwelling ecosystem offers rich opportunities for exploitation, but this does not exempt man from the responsibility of management. Clearly, there is not much to be done about the large-scale phenomena which generate

and control the upwelling systems, but detailed understanding of the eco-system may provide a basis for proper management of their fish resources. A drastic example of the interaction between man's activities and natural environmental events was demonstrated in the Peru upwelling system, when an El Niño phenomenon in 1972 coincided with a very high fishing pressure on the stock of anchovies. The result was that the anchovy harvest declined from an annual catch of about 12 million tonnes in 1970 to about 2 million tonnes in 1973,[20] and the consequence for fishermen and the fishing industry in the area can easily be imagined. The ability to predict phenomena like El Niño at an early stage may enable us to design proper measures which may prevent similar catastrophes in the future.

Inshore ecosystems

The term 'inshore ecosystems' is chosen for convenience to include a number of different ecosystems which have a strong interaction with terrestrial systems, and are similar in many physical, chemical and biological properties. Typical inshore ecosystems with wide geographic distribution and economic importance are estuaries, coastal lagoons and mangrove systems.

An *estuary* is defined as a 'semi-enclosed coastal body of water which has free connection to the open sea, and within which sea-water is measurably diluted with fresh water derived from land drainage'.[21] The degree of dilution as measured by the salinity of the estuarine water, varies both geographically and temporally within the estuary. The salinity may approach 0.5 per mille (or 0.05 per cent) (fresh water) near the river mouth, increasing outwards to more than 30 per mille (or 3 per cent) near the open sea. There are also considerable seasonal fluctuations depending on precipitation and variations in river flow, and daily fluctuations determined by the tidal flow. Few species have adapted to these environmental fluctuations, and the estuaries are therefore characterized by low species diversity. In estuaries at high latitudes the freshwater supply exceeds the evaporation at the surface, and the result is an outflow of low-salinity water near the surface and a deeper inflow of high-salinity water. This phenomenon is called *estuarine circulation*. In some estuaries in tropical and subtropical areas the opposite may occur. Evaporation exceeds the river input, resulting in the formation of water of very high salinity near the surface. This water is heavier than the deeper water, it sinks and flows outwards as a deep high-salinity current.

The *coastal lagoons* differ from estuaries in that the connection to the open sea is maintained through narrow channels, and some lagoons may even for periods of time be separated from the sea. Coastal lagoons in subtropical areas exhibit great seasonal fluctuations in salinity, related to variations in precipitation.

The most characteristic feature of the *mangrove system* is the mangrove trees, a number of different species of tree which are able to obtain moisture from the sea-water and nutrients from the sediments of shallow marine lagoons. The upper parts of the tree, the branches and leaves, are parts of terrestrial ecosystems as they provide food and habitat for insects and birds, whereas the lower parts, the branched roots, are important features of the marine mangrove ecosystem. Dead leaves from the canopy are important sources of organic matter for the marine mangrove system.

The inshore ecosystems are characterized by high productivity, particularly due to a very rich supply of inorganic nutrients. These are in part imported from land or from the sea, but a major part of the nutrient requirements of the inshore ecosystems is derived from regeneration within the ecosystems. All the inshore ecosystems are effective sediment traps, organic material brought into the systems by drainage from the land or produced within the ecosystem (including the leaves of the mangrove trees) being deposited in the sediments on the bottom. This leads to a very active degradation of the organic matter by micro-organisms, and subsequent release of inorganic nutrients from the sediments to the water due to the turbulence created by river flow or tidal action.

The primary production of the inshore ecosystems is high, and the major primary producers are different from those of the oceanic and coastal upwelling systems. The waters of the inshore systems are often characterized by high turbidity (oceanographers often refer to them as 'brown water systems' as distinct from the oceanic 'blue water systems'), and this prevents penetration of light and limits the growth of microalgae in the water masses.

The major producers in the inshore ecosystems are therefore plants that have the ability to utilize the water and the nutrients of the aquatic environment, and the light above the surface of the water. Such plants are eelgrasses and salt-marsh grasses, and, of course, in the mangrove ecosystems the mangrove tree itself. The primary production of these plants is high $(300-1,000 \text{ g C m}^{-2} \text{ y}^{-1})$.[22]

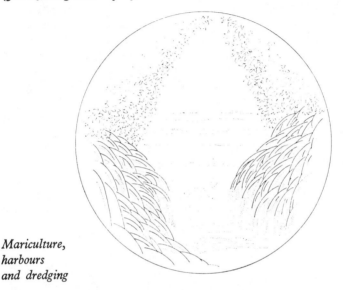

*Mariculture,
harbours
and dredging*

The plants of inshore ecosystems, in contrast to those of oceanic systems, are only to a small degree grazed by herbivores, most of the plant production going directly to decomposing micro-organisms, forming small fragments of organic material-detritus. The inshore ecosystems are therefore mainly detritus-based systems, in contrast to the phytoplankton-based oceanic and coastal upwelling systems.

The high organic production in the inshore ecosystems has resulted in rich fisheries, both in estuaries and in coastal lagoons, and these systems are also important for the mariculture of fish, molluscs and crustaceans. However, coastal systems have an additional ecological role to play which largely goes unnoticed. The detritus and the organisms that feed on it provide excellent food for larvae of many commercially important species of fish and crustaceans, and the larvae can also find shelter from predation among the seaweeds and grasses in the shallow estuaries, coastal lagoons and mangrove systems. The inshore ecosystems are therefore very important as nursery grounds for species which are subject to commercial exploitation both inside and outside the systems.

The inshore ecosystems have resources which are exploited by many different users. Estuaries and lagoons provide good harbours, but because of the high sedimentation, there is a constant need for dredging of ship canals. The shallow marshlands of estuaries and coastal lagoons are excellent for land reclamation, and large areas in different parts of the world have been turned into good agricultural land. The mangrove tree is an important source of wood in many countries, and mangroves are also cut for land-reclamation purposes. Such activities may have serious ecological consequences. Dredging operations lead to silt pollution, which damages the organic production in the natural communities and reduces the economic benefits of fisheries and mariculture. Land-reclamation practices in marshlands have much the same effect, but may in addition have serious consequences for birds that use the areas as nesting and feeding grounds. The destruction of the mangrove forests leads to shore erosion, and may destroy the mangrove systems as nursery grounds for economically important species of fish and crustaceans.

These are all legitimate uses of the ecosystem, but we must carefully regulate such practices to avoid irreparable ecological damage. Thorough research on the inshore ecosystems is therefore a basic requirement for the formation of proper policies to safeguard the ecological balance in the systems.

Ecosystem research and resource management

Traditional management schemes for commercial fisheries are based on the population dynamics of the particular species under consideration. A wealth of information about recruitment, growth and mortality of populations of commercial species has therefore been collected and analysed as a basis for political and economic decisions, and the fisheries research institutions in a number of countries have therefore enjoyed strong support from governments and fishermen's organizations. Management measures taken on the basis of advice from fishery scientists can point to spectacular successes, such as the revival of the Pacific halibut fishery,[23] but there are unfortunately equally spectacular failures. It is a sad fact that rich fish stocks such as the Atlanto-Scandian herring, the North Sea herring and mackerel have been decimated during the last two decades, in spite of the close attention to these fisheries by some of the most advanced laboratories in the world. Another drastic example is the collapse of the Peruvian anchovy fishery mentioned above. Clearly, bad management practices alone cannot be blamed for these developments, natural fluctuations and environmental anomalies such as the El Niño phenomenon certainly played their part.

The point remains, however, that proper management practices may require research into various components of the ecosystems of which the commercially important fish populations are integral parts.

The inadequacy of management on the basis of information about single species is now widely recognized, and fishery scientists are considering multispecies models which include the interactions between different species. Studies performed in various parts of the world have shown that a decline in the abundance of one species has resulted in the increase in other species. In an analysis of the fisheries in the North Sea, Jones and Richards[24] demonstrated that there was an increase in the production of groundfish following the reduction of the herring and mackerel stocks during the 1960s. Their explanation was that small fishes which constitute the diet of groundfish replaced herring and mackerel as the major predators on zooplankton, and thus directed more of the organic production of surface waters to the groundfish. Another example may be found in the increase in the stock of sardines after the decimation of the anchovy stock by the combined effect of overfishing and the El Niño phenomenon in the Peruvian upwelling system.[25]

Consideration of entire ecosystem dynamics could be an important element in the management of fish populations or coastal ecosystems having conflicting user interests. The question remains, however, whether the present state of knowledge about the dynamics of ecosystems and the methodology for prediction, i.e. mathematical models, is sufficiently advanced to serve as bases for management. Because of the high species diversity and the complexity of interactions among the components of ecosystems, we have only a superficial understanding of the behaviour of the systems. The mathematical models we use are mostly adapted from systems engineering, but they may not be applicable to marine ecosystems. Therefore, consideration of total ecosystems in the management of marine resources must be used with care and at the present time only as a supplement to the more traditional management methods, while the search for more ecological insight and better mathematical models continues.

International co-operation

Some of the more interesting and important, but also endangered ecosystems in the world are located in developing countries. These include the coral reefs, the upwelling ecosystems, the coastal lagoons and the mangrove systems. In order to ensure proper use of the biological resources of such systems and to protect them from environmental hazards, there is a need to develop management policies based on a thorough understanding of the dynamics of the ecosystems. It is therefore important that coastal states develop their marine science capabilities, and the matter is urgent because the growing world population calls for increasing the food production from marine ecosystems. Furthermore, the technology of resource exploitation is developing so rapidly that man is capable of disrupting the ecological balance of ecosystems. The Convention of the United Nations Conference on the Law of the Sea (UNCLOS) leaves areas within 200 miles of the coast under the jurisdiction of coastal states, and this gives states the responsibility of resource management which requires a marine-science capability.

The initiative and the major burden of developing a marine-science capability must be borne by the coastal states themselves, but the magnitude of the problem, particularly in developing countries, calls for assistance programmes to be developed by industrial countries and by international organizations. As a step in this direction, Unesco, through its Intergovernmental Oceanographic Commission (IOC), has launched a comprehensive plan for a major assistance programme to enhance the marine-science capability of developing countries.

Typical coastal ecosystems for which rational management strategies must be developed are found in a large number of coastal states, and it would be unnecessary and wasteful if the ecosystems research needed for management had to be repeated in all countries. Fortunately, the composition of important coastal ecosystems is identical in neighbouring countries and the ecological processes are more or less universal. The results of ecosystems research in one country will therefore be of benefit to other countries.

If networks of co-operating research institutions were developed, each participating country would share the collective knowledge of the ecosystems and thus be able to design better management plans. There are a number of examples of such international co-operation, particularly on a regional basis, in the study of ecosystems. The countries on the western coast of South America co-operate in the Estudio Regional del Fenómeno El Niño (ERFEN) through their regional association, Commisión Permanente del Pacífico Sur (CPPS) and in co-operation with IOC. Similar international collaboration is at present under development for the integrated study of mangrove ecosystems and coastal lagoons, and Unesco plays an important role in the organization of these activities.   ∎

Notes

1. E. Odum, 'The Emergence of Ecology as a New Integrative Discipline', *Science*, Vol. 195, 1977, pp. 1289–93.
2. E. Odum, *Fundamentals of Ecology*, Philadelphia, W. B. Saunders, 1971.
3. E. Fager, 'Communities of Organisms', *The Sea*, Vol. 2, pp. 415–37, London, John Wiley, 1963.
4. J. Ryther, 'Photosynthesis and Fish Production in the Sea', *Science*, Vol. 166, 1969, pp. 72–6.
5. C. Darwin, *The Structure and Distribution of Coral Reefs*, London, Smith, Elder & Co., 1842.
6. H. Odum and E. Odum, 'Trophic Structure and Productivity of a Windward Reef Coral Community on Eniwetock Atoll', *Ecological Monographs*, No. 25, 1955, pp. 291–320.
7. K. Mann, *Ecology of Coastal Waters. A Systems Approach*, Oxford, Blackwell Scientific Publications, 1982.
8. L. Muscatine and J. Porter, 'Reef Corals: Mutualistic Symbiosis Adapted to Nutrient-poor Environments', *Bioscience*, Vol. 27, 1977, pp. 454–60.
9. Mann, op. cit.
10. R. Margalef, *Perspectives in Ecological Theory*, Chicago, University of Chicago Press, 1968.
11. R. McArthur, 'Fluctuations of Animal Populations and a Measure of Community Stability', *Ecology*, Vol. 35, 1955, pp. 533–6.
12. R. May, *Stability and Complexity in Model Ecosystems*, Princeton, N.J., Princeton University Press, 1973.
13. R. Endean, '*Acanthaster planci* Infestations of Reefs of the Great Barrier Reef', *Proc. 3rd int. Symp. Coral Reefs*, Vol. 1, 1977.
14. R. Barber and R. Smith, 'Coastal Upwelling Ecosystems', in A. R. Longhurst (ed.), *Analysis of Marine Ecosystems*, London, Academic Press, 1981.
15. J. Walsh, 'A Carbon Budget for Overfishing Off Peru', *Nature*, Vol. 290, 1981, pp. 300–4.
16. T. Parsons, M. Takahashi and B. Hargrave, *Biological Oceanographic Processes*, Oxford, Pergamon Press, 1977.
17. Ryther, op. cit.
18. R. Margalef, 'What is An Upwelling Ecosystem?', in R. Boje and M. Tomczak (eds.), *Upwelling Ecosystems*, Berlin, Springer Verlag, 1978.
19. R. Margalef, 'Life Forms of Phytoplankton as Survival Alternatives in an Unstable Environment', *Oceanologica Acta*, Vol. 1, 1978, pp. 493–550.
20. Walsh, op. cit.

21. D. Pritchard, 'What is an Estuary: Physical Viewpoint', in G. M. Lauff (ed.),
    Estuaries, Washington, American Association for the Advancement of Science,
    1967.
22. Mann, op. cit.
23. T. Loftas, *The Last Resource*, London, Hamish Hamilton, 1969.
24. R. Jones and D. Richards, 'Some Observations on the Interrelationships
    Between the Major Fish Species in the North Sea', *ICES CM 1976*,
    Demersal Fish (Northern) Committee F: 33, 1976.
25. Walsh, op. cit.

To delve more deeply

KREMER, J.; NIXON, S. *A Coastal Marine Ecosystem: Simulation and Analysis.*
    Berlin, Springer-Verlag, 1978.
PLATT, T.; MANN, K.; ULANOWICZ, R. *Mathematical Models in Biological
    Oceanography.* Paris, Unesco, 1981.
POTTER, J. Management of the Tidal Thames. In: A. L. H. Gameson (ed.),
    Mathematical and Hydraulic Modelling of Estuaries Pollution. London, HMSO,
    1973.
WEBSTER, J. Hierarchical Organization of Ecosystems. In: E. Halfon (ed.),
    Theoretical Systems Ecology, pp. 119–29. New York, Academic Press, 1979.

PART 5.

THE FUTURE

For millennia, man has been sailing the coastal areas, fishing,
catching shellfish and extracting salt. Based on his experience, he
has 'opened up' the oceans by the use of scientific learning and
instruments, such as ships' clocks, for better navigation. Today,
nothing can happen in the sea without science having some part in it.
Science is the basis for the wise use of the ocean and its resources,
to discover what means are required to attain optimal balance between
exploitation and protection of very different marine environments.

Chapter 30

Marine science at the dawn of the year 2000

Eugen Seibold

Originally a geologist, the author obtained the Habilitation in
sedimentology in 1951, then taught geological science at Karlsruhe,
Tübingen and Kiel until 1979. Since 1980, Professor Seibold has been
president of the Deutsche Forschungsgemeinschaft. He is the author of
130 books and papers and, as an oceanographer, is well known for this
work as chief scientist on board R.V. Meteor, *in the Gulf (1965),*
near Portugal and Morocco (1967) and off West Africa (1971, 1975), as
chief scientist on board R.V. Valdivia, *West Africa, 1975, and as*
co-chief scientist on the Glomar Challenger *off West Africa and*
aboard R.V. Sonne *near north-western Australia in 1979. The author*
is past president of the International Union of Geological Sciences.
Address: Präsident der Deutschen Forschungsgemeinschaft, Kennedyallee
40, Postfach 205004, 5300-Bonn 2 (Federal Republic of Germany).

Let me begin by citing the importance of scientific results for decision-makers concerned with coastal management. They must be prepared to answer such questions as: Where are harbours, industry, tidal power stations and mariculture to be situated? Where should recreation areas be developed and what are their consequences for coastal regions? Science can help us resolve fairly these and other national (and often international) issues because science tries to be objective and because science speaks a truly worldwide language.

With the signing in 1982 of the Law of the Sea Convention by more than a hundred nations—and with its foreseeable ratification during the coming years—we enter a new historical epoch: a real oceanic decade or, more likely, an oceanic century. More marine science is thus urgently needed, and marine scientists are ready and willing to help. Together with engineers, researchers know that concerted action is needed at the national and world-wide levels in order to exploit the most modern technologies and to make the best use of the experience and skill found in universities, industry and governmental bodies. But what is the present state of the art? What are the major scientific problems and trends in marine research? What does the future portend?

I agree with Paul Valéry's remark, *Un regard sur la mer, c'est un regard sur le possible* ('looking at the sea is looking at what is possible'), and I interpret it directly, rather than as one of his many famous magical associations. Before continuing, however, let me say that within these few pages all the major problems in marine research and describe all their applications one cannot define, nor is it possible to predict all the developments in oceanography over the next two decades.

In 1981 a group of scientists were asked by the Intergovernmental Oceanographic Commission to produce a report, *Future Ocean Research* (FORE), in an attempt to answer some of the many questions. The researchers are aware that every scientist looks ahead and selects possibilities for future work: problems to be solved, methods to be used, regions to be visited and colleagues with whom to work. The scientists must, therefore, at least plan for the short term. Scientists having administrative responsibilities are aware, for example, that investment in a research vessel is based on the ship's life-span of about two decades, and realize that institutes, dedicated to the training of future generations, can be hoped to be active for at least half a century.

But the FORE group realized that no one is able to predict the unpredictable, the emergence of new ideas or instruments, the surprising effects of bringing together new creative groups. For these are the real driving forces, the very heart of basic science. Therefore, keeping these difficulties in mind, a forecast of the future evolution of ocean science and technology can be based only on the extrapolation of existing trends and awareness of the progress made during the last decades.

Physical oceanography

Physical oceanography has taken special advantage of the recent explosion in technical innovation and improvement. The dynamics of surface-water movement can be studied by satellite, using infra-red sensors; these indicate water temperatures and chlorophyll content. An almost continuous monitoring of upwelling waters off continental west coasts and in equatorial

---

Mini-glossary

Aquaculture	Growing useful organisms in fresh, brackish or saline waters
Bathymetry	Measurement of aquatic depths
Chemosynthesis	Synthesis by living organisms of organic compounds from simple, organic material
Cretaceous	Geological period estimated to be 65–141 million years old
Exclusive Economic Zone	National economic sovereignty extending to 200 miles offshore
Halocarbon	Carbon compound of one of the halogens (the elements bromine, chlorine, fluorine or iodine)
Hydropiston corer	Patented coring device for use in sedimentological research
Mariculture	Aquaculture in sea-water
Monsoon	Seasonal period of wind and heavy rainfall
Permian	Geological period estimated to be 245–290 million years old
Phytoplankton	Predominantly free-living population of microscopic algae, both uni- and multicellular, encountered in deep-sea and littoral environments
Plate tectonics	New concept of the structure of the earth's crust and upper mantle: mobile plates bearing both continents and sea floor
Sea beam	Patented radar-type scanning device
Tertiary	Geological period estimated to be 2–65 million years old
Tsunami	Japanese loan-word signifying huge oceanic wave of seismic origin, capable of causing extensive damage to life and property

---

regions has thereby become possible. It is known that such waters upwell from cooler depths richer in nutrients, thus favouring rich unicellular life when it reaches sunlight. With this phytoplankton the food chain begins, ending with fishes, and finally contributing to man's protein requirements.

Radar pulses emitted by satellites penetrate the clouds to detect wave motion from which wind direction and wind force can be worked out; this is an aid for shipping to select the most economical navigational routes (a computerized process, this is done in 'real time' to ensure maximal economy). Storm floods threatening coasts and estuaries can also be detected in this way. Similarly, radar serves as an altimeter to measure the sea's surface, where a remarkable resolution of 5–10 cm is possible.

As we are now able to eliminate seasonal tide and wind effects, we can reconstruct an 'average' sea-level influenced by gravity. This research offers a new dimension for exploration of the morphology of the sea floor, especially in remote areas like the Southern Ocean. The position of drift buoys which show the paths of surface currents, is monitored by satellite. Acoustic

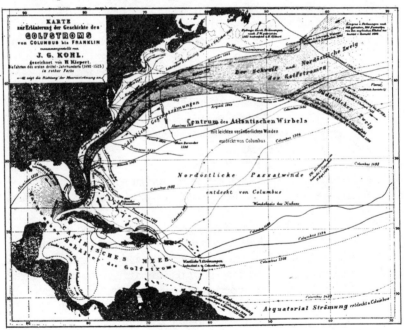

FIG. 1. (Simple) Gulf Stream as seen a century ago. (After J. G. Kohl, Bremen, 1868, courtesy of W. Krauss, Kiel.)

profiling deciphers the dynamics of deep-water masses. All these new methods produce an overwhelming flood of data, information which cannot be digested without larger and more sophisticated computers, or without better scientific strategies. There is a continuous feedback, therefore, between basic science, which brings forth new hypotheses and theories, and these new methods of investigation.

I would like to mention two areas of special progress in physical ocean-ography, concerning surface currents and air/sea interaction. At one time, the Gulf Stream was thought to be a broad continuous transport band of warm water heating northern Europe (Fig. 1). As more and increasingly complex counter-currents and eddies are detected, the latest results from drift buoys are reminiscent of the fascinating drawings of fountain waters by Leonardo da Vinci (Fig. 2). Even the cooling California current, formerly regarded by oceanographers as a 'broad, shallow, sluggish and not very exciting flow', now shows surprisingly baroque features. All these eddies transport plankton, fish eggs and pollutants, and they reduce gross current velocities.

A few more puzzles

Why, where and when are these currents formed? To what depth do they influence surface waters? Are they influenced by coastal or sea-bottom topography? Sharp, vertical boundaries of water masses have been detected: what are the origins and effects of these 'fronts'? As I have pointed out, many

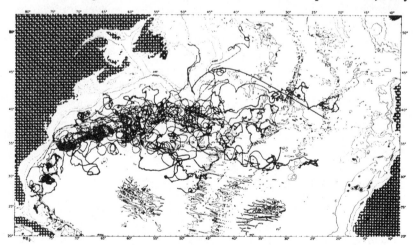

FIG. 2. (Complex) Gulf Stream, today. Trajectories are measured with aid of free-drifting buoys positioned by satellite. (After P. L. Richardson, *Journal of Physical Oceanography*, No. 11, 1981.)

more measurements from the air/sea interface are available than before. Air/sea interaction forms the heart of our global weather machine, and climate is the 'integral' or synergy of all weather.

Only when we understand all these factors better will we be able to improve weather prediction and, perhaps some day, climatic change. These consequences are important not only for sailors and fishermen but also for people on land. This is illustrated dramatically in the monsoon regions, where the date of onset alone can be the difference between starvation and agricultural abundance, as in large parts of India. To these ends, models and their refinements need to be improved as strategic tools for mastery of the many measurements involved, to establish further measurements for the checking of new theories, and, as I said above, larger computers are required.

One of the most astonishing results obtained in the last decade was the reconstruction of climatic factors (such as winter and summer sea-surface temperatures) of 18,000 years ago, that is, at the climax of the most recent ice age. Temperature indications stored as isotope ratios within plankton shells, buried in marine sediments dating from this epoch, are combined with modelling to form the basis of the global calculations needed. Such an approach is important as we try to look ahead to possible climatic changes in the future.

The greatest gaps in our present knowledge concern measurement and other data related to vertical mixing. Cold, dense water masses sink in the polar zones, especially around Antarctica. These areas remain huge white spots on our map of oceanic knowledge—chemical as well as biological and geological. Air/sea/ice interactions thus require additional investigation. As to the sinking water masses, their paths towards the equator are fairly well known; but where, how and how fast they mix with the waters above, and how they reach surface waters once again is not clear.

One last remark on the subject: in spite of the painstaking improvements in technology already mentioned, there remains a continuing need for research vessels because satellite measurements or oceanographic and climatic models have to be checked 'on the ground'. We are now in a better position, however, to plan expensive oceanographic expeditions than before, in order to investigate more systematically both typical and critical areas and processes.

Chemical oceanography

Chemical oceanography, also, has profited from improved, more exact and automated tools which often can be used directly on board ship. Thus we are able to optimize the position of observation stations established in support of expeditions. But despite this kind of progress, there remains a fundamental, generalized gap in our knowledge of the quantitative fluxes of inorganic and organic materials (including trace elements), various isotopes and anthropogenic substances (such as organic compounds), pesticides and some heavy metals. The nature of these fluxes is complicated by the fact that sea-water is not a simple liquid; rather, it contains many particulates both of inorganic and organic origin.

The fluxes may be very small and therefore dispose of huge surfaces for scavenging and transporting or releasing different ions, including metals. For example, we are alarmed by the continuing rise of carbon dioxide in the atmosphere, and at present we do not know what its climatic consequences might be. The oceans are, by far, the largest carbon dioxide reservoir because they contain 19,000 gigatonnes of CO_2 (1 gigatonne = 10^9 tonnes). The land (soil and plants) stores about 1,800 gigatonnes of CO_2, while the air contains only 700 gigatonnes. We must next determine the quantitative exchange of CO_2 between: (a) atmosphere and oceans; (b) different water masses; (c) sea-water and organisms; and (d) sea-water and carbonate sediments. These sediments cover about half of the deep-sea floor and dominate in regions with tropical reefs.

I have mentioned that even surface currents may be influenced by the sea floor. Here again there are interactions between the chemistry of sea-water and the ocean floor, whether these interactions be with sediments or interstitial waters. This is of practical interest in planning the disposal of radioactive and other wastes in oceanic sediments.

Vents, exotic biota and trace elements

The most exciting new scientific problems to arise recently result from the discovery of 'vents'—hydrothermal springs along the mid-oceanic ridges (Fig. 3). These hot springs lie at depths of about two kilometres, releasing sea-water which had previously penetrated the earth's crust. On its way down and through heating, the sea-water loses magnesium (Mg^{++}) ions and is enriched by potassium (K^+), calcium (Ca^{++}) and other ions. Some specialists believe, therefore, that this mechanism influences the global 'chemical budget' of the oceans.

Near these submarine vents are found quite unexpected biota: bacteria, mussels as large as 30 cm in diameter, worms almost two metres long, all higher organisms normally dependent on plants capable of photosynthesis. In the dark depths of the vents, however, chemosynthesis bridges the gap between the inorganic and organic worlds—a completely new field for chemical and biological research. We shall return to these vents as we discuss ore formation in the oceans.

Another surprising development in chemical oceanography is helping physical oceanographers. This is the contemporary use of trace elements, of isotopes of hydrogen and helium (^3H and ^3He) and of halocarbons (the

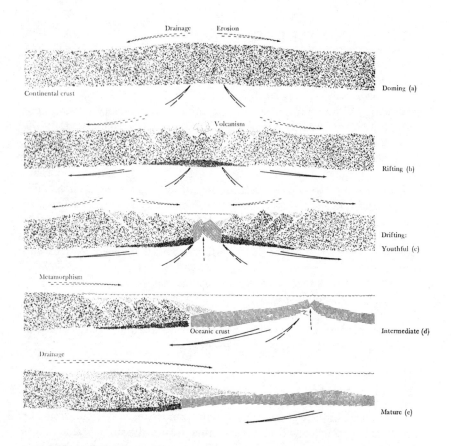

FIG. 3. Sea-floor spreading. An increased heat flow from the earth's mantle
produces 'updoming' of the continental crust (a). Hot mantle material penetrates
to the surface during drifting and causes volcanism (b). Both sides of these weak
linear zones drift apart from the spreading centre, at velocities of 1 to more
than 10 cm per year (c). A continuous supply of this basaltic material fills the gaps
and forms a new oceanic crust at the mid-oceanic ridges (d, right). Sea-water
penetrates into cracks of these basalts, is heated and precipitates metal sulphides,
oxides and hydroxides. The spreading oceanic crust cools and subsides. Water
depths therefore increase landwards (d); when these reach about 4.5 km, carbonate
particles from marine organisms are dissolved and accumulation rates drastically
reduced—a prerequisite for the growth and survival of manganese nodules. At
continental margins (e), sedimentation rates are higher by one order of magnitude
or more because of the material delivered from the continents. Additionally,
because of spreading, the ocean floor becomes older near the continents and more
time becomes available to accumulate sediments. Thick sediments are a prerequisite
for petroleum and gas formation; (e) is a schematic model of a 'passive' continental
margin. (After J. Curray.)

freons) as tracers of water masses, their movements and their mixing. We now understand far better than before the velocity of all these processes and their far-reaching consequences in terms of climate and pollution.

Biological oceanography

The biological sciences face the most complexities in regard to the oceans, for none of the 180,000 marine species identified to date live in isolation. They are either predators or prey, or else live in peaceful symbiosis with others. Most marine animals depend directly or indirectly on sea plants. (The exceptions are found near the vents mentioned above.) The very complexity of ecosystems leaves enormous gaps in our knowledge. How did ecosystems evolve? How did they react during earlier stages to the lack of oxygen in basin waters? (See Fig. 4.) How did the ecosystems react to the changing oceans during the ice ages? How are they distributed within different climatic zones in the oceans, from polar to tropical latitudes?

Antarctic marine ecosystems show many peculiarities, owing to their restricted food web, and thus are attracting many biologists. How do eddies and fronts affect the distribution of ecosystems in the Antarctic and elsewhere? What are the dimensions and causes of their fluctuations? In order to answer these questions, not only is careful and continuous stock-taking needed, we must learn much more about the relations between oceanic biology and chemistry, about physiological processes and even about organic behaviour. What a challenge, especially on the open seas! What a challenge, whether among the lesser-known ecosystems of high latitudes or of deep-sea bottoms or in fast-changing conditions near our coasts!

Modelling, palaeontology and mineral resources

Mathematical modelling has only just begun; a full, quantitative understanding of even simple ecosystems remains a long way off. In marine geology (as we shall see), a breakthrough has occurred during the past twenty years with confirmation of the theories related to spreading of the sea floor and plate tectonics. Confirmation of these theories has enabled us to integrate many isolated observations, many facts, in a unifying framework, as well as to formulate critical new theories to be confirmed or rejected. It is to be hoped that a general theory of ecosystems will emerge at the beginning of the next century.

Critical biologists have doubts as to whether all the conclusions reached by palaeontologists concerning palaeo-environments are valid. Changing environments may cause different kinds of adaptation, thus evolution, of species. This is a never-ending problem for researchers. But this and the

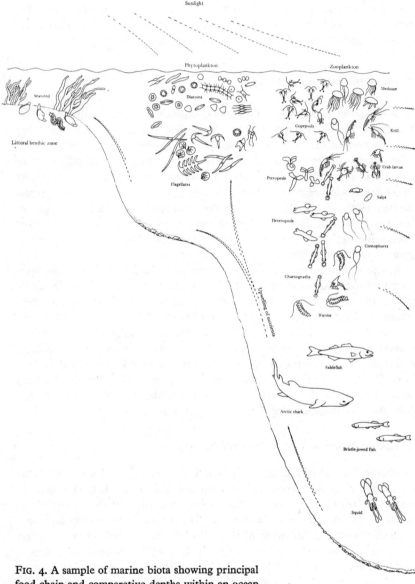

FIG. 4. A sample of marine biota showing principal
food chain and comparative depths within an ocean
ecosystem. The arrows show the fall or 'rain' of organic
detritus in the water, possibly the main transport of
food towards the depths. (After J. W. Hedgpeth, 1957,
J. D. Isaacs, 1969, and E. P. Odum, 1971. © Scientific
American, Inc., 1969.)

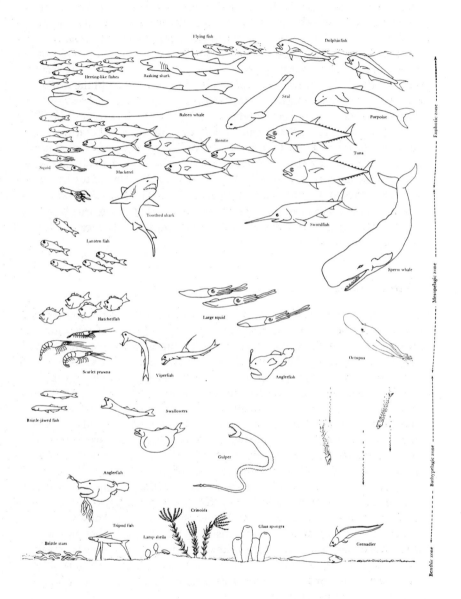

other questions I have raised are by no means purely academic. The foundation of all applications is basic science. Fishery and aquaculture in the seas, for example, stand to be improved by the development of new knowledge. Look at the case of what seems to be a largely untapped source of protein, the krill supply in Antarctica. How little we know of these krill! In the production of petroleum and natural gas, 30 per cent of the yield in 1981 came from offshore wells. The estimated value of this production, at least in the market economies, has been reckoned at $150,000 million. Perhaps the lesson here is that we should not forget that oil and gas originated in former marine life.

In 1980 the world's intake of fish, crustaceans and mussels from the sea (95 per cent of which came from the Exclusive Economic Zones, and 70 per cent of which was destined for human consumption) had a total value of nearly $15,000 million. This is only about a tenth of the value of the hydrocarbons indicated in the preceding paragraph. Mineral resources, however, are not renewable, whereas biological resources are. But this is true only if we use our living resources wisely and, once again, this prudence requires scientific knowledge.

Marine geology

Development of our knowledge of the earth's crust beneath the oceans could follow a specific strategy of research during the final decades of the century because of the status of the theories of sea-floor spreading and plate tectonics (Fig. 5). Many new metal concentrates have been found on the mid-oceanic ridges: oxides and hydroxides of iron and manganese, sulphides of copper and zinc and other non-ferrous metals, together with the strange biota discovered near the vents. These metallic compounds not only resemble those found in the Red Sea Deeps, in a similar tectonic setting (and currently the only compounds having potential economic value); they also represent a very interesting, ongoing experiment to understand fossil ore bodies found on land, thought also to be of hydrothermal origin.

Continuous deep-sea drilling activities aboard the *Glomar Challenger*, where I have conducted research activities, and especially use of the new Hydropiston corer, are beginning to discover more of the unknown. Stratigraphic chronological resolution, for instance, is now reasonably precise in the order of several centuries. And worldwide correlation is helping to explain better than before our concepts of palaeo-oceanography, palaeoclimatology and palaeobiogeography, as well as sea-level oscillations and their consequences for deep-sea currents.

Rare events, such as the possible impacts of celestial bodies with the earth at the epoch between the Cretaceous and Tertiary periods (with its drastic faunal and floral changes), or the desiccation of the Mediterranean, may

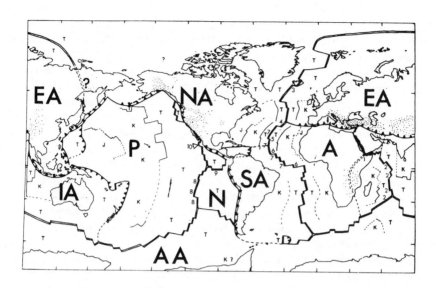

FIG. 5. The concept of plate tectonics: major plates of the outer, rigid crust of the earth—EA, Eurasia; NA, SA, the Americas; A, Africa; IA, Indo-Australia; AA, Antarctica; P, Pacific. The smaller letters J, K, T represent the ocean's crust formed during the Jurassic, Cretaceous or Tertiary periods. The mid-oceanic ridges and divergent plate boundaries are seen as double lines, while occurrences of metals are represented by arabic numerals. The spiked lines represent convergent plate boundaries. (After E. Seibold, *Bruun Memorial Lectures*, Paris, 1979.)

become new global time-markers. It is by utilizing these new tools during the coming decade that we shall surely gain new insights into organic evolution. One of the most exciting results so far is the confirmation of periodicities of 100,000, 40,000 and 20,000 years—cycles with great import for sea-level and climatic changes and possibly even for predicting such changes. Another significant development has been reconstruction of groundplans and vertical sections of the major oceanic basins since Permian times (about 200 million years ago). This is the base on which to reconstruct palaeocurrents, and it is important for the rational exploration of manganese nodules.

Origins of petroleum, gas and metallic nodules

Here again, many questions remain unanswered: Were all the 'passive margins' of continents formed in a similar way? Why and how fast did these subside? What is the nature of the very deep structures lying beneath these margins? The active continental margins reveal more and more paradoxes. Is it expansion and compression, or expansion or compression? Is it uplift or subsidence? Is it accretion and/or 'tectonic erosion'? It would seem that diversity is more widespread than similarity. What are the properties of the ocean crust's different layers? Whence come the 'stripes' of magnetic reversal? The most obvious problems now are (a) understanding the mechanisms and significance of the accretion of former oceanic plateaux at the active margins, such as round the Pacific Ocean, and (b) resolving the origin of the so-called back-arc basins, as in South-East Asia. These features play a special role in exploration for offshore petroleum and natural gas.

The first complete bathymetric chart of the oceans (scale 1 : 10,000,000) has just been published, but it still gives only a broad outline of the main features of the sea floor's morphology. New methods, using side-scanning of the sea floor by radar as in the case of the Sea Beam, produce directly aboard ship detailed morphological charts of remarkable accuracy. These are prerequisites for the selection of drilling sites, the use of submersibles, the study of submarine canyons, etc.

By far the most important marine mineral resources offshore are petroleum and natural gas. In applying the theory of the spreading of the sea floor to these resources, the future scenario is pessimistic; 80–90 per cent of the ocean floor will never produce hydrocarbons economically because the sedimentary cover is not thick enough for organic matter to mature sufficiently at high temperatures. This is because of the youth of vast, central areas of the ocean's floor and reduced rates of accumulation of sediments in the deep sea. The best prospects, therefore, are restricted to the continental margins.

Manganese nodules and especially mid-oceanic ridge ores are not easily exploited, in spite of successful pilot operations conducted to obtain nodules bearing more than 2 per cent of combinations of cobalt, nickel and copper from the depths of the equatorial Pacific Ocean. Economical exploitation of marine mineral sources is currently restricted to placer deposits, that is, to mechanical concentrates of minerals containing tin, zirconium, titanium or iron at depths of less than 40 m. Aggregates such as sand and gravel, if situated not too far from industrial sites, can be expected to have continuing economic importance.

Summary and general conclusions

In the FORE report mentioned above, I raised some topics of a more general nature. A few of these follow.

No single publication can give a full impression of the many facets of the sea, with all the challenges and adventures implicit for both man and science, extending from rocky and sandy coasts to the endless and stormy southern oceans.

One needs to live and work aboard ship for a while in tropical or polar seas; one has to dive among coral reefs; one should walk the soft, tidal flats or penetrate the labyrinth of mangroves, and breathe the wind of the sea and smell its organisms in order really to understand the fascination of marine research.

Everyone is aware that the oceans cover vast areas, so that even today little is known from offshore, especially in remote or hostile regions like the Arctic or the south-eastern Pacific. And the marine realm constantly presents new discoveries, such as the mid-oceanic hot vents and their peculiar biota.

Because of limited manpower, shipboard time and finances, the huge stretches of ocean need the elaboration of strategies for the selection of research projects. Models of circulation, future models of ecosystems, concepts such as global tectonics are examples of strategies. New techniques are required for simultaneously obtaining a variety of data from large areas of the marine environment. Examples of this information include sea-surface temperatures, wave observations from satellites, airborne magnetic measurements, multidimensional data from the sea bottom via side-scanning radar, Sea Beam, and continuous seismic reporting.

In order for large oceanic areas to be studied with a 'big science' approach, further international co-operation is needed. Many exciting discoveries resulted from enterprises such as the International Geophysical Year, the International Indian Ocean Expedition and the Deep-sea Drilling Project. The vast ice ocean of Antarctica cannot be investigated without such co-operation. And all these examples show that marine research is expensive, so that the financial aspects alone beg continuing co-operation between large and small countries, between industrialized nations and those in varying stages of development. The world ocean has unity, implying interaction on a global scale; the seas cannot be divided into small compartments as was medieval Europe.

Differentiation and interdisciplinarity

Oceanic processes are governed by the fundamental laws of physics, chemistry and biology. But as we are accustomed to distinguishing between industrialized and developing countries (looking at them as groups whose constituents are in themselves different), we must keep in mind the many differentiations applying to the ocean. Climate causes the most obvious differences; exposure to polar seas, both at the surface and in deep water, is a comparable factor. Regional and local environments result from complex interactions, just as connections with adjacent seas dictate salinity and oxygen content. Processes taking place at the boundary between atmosphere and sea, between the sea and its floor, and especially along the coasts where all these processes meet, these are all influenced by a great number of factors. Taken altogether, the ocean is more than the sum of its parts. That is why ocean research always needs an interdisciplinary approach.

New efforts in transdisciplinary teaching are therefore necessary, especially in the developing countries. All nations and international scientific organizations should also improve the dialogue between oceanologists and society in general, thus assuring a ripe and fertile scientific climate round the globe (and particularly among the developing nations).

Many of the industrializing countries need more and better scientific and technical manpower, facilities and equipment, as well as services and a sufficient budget. All these elements can help the developing nations to establish their own expertise in order to be able to judge independently their own situations and meet the many new responsibilities evolving from the rights deriving from the Law of the Sea. We must appeal to such a global partnership because the sea is one of the common heritages of mankind—much as science is, too.

The sea, teacher as well as provider

The sea, I should add, is a good teacher. Scales of time and space are involved: waves vary in length from centimetres to thousands of kilometres (as in the case of tsunami) and in periods of time from seconds to months. Surface circulation covers hundreds or thousands of kilometres, over weeks and even years; periods of deep circulation can last thousands of years. Cyclical temporal parameters for ocean climates extend from 10,000 to 100,000 years, whereas changes in the morphology of ocean basins require millions of years.

Although the average physicist or chemist may not learn much from ocean research to improve his knowledge of a specific subdiscipline, oceanic investigation depends heavily on progress in the basic fields of science; the same can be said for many aspects of modern biology. Yet the very difficult study of ecosystems seems to be more easily accomplished on (or in) the sea, despite dramatic changes in temperature and other meteorological factors, and the need to avoid problems of humidity.

Geologists benefit from contemporary marine conditions by increasing their understanding of the environment, provenance and origin of fossil sediments and mineral resources. This knowledge in turn helps them to reconstruct reasonably accurately the earth's history. And from these findings, geologists can help other scientists to understand the present configuration of the oceans, their climates and their marine life.

The sea presents a dangerous environment to many, so marine research has an obligation to give warnings of (if not forecast) tsunami, volcanic eruptions, icebergs, storm floods and the presence of sharks. But the oceans are also promising environments. Fishing, aquaculture, the exploitation of mineral resources on the sea floor and beneath, the dumping of all sorts of waste—these are activities requiring constant scientific consideration.

The ocean has a ceaseless impact on the increasingly populated continents; it is a most effective weather machine, as I have said, influencing harvests, river floods, traffic conditions and the climate in general. Exploration forecasts, and even hints at such exploration, result from different methods of research. Investigation planning normally begins with descriptions of fact, then observations leading to consideration of processes, with the aim of understanding these in a quantitative manner. Often a general concept, or model, marks the beginning of research and then serves as the basis for forecasts if the model's postulates are supported by field or laboratory experiments.

So in order to use the ocean wisely, we need first to understand it. We need to ask how we should maintain its renewable biological resources; specifically, how we can conserve the genetic potential of the many marine species identified so far, and how we should conserve ecological integrity? In regard to waste disposal, what is the ultimate compatibility of the oceans with the many different kinds of pollutants (both artificial and natural) transported from the land by rivers and winds? All these and many more questions imply more and better research in the future because, at least since the time of Sir Francis Bacon, we have learned that 'Nature, to be commanded, must be obeyed.' ■

To delve more deeply

BEHRMAN, D. *Assault on the Largest Unknown.* Paris, Unesco, 1981.
Climate—Chemistry and Physics of the Atmosphere. Theme of entire issue of *Impact of Science on Society,* Vol. 32, No. 3, 1982.
MAULA, E. *The Archipelago as a Focus for Interdisciplinary Research.* Karachi/Helsinki, Hamdard Foundation Pakistan/Interdisciplinary Academy, 1981.
VANDERPOOL, C. Marine Science and the Law of the Sea. *Social Studies of Science,* Vol. 13, No. 1, 1983.

*An American specialist dealing with new and renewable forms of energy
passes in review some of the realistic possibilities as we look about to
broaden our energy base for the mid-term future. The author states a few
caveats, then describes the potential represented by photovoltaics,
industrial alcohol, by-product hot water, new materials and long-range
energy transfer. All these are related to the sea as source or transfer
medium.*

Chapter 31

Exploiting the ocean's
energy, 1985-2005

David W. Doyle

*An engineer with long experience in the field of the energy industries,
the author of this analytical projection is currently vice-president,
Pacific Operations, United Energy Corporation, with headquarters in
Foster City, California. Mr Doyle can be reached at 65 Hanapepe Loop,
Honolulu HI 96825 (United States of America).*

Background to a scenario

This chapter touches on some of the ocean energy systems most likely to be brought to the stage of commercialization within the next twenty years. This is only a set of guesses, educated speculation but still merely guesses—for no one can predict where the recovery of energy from, or associated with, the ocean will be taken by developing and new technology over a period of five years, let alone twenty years.

We are also presenting a point of view derived from the concept of integrated energy/food/feed farming systems; these are only now beginning to be developed on the United States' mainland. Our company concentrates on the establishment of integrated systems producing electricity and hot water (or steam) from solar energy, as well as industrial or fuel alcohol from feedstocks that include algae. Food, feed and edible fibres are derived from these same feedstocks.

Land-based integrated energy farms (including their food and feed aspects) can be replicated in many ocean areas, particularly along unused or only marginally useful coastal zones. But there is also a potential for significant exploitation of ocean energy resources by using floating platforms tethered to the ocean floor. The capital required to build such platforms would be equivalent to the value of land in coastal areas that would then become (or remain) available for other development.

Some general reservations

On the basis of research and development work already accomplished or scheduled by United Energy Corporation (and some others working in our field), it seems realistic to expect the following scenario to be put into practice by the private sector—regardless of government assistance or the lack of it. We assume that systems costs will diminish dramatically over the next few years, thus making what we say possible on a competitive level with fossil and nuclear fuels. It is also assumed that there will be few areas where governments actively oppose the systems that we believe will be established.

With the possible exception of the production of some edible (dietary) fibres, the systems we foresee do not require agricultural land or standard agricultural techniques. Everything involved can be produced in water, or be suspended in it or floating on it. For each technology, we present some of the problems now being faced and some of the solutions being sought.

Photovoltaics

The most significant barrier to widespread use of photovoltaics (PV) on land or at sea, in regions having adequate sunlight, continues to be the high capital cost of manufacture. A close look at what has happened in the past ten years and what is now happening reveals that the system cost of an 'installed watt' has been reduced from more than US\$100/W_p (for 'peak watt') fifteen years ago less than \$8/W_p in 1984. Several American firms believe that, if sales continue at their current level, the cost can be lowered to \$3/W_p by 1988. The next large reduction—other than reductions attributable to new technological breakthroughs—will probably occur as a result of the robotization of manufacture and assembly.

Other obstacles include corrosion, degradation (including damage caused by ultra-violet radiation), and inadequate or poor cost-effective storage and

Photovoltaic and marine thermal energy conversion

Photovoltaic effect—the direct conversion of sunlight into electricity, whereby light (photons) falling on, for example, silicon cells liberates electrons, generating electromotive force that in turn creates a current.

Concentrating photovoltaic system—one in which the systemic efficiency of the PV effect is combined with photo-thermal energy.

OTEC (ocean thermal energy conversion)—a Carnot-cycle heat engine exploiting the temperature difference between the tropical ocean's warm, upper layers and the deep, colder layers at 650-1,000 m below the surface. Arsène Arsonval, the French physicist, first suggested the idea in 1881; Georges Claude, his student who was to make neon gas luminesce, made attempts during the Second World War to use OTEC. Only today's level of technological development has made such thermal conversion economic.

back-up (redundancy) systems. Work now in progress, however, indicates rapid movement towards removal of all these barriers. For ocean-borne systems, there are some special problems. Corrosion becomes a much greater threat. Tethering is a complex problem, where allowance must be made for tidal movements. The advent of high winds and high seas, the need for special tracking systems: these are all of obvious concern.

And yet, we can see how all these problems can be resolved within the limits set by economics and system life. The result is that it should become economic to establish the same kind of floating PV concentrating systems as we have installed in California in order to provide power to utilities, in coastal waters including relatively protected bays, salt marshes, estuaries, deltas, fjords, lochs and so on.

Protection against wave action has become a possibility, using floating booms that contain flexible containers in which algae can be grown and harvested. Stable temperatures and normal wave action (producing light flashing through turbulence) would contribute to growth, and concurrently the use of wave action to cause this turbulence in algae containers can dampen waves. If current work on use of non-corrosive materials is successful, the equipment used to contain and enhance the growth of algae could also produce energy derived from wave action.

Alcohol

Concentrating PV systems produce hot water from their cooling circuits. Our firm uses such hot water for a variety of purposes, including the distillation of alcohol intended for use as fuel. Distillation occurs, because of a vacuum system, at precisely the outlet temperature of the PV system—60 to 65.5°C (140-150°F). Using algae of a variety now under test by our laboratories, the resulting by-product is a high-protein residue suitable as food for humans or, for animal, bird and marine life.

Work is proceeding on systems to produce power from renewable sources when the sun is not shining; the main contender among these sources is alcohol. A concentrating PV system can generate, in effect, its own back-up or redundancy power by storing alcohol made when the sun heats the PV cells.

In cases where wave action and the danger of storms is sufficiently present, additional protection from wave action can be obtained by floating the combined PV/alcohol/algae complex on materials sturdier than the 10 cm boat-dock foam now used and, if necessary, by building concentric or multiple booms. In many cases, economic considerations will permit the construction of jetties or causeways in order to protect large systems in locations where nearby land is valuable.

There are no technological barriers to growing algae commercially in or on the oceans, where climatic factors allow this, as a feedstock for alcohol. The protein by-products will provide, in most cases, significant additional revenue. CO_2 generated by the fermentation process is recycled to the algae in order to increase growth. Even in cool waters, heat detoured through algae containers (after the heat has been used for distillation) can sustain growth. All that is required is adequate direct-beam sunlight to ensure that PV concentrating systems will work effectively.

Hot water; heat

Hot water coming from the cooling circuits of PV systems can be used for several purposes other than for making alcohol. It can be used to raise the temperature of the warm side of the ocean thermal energy conversion (OTEC) systems. It can be used to warm the nutrient-rich, bacterially clean bottom-water brought up by OTEC systems, thereby permitting growth of marine life at relatively warmed temperatures in a greatly improved medium.

Hot water can also be used to increase the warm-layer temperatures in salt solar ponds. And, of course, it can be used too as the heat source for all sorts of industrial processes: for cleaning, as hot water, and so on.

It is worth noting that OTEC systems appear increasingly practical in the light of encouraging results in the cases of the anti-fouling and anti-corrosion work done at the Hawaii Natural Energy Laboratory situated at Keahole, Hawaii. (See the Laboratory's annual reports for 1983 and 1984.)

Materials

Technological improvements in materials designed to reduce damage from ultra-violet light, from corrosion and biofouling, and intended to increase the efficiency of heat exchangers, all combine to indicate that ocean-energy systems of several types will come into commercial operation within the next twenty years—particularly if the price of fossil fuels should rise substantially. But even if this should not occur, such systems will be brought into use on a growing scale in areas having the required renewable resources or climatic conditions, and areas that suffer from the burden of importing fuels or other energy forms from outside their borders.

Meanwhile, the trend in photovoltaics is towards augmented cell efficiencies (gallium arsenide should add more than 20 per cent to systemic electrical efficiencies), while improved thermal conversion systems are already bringing a secondary advantage to concentrating systems: efficiencies exceeding 40 per cent of the total sunlight available will soon be commonplace. This is a substantial commercial advantage, over and above the reduced cell sizes permitted by concentrating sunlight.

A higher quality of heat is also derived when solar cell temperatures higher than 121 °C(250 °F) are attainable without losing electrical efficiency in the cell. The tighter grids necessary for such concentrations are already in production.

Long-range transfer

There remain few significant barriers to long-range transfer (from space to ocean, for example) of power on a commercial basis. At least one company in the United States is planning to conduct a test of such power transfer, initially from a geosynchronous orbit to an uninhabited Pacific Island. Longer term plans include photovoltaic collection in orbits having 24-hour sunlight, as well as transfer to land- and ocean-borne receiving stations by means such as microwave beams. Systems of this kind will not really use ocean energy; they will use the oceans and marginally useful coastal waters as bearing surfaces, where there is little or no risk to human or other life, or to property.

Wind, wave and tidal energy

Improvements in materials, coupled with continuing work to combat biofouling and corrosion, will make commercial exploitation of wind, wave and tidal energy practical in the long term—very probably within the next twenty years. Ocean-borne 'wind farm' platforms, such as those proposed for use in the English Channel (as well as those that we foresee for the photovoltaics) should be commercially practical as the value of land and the price of hydrocarbon fuels rise. The harnessing of tidal energy is, of course, site-specific on the commercial level. We believe, however, that the exploitation of tidal power will develop further during the next two decades.

OTEC and desalination

Work being conducted on OTEC cycles (such as that in Hawaii, already mentioned) by the United States, Japan and France and commercial-scale desalination at costs of less than $0.50/thousand litres is very encouraging. United Energy Corporation, for example, plans to build a small ocean desalination system in 1986-7, which will allow us to continue and enlarge current developmental work. One of the options being studied is similar to an open-cycle OTEC system, with the PV cooling circuits adding heat to the warm side.

Improved heat-exchanger technology and better biofouling and corrosion protection combine to make both OTEC and desalination increasingly attractive options for developed as well as developing countries. Problems still encountered with OTEC include the handling and installation of large-diameter pipes, and of course ocean-floor configuration, as well as the relatively restrained number of commercially sound potential sites. But the availability for use of a large range of energy systems for desalination— including PV, solar thermal, wind, wave and tidal energy, as well as the use of ocean-produced alcohol—will make it possible to select the most cost-effective commercial systems for each site and for conditions pertaining to the site. This is a wealth not yet available to us.

Hybrid systems

It is likely that work now in progress intended to join together various ocean-energy systems, and to 'cascade' heat from function to lower quality function, will result in some interesting and practical synergistic alliances. I have mentioned some of these. To recapitulate, such systems will include:

- Production of electricity and fresh water

- Production of electricity and alcohol, food and feed

- Production of electrical and mechanical energy

- Production of energy and damping of waves

- Linkage of PV with thermal systems

- Linkage of PV with OTEC and (or) desalination

In other words, we are previewing the introduction of integrated energy/ food/feed ocean farming systems on a commercial scale. We foresee this as being entirely within our grasp in the next two decades.   ∎

Growing concern about shortages of raw materials is concentrating attention on the sea's natural resources and the prospects for their use. Perhaps not surprisingly, increasing attention is being paid by states to the future role of the ocean in international affairs. Nations look to naval power to secure and defend their perceived interests at sea. Inevitably, navies are expanding in almost all regions, and naval arms races are becoming more intensive than ever before. Navies are seen as enhancing a country's status.

Chapter 32

Military use of the oceans

Frank Barnaby

Dr Barnaby is a physicist, peace researcher and prolific author. He worked for the United Kingdom's Atomic Energy Authority from 1950 to 1957, the Medical Research Council from 1957 to 1968, as executive secretary of the Pugwash Conferences on Science and World Affairs (1968–70), and later directed the Stockholm International Peace Research Institute (SIPRI) for almost eleven years. The author is currently professor of peace studies at the Free University, Amsterdam. His address is: Brandreth, Chilbolton, Stockbridge, Hants SO20 6AW (United Kingdom).

Traditionally, naval power has been used to influence international affairs, both globally and regionally. It has not only been a tool for military action, but has more frequently been used for political coercion. With the rapidly growing interest in exploiting the ocean's resources, navies will be increasingly used to police and defend economic zones.

The superpowers will continue to use their navies to extend their influence globally, and we must expect that the superpower's naval rivalry in the major oceans will intensify. Dominating all the military uses of the oceans are those activities related to strategic nuclear war: strategic nuclear submarines carry large fractions of the American and Soviet arsenals. The superpowers are devoting large resources to developing effective anti-submarine warfare techniques, and success in this field will considerably enhance perceptions of a first-strike capability.

Given all these trends, there is little doubt that future naval developments will have serious repercussions on international security.

Navies of the superpowers

Any survey of the world's navies will be dominated by the enormous naval forces of the United States of America and the USSR, but it will also show a rapid build-up of light naval forces worldwide. In fact, the steep rise in the number and quality of light naval forces is a major reason for the global militarization of the oceans. A modern patrol-boat, for example, can carry missiles of considerable fire-power. The acquisition of 'light' naval vessels, therefore, allows new and small navies to expand rapidly into relatively powerful ones.

Each superpower publishes its assessment of the other's military forces. According to *Soviet Military Power* (an American publication),[1] the Soviet navy has two main missions: (a) to protect the seaward approaches of the USSR and its allies, and (b) to destroy naval forces of the North Atlantic Treaty Organization which could threaten Soviet military operations. The first mission requires naval forces which are larger in number, smaller in size, more designed for a single task and less capable in terms of weapons and endurance than the forces required for the second. Estimates available of Soviet naval strength support this view of the USSR's naval missions.

The Soviet navy operates four fleets ranging the Pacific Ocean, the Baltic, the Mediterranean, the Indian Ocean, the South China Sea and the South Atlantic.[2] Figure 1 shows the principal constituents of these fleets.

The Soviet Union is still building seven classes of surface warships, five classes of submarines, and four types of naval aircraft. These include a second nuclear-powered cruiser, and three classes of nuclear attack submarines. One of the nuclear submarines has a titanium hull which gives it a speed of 40 knots, the world's fastest. The Soviet Union is believed to be planning the construction of a large, nuclear-powered aircraft carrier, to enter service in the late 1980s.

The most modern aircraft being built for the Soviet Naval Air Force is the supersonic, variable-geometry-wing Backfire capable of carrying air-to-surface anti-ship cruise missiles having speeds of about three times the speed of sound.

The Soviet assessment of the military forces of the United States is published in *Whence the Threat to Peace*.[3] The United States navy, it says, is assigned a special role not only in war but in carrying out American global policies, as an 'instrument for demonstrating force and direct military intervention'. Figure 1 shows the details of this Soviet assessment of American naval power.

The United States has built very large surface warships, the biggest being the 91,400-ton Nimitz-class aircraft carriers. The biggest Soviet surface warship is the 37,000-ton carrier *Kiev*. On the other hand, the Soviet Typhoon strategic nuclear submarine displaces about 25,000 tons when submerged, significantly more than the American Trident strategic nuclear submarine, which displaces about 18,700 tons.

If the navies of the European allies of the United States and the USSR are added to those of the superpowers, NATO naval forces have a significant numerical advantage over those of the Warsaw Pact, for all categories of major surface combatants (frigates and above). Stockholm International Peace Research Institute figures[4] show that Western superiority is greatest in the category of aircraft and helicopter carriers (about fifteen to one) and in the category of frigates (about five to one). For all major surface warships it is just over two to one. Warsaw Pact naval forces are numerically superior in light vessels (corvettes and fast patrol-boats) and in mine-warfare forces.

The Warsaw Pact navies have more submarines than have NATO navies—about a hundred more. A further large difference exists between NATO and the Warsaw Pact in the numbers of modern submarines carrying ballistic missiles—NATO has forty-two against the Warsaw Pact's sixty-two. But, at any one time, only about 15 per cent of the latter are out of port, compared with over 50 per cent of the NATO submarines.

As we have seen, naval forces operate large numbers of aircraft and helicopters. The two alliances operate more than thirty different kinds of fixed-wing naval aircraft, some in several versions, and about twenty types of helicopter. NATO naval air power is considerably superior to that of the Warsaw Pact, especially in sea-based aircraft. The Soviet Union does, however, operate a fleet of land-based long- and medium-ranged bombers, assigned a maritime role and armed with various types of anti-ship missiles. The Backfire bomber is becoming the mainstay of this bomber fleet.

The role of naval air power is steadily increasing. The USSR is expanding its fleet of aircraft and helicopter carriers and almost all NATO countries

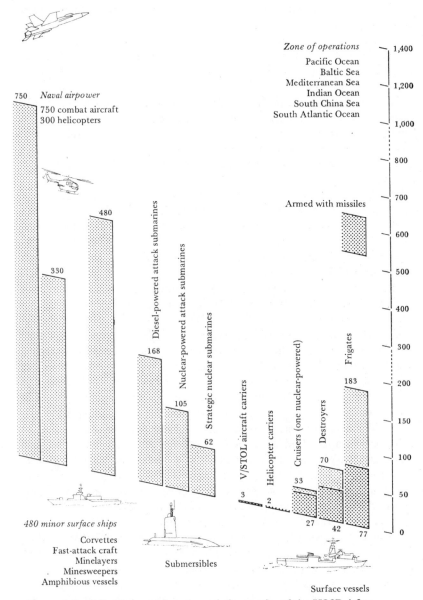

750   *Naval airpower*

750 combat aircraft
300 helicopters

480

330

Diesel-powered attack submarines

Nuclear-powered attack submarines

Strategic nuclear submarines

168

105

62

V/STOL aircraft carriers

Helicopter carriers

Cruisers (one nuclear-powered)

Destroyers

Frigates

3

2

33

70

183

27

42

77

Zone of operations
Pacific Ocean
Baltic Sea
Mediterranean Sea
Indian Ocean
South China Sea
South Atlantic Ocean

Armed with missiles

1,400

1,200

1,000

800

700

600

500

400

300

200

150

100

50

0

480 minor surface ships
Corvettes
Fast-attack craft
Minelayers
Minesweepers
Amphibious vessels

Submersibles

Surface vessels

FIG. I. (*above*) Combat vessels and naval air capacity of the USSR (after
International Institute of Strategic Studies, *The Military Balance, 1982–1983,*
London, 1982); (*opposite*) Naval capacity of the United States, according to Soviet
sources (After *Whence the Threat to Peace,* Moscow, Military Publishing House,
1982).

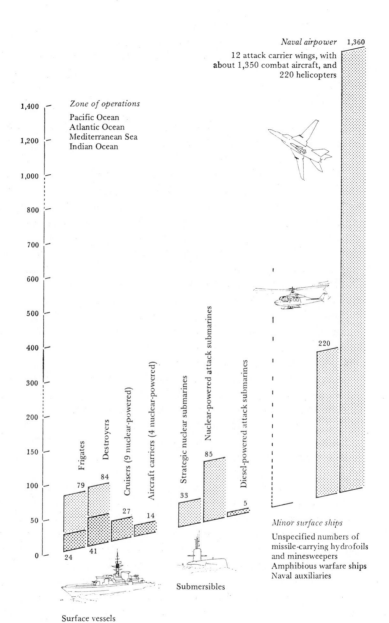

Naval airpower 1,360

12 attack carrier wings, with
about 1,350 combat aircraft, and
220 helicopters

Zone of operations

Pacific Ocean
Atlantic Ocean
Mediterranean Sea
Indian Ocean

1,400
1,200
1,000
800
700
600
500
400
300
200
150
100
50
0

Frigates

Destroyers

Cruisers (9 nuclear-powered)

Aircraft carriers (4 nuclear-powered)

Strategic nuclear submarines

Nuclear-powered attack submarines

Diesel-powered attack submarines

79

84

27

14

33

85

5

220

24

41

Submersibles

Surface vessels

Minor surface ships

Unspecified numbers of
missile-carrying hydrofoils
and minesweepers
Amphibious warfare ships
Naval auxiliaries

are modernizing their naval air forces. It can be expected that NATO's naval air superiority will, in general, be maintained and even increased.

Qualitative advances

Although numbers of ships are usually used in assessing naval 'balances', a more important feature is qualitative, i.e. the effectiveness of naval units in combat. This will be determined to a large extent by technology. In an assessment of naval technologies, SIPRI[4] has shown that naval weapon systems are undergoing rapid changes because of developments in electronics, propulsion plants, engines and construction materials. One particularly important use of naval technology is to offset the rapidly increasing costs of naval vessels. The cost factor is limiting the size and number of naval vessels that even the richest industrialized countries can afford. The cost, for example, of an American nuclear-powered aircraft carrier exceeds $2,000 million and a modern destroyer may cost about $500 million; even a frigate can cost nearly $200 million. But a missile-armed fast patrol-boat costs about one-fifth that of a missile frigate. Even though it is much smaller in size, the performance and fire-power of a modern fast patrol-boat has been increased to the former level of very much larger ships such as destroyers. Fast patrol-boats are rapidly evolving into heavily armed, multi-purpose warships—a trend shown by, for example, the French Combattante III, the Israeli Reshef class, and the Swedish Spica II. Modern fast patrol-boats operate all types of weapon systems—guns, torpedoes and missiles. Advances in automating light anti-aircraft guns and surface-to-air missiles have provided fast patrol-boats with effective defence against air attack. Fast patrol-boats are a good example of how technology is used by designers to produce ships at less cost.

Another example of the naval exploitation of new technology is the increasing interest in hydrofoils. These vessels can operate at speeds of up to seventy knots, even in rough seas. And, because roll and pitch angles are very small, weapons can be handled with greater accuracy. Although current hydrofoils are small they are still powerful multi-purpose weapon platforms. For example, the Italian PHM carries an automatic dual-purpose heavy gun, two ship-to-ship missiles and an electronic fire-control system, even though it displaces only sixty tons. The bigger Soviet Sarancha-class hydrofoil, of 235-ton displacement, carries four ship-to-ship missiles, two surface-to-air missiles, and a 23-mm six-barrelled gun. The United States plans to build hydrofoils of 1,000 or 2,000 tons' displacement with ocean-going capabilities. Hydrofoils will greatly increase the potential of light naval forces, a fact recognized by, for example, China which operates over 120 hydrofoils.

The main aim of a ship designer is to equip his ship with the most powerful weapons possible. Great care is taken to economize on space and weight by the use of special materials. For example, all-aluminium hulls have been developed for frigates, fast patrol-boats, hydrofoils, etc. Non-magnetic plastic hulls are used on minesweepers and a great effort is being made to develop better steels, particularly for submarine construction. The USSR has constructed a submarine hull from titanium for deeper diving and faster speeds.

Naval missiles

The most remarkable developments in naval weapons are those affecting guided missiles. New propulsion, fuel, guidance, launching and warhead technologies are revolutionizing anti-ship missiles. The success of the Exocet in the Falklands war dramatically brought home how vulnerable modern warships are in the face of anti-ship missiles. Many now believe that large warships are obsolete, except to project military power abroad in peace-time and to threaten smaller countries.

Modern anti-ship missiles are exemplified by the American Harpoon and Tomahawk missiles. The Harpoon is already deployed in large numbers on surface ships, submarines and aircraft. The Tomahawk is in the final stages of development and testing. Both missiles can be fired from the standard torpedo tubes of submarines.

In its basic version, the Harpoon is about 4.5 m in length and weighs over 660 kg. It has a high-explosive warhead of 225 kg, of the blast-penetrating type, but can also be fitted with a nuclear warhead. The Harpoon has a range of about 100 km, well over the horizon. The location of the target is fed into an inertial platform before the missile is fired. The on-board computer-controlled inertial guidance system steers the missile towards the target even if it is fired perpendicularly to the correct direction.[5] A radar altimeter is used to fly the missile just above the surface of the sea; when it gets near its target an active radar seeker finds and locks on to the target. The seeker commands the missile to gain height to out-manœuvre the target's defences and then dive down on to it from above. A solid-fuel-boost motor accelerates the Harpoon to a speed of about 75 per cent the speed of sound in less than three seconds; the missile is then propelled by a turbojet engine. Once fired, the Harpoon need have no contact with its parent ship. Information about targets at the extreme range is obtained using over-the-horizon radar and fed into the missile's computer before it is fired.

The Harpoon is essentially a cruise missile, as is the Tomahawk. The Tomahawk, the naval tactical version of the air-launched cruise missile, is a very versatile weapon; it can be fitted with a high-explosive or nuclear warhead and fired from surface ships or submarines, as an anti-ship, anti-shore or anti-submarine weapon. The Tomahawk is about 6.1 m long, has a diameter of about 0.55 m, weighs about 1,200 kg, and can carry a 475-kg armour-piercing warhead. Like the Harpoon it is first propelled by a solid-fuel-boost engine but uses a turbofan engine for cruising. The missile can be fired from a submerged submarine and has a range of up to 500 km, travelling at about 70 per cent the speed of sound. An advanced Harpoon system is used to guide the Tomahawk and it too skims about 10–15 m above the surface of the sea.

The Soviet navy operates a number of types of surface-to-surface anti-ship cruise missiles but probably none is able to fly at very low altitudes (to escape detection). Soviet naval bombers are often armed with air-to-surface anti-ship missiles. The Backfire bomber, for example, carries the AS-6, believed to be capable of 200-km range at high altitude, supersonic, and inertially guided flight, and to use a radar-homing device to dive on to its target. Many other countries operate anti-ship missiles—such as the British Sea Skua, the French Exocet, the Norwegian Penguin, the French/Italian Otomat, and so on.

Effective anti-ship missiles certainly give navies a considerable offensive capability. The threat from anti-ship missiles has stimulated efforts to develop a wide range of highly automatic, quick-reaction ship-defence systems and electronic counter-measures. An example of such a ship-defence system is the American Aegis, which consists of a modern multi-function, phased-array radar for target detection and tracking, numbers of radars for target illumination, several computerized systems for control of different weapons (missiles and guns) and an automatic multi-purpose launcher. Quick-reaction missiles which may operate with this sophisticated fire-control system include: the American Standard SM, a ship-to-air missile, with a range of over 100 km, mid-course command guidance, and a new digital on-board computer; and the British Sea Dart, which is designed to engage a number of low- and high-flying attacking aircraft and missiles at ranges of 30 km or so. The development of ever-more sophisticated anti-ship missiles and systems to defend ships against them is stimulating an intensive superpower naval arms race.

Anti-submarine warfare

One component of this arms race is the efforts being made by both super-powers to develop effective anti-submarine warfare (ASW) techniques. If strategic ASW succeeds in giving one side or the other the perception that it can greatly limit the damage the other side can do in a retaliatory strike with its submarine-launched ballistic missiles, then this perception may well contribute to the belief that a first nuclear strike may be effective. Of all naval developments those in strategic ASW are potentially the greatest threat to world security, in that they may significantly increase the probability of a nuclear world war.

The very large effort being made by the superpowers to improve ASW techniques to detect and destroy the other side's submarines will almost certainly lead, in time, to success. Even in the absence of a technological breakthrough—which cannot, of course, be discounted—steady progress in limiting the damage that can be done by enemy strategic nuclear submarines must be expected.

In ASW, detection is the critical element. Detection methods are being improved by increasing the sensitivity of sensors, improving the integration between various sensing systems, and better computer processing of data from sensors. Advances in microelectronics are rapidly improving these elements.

All types of ASW sensors are being improved: (a) electronic, based on radar, infra-red, lasers and optics; (b) acoustic, including active and passive sonar; and (c) magnetic, in which the magnetic disturbance caused by the presence of a submarine is measured. Airborne, space-borne, ocean-surface and sea-bottom sensors are being increasingly integrated into detection systems and, therefore, made more effective. ASW aircraft, surface-ships, and hunter-killer submarines are also being made increasingly complementary. Each system has special characteristics and the integration of those that complement each other considerably improves the overall effectiveness.

American ASW activities are worldwide and continuous, involving a total system of extreme complexity and a network of foreign bases and facilities. A typical American ASW task force uses an ASW aircraft carrier carrying specialized aircraft, destroyers equipped with ASW helicopters, and nuclear-powered attack hunter-killer submarines. The task force works with long-range land-based aircraft and receives information from unmanned surveillance systems, including sea-bottom arrays and space-based systems. The job of the task force is to hunt down a single enemy strategic nuclear submarine. Once detected the submarine would be relatively easy to destroy.

Soviet ASW is based mainly on naval helicopter carriers and long-range, land-based aircraft. The Soviet helicopter carriers in service are mainly intended for fleet defence against submarine attack, or, in other words, tactical ASW. Search helicopters carry very sophisticated electronic equipment to detect and track enemy submarines. Armed helicopters would drop weapons to destroy the enemy submarines. The USSR also operates a number of cruisers and destroyers for ASW activities.

The USSR uses several types of long-range aircraft, equipped with the most modern high-resolution radar and magnetic-anomaly detection equipment, to hunt down enemy strategic nuclear submarines. Soviet ASW activities tend to have a shorter range than those of the United States and are largely confined to areas close to Soviet territory or fleets. As time goes by, though, one must expect the USSR to extend the range of its ASW activities.

The most effective single ASW weapon system is the hunter-killer submarine—a nuclear submarine equipped with sonar and other ASW sensors, underwater communication systems, and a computer to analyse data from the sensors and to fire ASW weapons. The United States and the USSR each operate hunter-killer fleets, a hundred or so strong.

Submarines can be destroyed with torpedoes, depth charges or missiles. A typical ASW weapon is the American Captor, or encapsulated torpedo, which is a torpedo inserted into a mine-casing to allow it to be stored in deep water for a long time. The Captor has an acoustic detection system and a small computer which activates the launching mechanism when an enemy submarine is detected and identified in its vicinity. The weapon lies on the ocean bottom for a long time, waiting for a target, and is ideal for sealing off narrow straits to create an ASW barrier. The torpedo's sensor can distinguish between surface vessels and submarines, and has a range of about ten kilometres. If it misses its target on the first attempt, it can turn and try again.

American hunter-killers carry the SUBROC ASW missile, which, equipped with a nuclear warhead, is launched from a torpedo tube, rises to the surface, flies as a missile over a range of about fifty kilometres, re-enters the ocean near the enemy submarine and explodes as a nuclear depth charge.[6] Other examples of ASW missiles include the Australian Ikara weapon, the Soviet SSN-14 of thirty kilometres range, and the French Malafon of thirteen kilometres range.

The British have a sophisticated ASW capability, including weapons with nuclear warheads. It is believed that the naval task force sent to the Falklands had nuclear weapons on board, including ASW depth charges.

British ships sunk during the war may have nuclear weapons on them. The dispatch of nuclear weapons to the South Atlantic is against the spirit, if not the letter, of the Treaty of Tlatelolco, intended to make Latin America a nuclear-weapon-free zone.

Smaller navies

There has been, over the past ten years or so, a sharp increase in the number of small naval vessels, displacing less than, say, 1,000 tons, operated by the navies of the smaller countries. Currently, about fifty countries (out of 122 with direct access to the sea) are operating a total of more than 5,500 patrol vessels. These vessels vary from missile- and torpedo-carrying fast patrol-boats to slow machine-gun-armed patrol-boats. Some of these missile-carrying light naval vessels have considerable fire-power.

The proliferation of light naval forces to many Third World countries reflects the level of conflict there and is part of the general escalation of the global arms trade. An increasing number of countries are investing in patrol vessels of various kinds to protect their extended interests at sea. The extensions of territorial waters by certain countries and the establishment of New Economic Zones by the Law of the Sea Treaty mean that states are keen to secure their interests at sea by naval force. Modern fast patrol-boats are seen as the best vessels for the purpose because thay have a strike potential sufficient to deter much larger naval vessels. They are also relatively cheap. We must, therefore, expect the increase in light naval forces to continue.

Countries like Argentina, Brazil, Indonesia, Japan, Mexico, Norway, and Peru have developed light naval forces to protect their extended interests at sea.[7] This list will obviously grow. The establishment of New Economic Zones, together with military conflicts and regional tensions, will be the main factors in generating and escalating worldwide naval arms races.

Strategic nuclear submarines

Of all the naval vessels using the oceans the most destructive by far is the strategic nuclear submarine. The Soviet and American navies operate a total of ninety-five modern strategic nuclear submarines—the USSR has sixty-two and the United States has thirty-three. France has five strategic nuclear submarines and the United Kingdom has four. The ballistic missiles carried by these submarines are normally targeted on the enemy's cities and industrial areas to provide the assured destruction on which nuclear deterrence currently depends. A single American strategic nuclear submarine, for example, carries about 200 warheads, enough to destroy every Soviet city with a population of more than 150,000. American cities are hostages to Soviet strategic nuclear submarines to a similar extent. Just four strategic nuclear submarines, out of the 104, on appropriate stations could destroy most of the major cities in the Northern Hemisphere.

The most modern operational class of Soviet ballistic-missile submarines is the Delta class, carrying twelve or sixteen missile tubes, the missiles having ranges of about 8,000 kilometres. The SS-N-18, the missile carried by Delta-class submarines, is the first Soviet submarine-launched ballistic missile (SLBM) to carry multiple independently targetable re-entry vehicles (MIRVs). These missiles could hit most targets in the United States from Soviet home waters.

In 1980, the USSR launched its very large strategic nuclear submarine, the 25,000-ton *Typhoon*. This 160-metre-long boat carries twenty SLBMs. It will become operational in the mid-1980s and be equipped with a new, more accurate ballistic missile, the SS-NX-20. This SLBM will probably carry twelve warheads over a range of about 8,000 kilometres.

At the end of 1982, the USSR had deployed 937 SLBMs, 260 of them MIRVed, in its sixty-two nuclear strategic submarines. These SLBMs are capable of delivering about 2,800 nuclear warheads with an explosive power equivalent to that of about 880 million tons (880 Mt) of TNT. (In all history man has used about 20 million tons of high explosive!) Soviet strategic nuclear forces can deliver a total of about 8,500 warheads and the SLBMs carry about 33 per cent of the total.

The USSR had, by the end of 1982, deployed 240 SS-N-18s, each equipped with seven MIRVs having an explosive capacity equivalent to 200,000 tons (200 kt). The SS-N-18s are carried on fifteen Delta-class submarines. The other main Soviet SLBM is the SS-N-8, with a range of about 8,000 km and a single 1-Mt warhead; 289 of these are deployed, mainly on twenty-two Delta-class submarines. The USSR also operates twenty-four Yankee-class strategic submarines, each carrying sixteen SS-N-6 SLBMs, a 3,000-km range missile carrying either a 700-kt warhead or two 350-kt warheads.

Some strategic plans

The United States now operates two types of SLBM—the Poseidon and Trident-I. The Poseidon carries on average nine MIRVs, each with a yield of 40 kt; the Trident-I carries on average eight MIRVed warheads, each with a yield of 100 kt.

Two new Trident ballistic-missile submarines are now operational. They are approximately twice as large as the Poseidon missile submarines which they are replacing. Each Trident submarine carries twenty-four Trident SLBMs, with ranges of about 7,500 kilometres. Seven more Trident submarines are being built, which should become operational at the rate of about one a year.

Trident-I SLBMs are also being retrofitted into Poseidon submarines. So far, twelve such submarines are in service. Nineteen other Poseidon submarines are operational; each carries sixteen Poseidon missiles with ranges of 4,500 km. The American strategic nuclear submarine fleet can deliver 544 SLBMs, carrying a total of about 4,600 warheads with a total explosive power of around 300 Mt. Nearly 50 per cent of American strategic nuclear warheads are carried on submarines.

The American strategic plans call for the development and deployment of the Trident-II SLBM. This new missile will have a range of about 11,000 km and carry up to fourteen warheads each with a yield of 150 kt.[8] The missiles will be much more accurate than the ones they replace. The submarines themselves will have more accurate navigational techniques, some based on satellite systems. Finally, SLBM warheads may be fitted with terminal guidance, in which a radar altimeter or laser will search the area under the target after the warhead has re-entered the atmosphere and guide the warhead very accurately on to its target.

Ballistic accuracy

Accuracy of missile delivery is a key element in the current changes in superpower nuclear strategies from nuclear deterrence by mutual assured destruction to fighting a nuclear war. The more the superpowers adapt to war-fighting doctrines the greater will be the probability of a nuclear war, because the idea that such a war is 'fightable and winnable' will gain ground. At the moment, land-based intercontinental ballistic missiles are accurate enough to destroy enemy land-based ballistic missiles in their hardened silos and are, therefore, nuclear war-fighting weapons. SLBMs are not yet accurate enough for this purpose, and still provide an element of nuclear deterrence by mutual assured destruction.

The accuracy of a nuclear warhead is measured by its circular error probability (CEP), the radius of the circle centred on the target within which half of a large number of warheads fired at the target will fall. The CEP of the Trident-I SLBM is probably about 500 m at maximum range; that of the Poseidon is about 550 m. The improved accuracy obtained from the deployment of mid-course guidance techniques, coupled with the more accurate navigation of ballistic-missile submarines will steadily reduce the CEPs of SLBMs. The deployment of missiles with terminal guidance could give CEPs of a few tens of metres. SLBMs will then be so accurate as to cease to be only deterrence weapons aimed at enemy cities and become nuclear-war fighting weapons. Superpower nuclear policies will then be pure nuclear war fighting. Nuclear deterrence by mutual assured destruction will be a dead letter.

Soviet SLBMs are significantly less accurate than their American counterparts. Operational Soviet SLBMs are thought to have CEPs exceeding 1,000 metres,[9] but the accuracy of the missiles will steadily improve. The Soviets are said to have already tested a new SLBM, the SS-NX-17, with a CEP of about 500 metres.

The naval arms race: costly, threatening

The military use of the oceans is increasing with disturbing rapidity. In fact, intensifying naval arms races are leading to the total militarization of the oceans.

There is a Treaty on the Prohibition of the Emplacement of Nuclear Weapons and Other Weapons of Mass Destruction on the Sea Bed and the Ocean Floor and in the Subsoil Thereof, which entered into force in May 1972. The SALT-II treaty prohibits the development, testing or deployment by the United States and the USSR of fixed ballistic or cruise missile launchers for emplacement on the ocean floor, on the sea bed, or on the beds of internal waters and inland waters, which move only in contact with the ocean floor, the sea bed, or the beds of internal waters and inland waters, or missiles for such launchers. The SALT-II treaty has yet to be ratified by the United States Senate but both the USSR and the United States say they will stick to the requirements of the treaty; but no treaty significantly limits the naval activities of the powers or controls naval arms races.

The rising costs of warships and their increasing vulnerability are tending to reduce the acquisition of large warships in favour of smaller and cheaper vessels. But, because of the technological advances in ship construction and naval weapons, the fire-power of ships is not decreasing correspondingly. The naval arms race has, in fact, become a race for quality.

An increasing number of developing countries are participating in the global naval expansion. This is leading to a rapid increase in light naval forces. A major factor stimulating this expansion is the perceived need to police and defend New Economic Zones. The expansion is being encouraged by the aggressive efforts being made by industrialized countries to export naval vessels and naval weapons.

The escalating naval arms race is very expensive and absorbs large financial and skilled manpower resources, which the world can ill afford. Moreover, naval arms races are aggravating tensions in a number of regions. Most serious of all, the strategic nuclear forces of the superpowers are being developed into nuclear war fighting forces. If ASW techniques become effective they may give rise to perceptions of a first-strike capability. This superpower naval activity is the greatest threat to world security. The naval arms race may, if it is not controlled in time, be the major contributor to a nuclear world war.   ■

Notes

1. *Soviet Military Power* (2nd ed.), Washington, D.C., Government Printing Office, March 1983.
2. *The Military Balance, 1982–1983*, London, International Institute of Strategic Studies, 1982.
3. *Whence the Threat to Peace*, Moscow, Military Publishing House, 1982.
4. SIPRI, *World Armaments and Disarmament* (SIPRI Yearbook 1979), London, Taylor & Francis, 1979.
5. W. Gunston, *Encyclopaedia of Rockets and Missiles*, London, Salamander, 1979.
6. SIPRI, op. cit.
7. Ibid.
8. SIPRI, *World Armaments and Disarmament* (SIPRI Yearbook 1982), London, Taylor & Francis, 1982.
9. *Ambio*, Vol. XI, Nos. 2–3, 1982.

Three young researchers at the University of Manchester's Programme of Policy Research in Engineering, Science and Technology undertake a long-term forecast, 'a thankless task where it is easy to be proved wrong and difficult to be proved correct'. Anticipating the future of the marine technologies is complicated by the fact that the ocean is one of our last frontiers on earth, of great potential, but for which we have few guidelines in terms of the numerous possibilities to be realized.

Chapter 33

Using the seas during the twenty-first century

Glyn Ford, Luke Georghiou and Hugh Cameron

James Glyn Ford is co-ordinator, Marine Resources Project, Programme of Policy Research and Engineering (PREST), University of Manchester, and a member of the British Labour Party's Science and Technology Committee. Hugh Cameron is a research associate on the same project, while Luke Georghiou co-ordinates another PREST project on technological innovation. They can be addressed as follows: PREST Marine Resources Project, University of Manchester, Oxford Road, Manchester M139 PL (United Kingdom).

Past and present

To put the problem of foretelling oceanic research and technology in perspective, imagine forecasting the marine technology of our own decade from some time in the early 1880s. At that time the *Challenger* expedition (1873–76), the world's first oceanographic expedition, had returned; but final analysis of the materials collected would take two further decades. The steamship age was well established; but the principal marine power sources of the present century, the steam turbine and the diesel engine, would not be developed before the 1890s and nuclear power was undreamed of. Mechanization of fisheries had just begun, but the use of technical aids such as echo-sounders was half a century away. Offshore petroleum, if it had been known about, would be seen as little more than a curious facet of natural history. Submarines would not be truly feasible before the end of the century.

The list could continue but the point is that long-range technological forecasts will generally fail because they miss the novel, surprise factors. The danger of this is greatest with linear extrapolations of contemporary trends. To give an example, the preoccupation of nineteenth-century naval architects with speed dominated ship design for many years, but the particular demand they were satisfying came to be better supplied by air transport and preservation of perishable cargoes. New factors such as scale, fuel and labour economy became the objectives of the designers.

If the future is to be examined without falling into this trap, it is necessary to integrate consideration of the way new technology develops with that of the economic and social environment in which it develops. For the present purpose we must consider, on the one hand, the potential resources of the ocean and the demands society will make of them and, on the other, the supply of technology to meet those demands. These will not necessarily coincide. To illustrate this, we can look at our example of the submarine.

The military value of being able to approach enemy shipping underwater had long been recognized, and repeated attempts had been made to create such a device during the Napoleonic Wars and the American Civil War. These early 'submarines' relied on human effort for motive power, since steam engines were unable to operate underwater without rapidly exhausting the air supply. Ideas abounded, not least the fictional *Nautilus* of Jules Verne. Nevertheless, the demand could not be satisfied until another technology, the electric motor, had been developed and applied to run from batteries charged from diesel engines operated when the submersible was on the surface.

Hence we see that the initial conception of a technological problem often anticipates by decades the availability of adequate technological knowledge,[1] though it may stimulate the development of the necessary technology. This analysis will be based on the confluence of the two streams of development, demand and supply. In the first, the potential of marine resources and the specific technologies conceived to realize them is assessed, while the second considers the underlying technological trajectories[2] which may pervade developments in the next fifty years through the supply of basic technologies applicable on a wide scale.

Before embarking on the forecast itself, it is necessary to sketch the broad social and economic assumptions upon which it is based. Firstly, our prime concern is with peaceful uses of the ocean and we will not consider military technology though we recognize that 'spin-offs' may occur. With this exclusion, it may be assumed that the main motive for exploiting the oceans is broadly economic: that is, the satisfaction of human needs by the use of oceanic resources. The sea is generally a hostile environment and therefore it is only where the sea offers an advantageous alternative source of supply, or a unique use (compared to land) that it will be exploited. More fundamentally, it is assumed that a commitment to technological progress will remain and that the Third World will increasingly share in both the resources available and the large-scale high technology to exploit them.

The demand for ocean resources

The oceans are a resource bank that could provide a variety of mineral, energy, biological and 'space' resources. For exploitation to take place, three conditions must be satisfied. Firstly, the resource must be accessible, taking into account geographical, geological, legal, political and social factors. Secondly, the technology must be available to exploit the resource. And lastly, exploitation must be economic; that is, there must be a demand for the resource at the price the oceans can supply it. Over the next half century or so, these conditions could be met, creating new marine industries in the following areas.

Minerals

The most obvious untapped ocean resource is seawater itself. The mineral wealth it contains is almost as incomprehensible as it is unexploitable. Although almost since the origins of civilization the production of sea salt has taken place, little else has been recovered from a 'liquor' that contains traces of essentially all naturally occurring elements. The reason is that these traces are so dilute that it seemed impossible to think of recovering them economically, as, for example, in Fritz Haber's attempts to recover gold from seawater after the First World War. However, it is possible to extract minerals from seawater economically if the right conditions occur, involving a combination of characteristics such as: value of the mineral, concentration in seawater, ease of extraction and availability of alternatives considering geological, geographical, technological, economic and political dimensions. At present a limited number of materials are recovered, including bromine and the major part of the world's magnesium.[3]

The future looks promising. Recent advances in the chemistry of chelating—or bonding so as to form ring structures in—ion-exchange resins offer the possibility (in the long term) of producing very effective reagents that can be woven into cloth that will selectively and effectively strip various elements out of seawater. It is likely that uranium, vanadium and molybdenum offer the most promise. The first of these, because it is a fuel for nuclear power stations the price of which has a relatively small impact on the overall economics of electricity production, must emerge as the prime target. Its strategic importance also means that a premium will be available to ensure guaranteed sources of supply. Production may be parasitic upon some other ocean activity, however, as the economics of treating the necessary volumes of water will give a major advantage to plants in conjunction with tidal power stations or similar activities.[4] The other possibility might be to exploit those few locations where seawater is extra-rich, that is, the Dead Sea or the Gulf of Kara Bogaz in the Caspian.

Apart from seawater itself, the most well-known form of mineral wealth found in the oceans is manganese nodules. These are potato-sized concretions that litter the floor of the deepest part of the oceans, well away from the shore. Some 'commercial' grades contain, apart from manganese, significant amounts of nickel (1.3 per cent), cobalt (0.2 per cent) and copper (1.1 per cent). While short-term land-based supplies of all these metals seem adequate, in the longer term, this is far less certain. Nickel-sulphide ores are becoming expensive to recover and refine, while the alternative, nickel laterites, requires vast capital investments little different to the thousand million dollars required for a nodule-mining operation. The prospect is that, within the next thirty years, nodule ships may be grazing the deep oceans, each sucking millions of tonnes per annum of nodules up the three or four kilometres from the ocean floor. There already exists the proof of concept for such operations.[5]

Less well known are the polymetallic sulphides that are found beneath the deep-sea floor along medium to fast spreading centres, to use the terminology of plate tectonics (see Fig. 1) for occurrences in the East Pacific. These massive deposits, formed by hydrothermal activity engendered by the large heat-flows along the spreading centres, contain significant amounts of zinc, silver and copper. Samples have shown zinc contents of 30 per cent and more. Because, unlike manganese nodules, these deposits occur as a more traditional ore body, their exploitation at depths of two kilometres or more beneath the sea surface will be extraordinarily difficult. It is possible that either the hot, supersaturated liquids carrying the metals in solution can be tapped as they come to the surface of the sea floor or, more likely, the unique

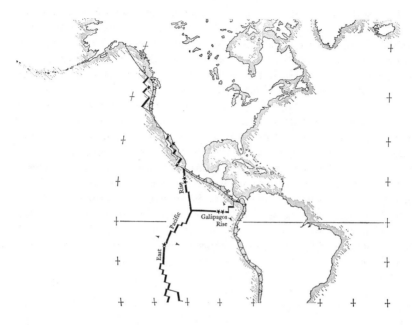

FIG. I. Hydrothermal and polymetallic sulphide deposits.

manifestation of this resource in the Red Sea[6] in the form of unconsolidated mud will be exploited by pumping to the surface.

Radiolarian oozes (abyssal siliceous deposits) which are 100 per cent pure silica, are another deep-ocean source of pure material for ceramics. The technology involved will be similar to that used for nodule-mining, albeit at a much smaller scale. The same is true of phosphorite nodules, a potential fertilizer, found hundreds of metres deep on continental shelves.[7]

Energy

The oceans offer a number of potential energy sources. Offshore oil and gas activity will increase but, because of the mode of formation, it is unlikely that any significant amounts will be found off the continental rise in the deep ocean. Undersea coal exploitation will be either an extension of land technology using industrial islands (see below), or it will use these for coal gasification techniques that have yet to be developed. In contrast to these, a number of ocean energy technologies will use the seas themselves. These are tidal energy: two plants are already in operation, in the La Rance estuary, France, and Kislaya Inlet, Soviet Union; wave energy, with a variety of uses ranging from small-scale power for offshore operations (through the use of

systems like the Japanese *kaimei)* to proposals for arrays that will provide
power in quantities similar to those provided by conventional power-
stations. Finally, there is ocean thermal energy conversion (OTEC) that will
exploit the temperature differences between the surface and bottom waters of
the ocean, and for which the world's first commercial plant is under
construction on the island of Nauru in the Pacific.[8]

None of these will be the sole ocean energy technology, for each has its own
geographical peculiarities. OTEC is only viable where the temperature
difference between the surface waters and the water at 500–1,000 m is greater
than 20 °C. This essentially limits OTEC to the tropics (see Fig. 2). The
technology involved will require some massive scaling-up of heat-exchanger
technology, plus, for the larger plants, the construction and deployment of a
pipe of dimensions never built before, to bring up the cold water from depth.
Initially OTEC will probably be a technology for small tropical islands,
often with land-based plant. Large floating stations of a capacity of 100 MW
of electricity feeding into national power grids will probably come well
behind these.[9]

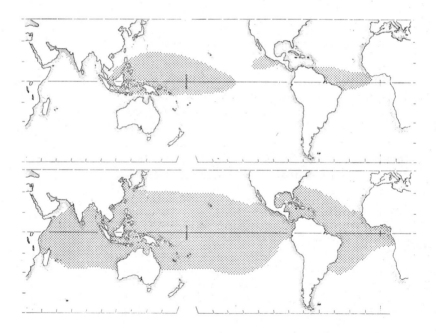

FIG. 2. Worldwide locations of OTEC thermal sources: (a) between surface and
depth of 500 metres; (b) between surface and depth of 1,000 metres. (After
Department of Energy, Washington, D.C.)

Large-scale wave energy operations are limited to coastlines with a large fetch, the sweep over open water along which the wind blows. These are usually to be found in the mid-latitudes. Various technical options are under development with no design yet dominant.

Tidal energy is not destined to become a common form of energy production. It is a technology requiring a tidal range of an order of magnitude seven to ten metres above that found in the open ocean. Sites with such a range are relatively rare; they occur only in estuaries where tidal inflow is funnelled to an unusual degree. Large-scale projects will require a massive civil engineering commitment. The Severn estuary in England, the Bay of Fundy in North America and the Gulf of San Jose in Argentina are all potential sites.[10]

More long-term, exotic ocean energy technologies have been suggested that will tap ocean currents like the Gulf Stream. These are not likely to be taken up, however, until the potential of the technologies mentioned above has been explored and exhausted.

Biological resources

The demand for food resources will inevitably continue to increase, and the seas offer a significant potential contribution towards meeting this requirement. However, the improvement of fisheries technology during this century has little more to offer because, for many species, the maximum sustainable yield has been reached or exceeded. Major expansion is possible only through improved resource management and the exploitation of some unused potential in the Southern Hemisphere.

The future exploitation of biological resources will be dependent upon such a managed approach. This should come through large-scale mariculture. OTEC plants produce as a 'waste product' large volumes of cold but nutrient-rich and disease-free bottom water; this cold water offers an ideal cultivating medium for mariculture.

It has even been suggested that the benefit from mariculture could exceed that from energy production. On a larger scale, one can anticipate that natural semi-enclosed basins could be managed to produce a wealth of fish resources. This is illustrated by a scheme put forward by the Japan Society of Industrial Machinery Manufacturers for a large-scale marine farm project for Saeki Bay near Fukuoka in Japan.[11]

The seas also contain a vast number of organisms that have so far received only the most cursory scientific investigation. Among these it is inevitable that chemicals and drugs will be discovered, products that can be exploited by the pharmaceutical industry.[12] Bearing in mind the vast changes that biotechnology promises to bring to chemical processing, it is unlikely that the seas will not make their contribution in this regard.

'Space' resources

The world is increasingly crowded, and land, at least in some regions of the world, is becoming a commodity in short supply. At the same time, the environmental movement has made people increasingly aware of the problems of pollution. These factors have created the desire to isolate risky activities, materials and wastes from human settlements. One outcome of this is that the seas seem to offer the 'space' for hazardous industrial or other activities that require large areas of space and the disposal of dangerous waste, including nuclear waste.

Many of the proposed ventures can be stimulated by the construction of artificial (industrial) islands that will enable current industries to relocate offshore and new industries to start up. Proposals have been made to site nuclear power-stations offshore, as well as installations for oil terminals, airports and liquefied natural gas storage.[13] The technology is already available to do this in shallow water. More speculatively, industrial islands have also been suggested for massive leisure complexes and as sites for the production of wind or other forms of energy. In the case of wind energy, there seems no reason why this could not take place in shallow water without the construction of an artificial island.

In the longer term, particularly in parts of South-East Asia and Japan where land prices are astronomically high, artificial islands could be used for residential purposes. The Japanese have already successfully built a combined industrial/residential island at Kobe.

Waste disposal at sea has been going on in an ad hoc manner for centuries, while low-level nuclear waste has been dumped at sea by a number of countries over the past decades. This dispersal of waste into the oceans has become an increasingly controversial subject, with regional groups of states campaigning for nuclear-free Pacific and Indian Oceans. The future containment of, at least, nuclear waste, is likely: high-level nuclear waste may undergo 'engineered emplacement' in the deep-sea bed using a variety of new technologies. The most extreme of these would entail platforms and sophisticated positioning equipment to place the waste 'packages' in holes drilled as deep as a kilometre below the sea bed, where continuous monitoring will then take place.[14]

The next century should see the rich mineral resources of the Antarctic opened to exploitation, which will mean increased use of the oceans for transport of bulk raw materials in conditions that should stretch current ship technology to the limit. New methods of carrying non-perishable cargo will begin to be used once the energy saving is estimated to justify the cost of developing new transport vessels.

The supply of marine technology

What we have said above implies that the sources of design and engineering will have to provide the various elements of the technologies necessary. The pace of technical advance will determine the time-scale upon which resources will become available. The nature of the marine environment suggests that three aspects of technical development will be of fundamental importance in expanding use of the oceans in the areas outlined above: materials, information and control.

Materials

Whatever stands or floats in the ocean is subject to major physical forces: waves at and close to the surface, and pressure far beneath. In addition, the corrosive nature of seawater creates problems with steel, at present (along with cement) one of the two dominant construction materials. It is difficult to see their replacement before the next century, although corrosion control will improve and more advanced versions of these materials will be developed, particularly the use of composites to achieve lightness and strength. Countervailing trends are already in operation: the resistance to corrosion and high strength/weight ratio makes titanium, and, to a lesser extent, aluminium ideal for some ocean applications. Titanium is now used for piping in desalination plants and in the offshore oil industry and is favoured for the construction of heat-exchangers in OTEC plants. Yet, apart from fabrication problems, the very high energy cost of titanium production, when compared to that of steel, will mean that rising energy costs will limit its use to a small number of components where titanium's advantageous properties make it indispensable. And if a direct refinement method is found for aluminium, the use of this metal could dramatically increase.

More radical prospects for new materials are found in synthetic carbon and fibres. These might well migrate from special, high-technology application to large-scale use in construction. Materials substitution is a slow process, given the cautious nature of the engineering profession, but on this time-scale and with the new demands it could be significant. The materials are available to allow for operation at much greater depths while higher strength/weight ratios can make even seemingly expensive materials economical in uses such as wave-power generation.

Information

The oceans are part of the current information revolution. All marine activities are dependent upon environmental data as to weather and sea conditions, while ocean exploration requires the systematic handling of large volumes of information. Two related developments have to be taken into account: the gathering of information and its subsequent processing and application. Satellite navigation is already available and satellites can also monitor surface phenomena, although it seems unlikely that they will be able to monitor the deep-sea bed. In this area, acoustic techniques will continue to be improved with increased range and resolution.[14] Lasers or extremely low-frequency electromagnetic waves offer the prospect of improved underwater communications such as air-to-submarine links.

While remote sensing has been advanced by offshore mineral exploration, the data still cannot be handled in the optimal manner. Real-time computing of marine data on a wide scale would permit this and make far better weather and sea-condition predictions possible, together with advance warnings of earthquakes and tsunami.

Control

It seems paradoxical that one feature of progress in ocean technology is the removal of humans from the sea. However, with the exception of leisure activities, man is at sea only for what can be brought to land, and increased automation is welcomed when it frees individuals from dangerous or unpleasant tasks. Already ships operate with crews a mere fraction of the size of those in the first half of this century. What is more important is that the future technological trajectories outlined above mean that ocean exploitation will occur increasingly in extreme environmental conditions, unsuitable for humans. The polar seas and the deep oceans, where resources such as nodules and sulphides are found, will require unmanned undersea operations. The cost of creating human habitats in these conditions can be short-circuited by the use of remote control. The most advanced nodule recovery concept, now being researched by the French AFERNOD group, involves the use of remote-controlled shuttles travelling to the sea bed to collect the nodules.

As operations in the North Sea have shown, in certain situations man is indispensable. At a price, technology can create the conditions to allow man's presence, even perhaps enabling oxygen to be extracted directly from the water by equipment carried by divers or submarines.

Other factors

Many more technical advances will contribute to the development of the oceans to their limits. New forms of power, chemicals and modelling will all contribute, affecting areas as diverse as biofouling (the fouling of power units and materials by plant and animal growths in the aquatic medium) and mariculture, *in situ* analysis and platform stability.

Land-based trends may be taken to the oceans. For land-based minerals, there has been a steady concentration of activities closer to the mine site, eliminating transport costs and creating 'vertically integrated' operations.

Once new developments begin to be used in parallel, then one can imagine nodule-mining operations occurring on floating islands powered by satellite OTEC plants, with processing that uses chelating ion-exchange resins to strip metals out of solution. This emphasizes the point that the new ocean technologies should be complementary in their development.

The path to the future

There is a danger in trying to forecast the future in one particular area—that one may ignore the influence of other areas. Just a new marine technologies will be developed, so may even more attractive land-based alternatives. The demand for the marine products may also be superseded by technological change. Non-ferrous metals for the steel industry (manganese, cobalt and nickel) would have only minor application, for example, if there were widespread substitution for steel. Already, sectors of copper demand are being threatened by the use of fibre optics in communications. A further point is that, on occasion, the ability to exploit resources exists without sufficient demand for what is produced. The problems of the Anglo-French Concorde aircraft illustrate this pitfall.

A wider context which must also be recognized, through not dwelt on at length here, is the importance of the social and political dimensions of the environment in which a technology develops. The difficulties encountered over manganese nodule development emphasize that the international character of the oceans be accepted in order to ensure agreement and co-operation between all nations for prolonged, stable and fair development.

Forecasts such as this have no value beyond that of curiosity unless their content is challenged, debated and ultimately becomes a factor in policy decisions. Much of the future is random but it is not too early to begin formulating serious policies to develop desired 'ocean futures'. As we witness the difficulties many new technologies face when they are first introduced, the importance of persistence is clear. The most vulnerable part of this development is the initial period. For some of the concepts considered (such as OTEC), the momentum has begun to build up. Others, such as sea-bed minerals, are in danger of losing impetus because of political disagreements. A Third category (for example, the extraction of a number of seawater metals) is still at the small-scale, experimental stage. There may be also a fourth category, whose development may not have begun and which has not yet been perceived as marine technology. Only the future will tell what this could be. ■

Notes

1. Other examples of this can be found in N. Rosenberg, 'Science, Invention and Economic Growth', *The Economic Journal*, London, No. 333, 1974.
2. The concept of technological trajectories is defined in R. Nelson and S. Winter, 'In Search of Useful Theory of Innovation', *Research Policy*, No. 6, 1977.
3. W. McIlhenny, 'Extraction of Economic Inorganic Materials from Seawater', in J. P. Riley and G. Skirrow, *Chemical Oceanography*, Vol. 4, London, Academic Press, n.d.
4. H. Cameron, F. Vernon and L. Georghiou, *'Prospects for the Extraction of Uranium from Seawater'*, *Institute of Chemical Engineers Symposium*, series No. 78, 1982.
5. T. Marjoram et al., 'Manganese Nodules and Marine Technology', *Resources Policy*, Vol. 7, No. 1, 1981.
6. Z. Nawab and K. Luck, 'Test Mining of Metalliferous Muds from the Red Sea Bottom', *Meerestechnik*, Vol. 10, No. 6, 1979.
7. P. Guest and G. Ford, 'Deep Sea Nodules Help the Farmer', *New Scientist*, 1 July 1982.
8. Preparatory Committee for the United Nations Conferences on New and Renewable Sources of Energy, *Report of the Synthesis Group*, United Nations, 10 March 1981 (doc. A/CONF.100/PC/41).
9. F. Ford, C. Niblett and L. Walker, 'Ocean Thermal Energy Conversion', *IEE Proceedings*, Vol. 130, Part A, 1983.
10. A. Brin, *Energy and the Oceans*, Guildford, Westbury House, 1981.
11. 'Visionary, Large-scale Marine Farm Project', *Technocrat*, Vol. 15, No. 8, August 1982.
12. J. C. Braekman and D. Daloze, 'Les médicaments de la mer', *La Recherche*, No. 143, April 1983.
13. *Industrial Islands*, Conference Proceedings, London, Institute of Mechanical Engineers, 17–19 November 1981.
14. D. Talbert, *Sub-seabed Disposal Programme Annual Report. January–December 1979*, Vol. 1, *Summary and Status*, Albuquerque, Sandia National Laboratories.

To delve more deeply

LEWIS, B. The Process of Formation of Ocean Crust. *Science*, Vol. 220, No. 4593, 8 April 1983.

Since the beginning of man's cultural expression, the sea has evoked in the imagination reveries of the world ocean that have been expressed in words, pictures and music. Although many animal species have originated in or been intimately associated with the aquatic environment, it is man only who has left ineffaceable traces of how the brain interprets the sea and its effect on life, on contemplation, on our view of the world.

Chapter 34

The sea and the dreams of man

Elisabeth Mann Borgese

Elisabeth Mann Borgese, who was born in Munich, has long been concerned with the sea; she is the author of The Ocean Regime *(1968),* Pacem in Maribus *(1972),* The Drama of the Oceans *(1976),* Seafarm *(1980),* The Mines of Neptune *(1983), and numerous articles on the Law of the Sea and on ocean management. She has been associated with the Center for the Study of Democratic Institutions in California since 1965, the International Ocean Institute, Malta, and she is currently professor of political science in Halifax, Nova Scotia. Her address is: Dalhousie University (Department of Political Science), Halifax B3H 4H6 (Canada).*

My study overlooks the sea: more precisely, one of the picturesque coves
characterizing the indented Nova Scotia coast. Masses of ice are drifting in
and out with the tides, reflecting the sunset in pastel pale pink and gold. The
pines ashore are dwarfed by salt and wind and cold. The homes nearby,
simple wooden structures, have a somewhat improvised, temporary aspect.

Ever since I saw it first at the age of five, led by my father's hand, I have
been under a spell of love of the sea. I have dedicated the past sixteen years
to the sea and its role in the economic, political, legal and other social evolution of the world community, dreaming about the oceans and the future.
And yet the title that has been suggested to me for these pages stuns me,
confronts me with 'why': Why on earth am I doing what I am doing? The
answers are bafflingly complex, involving levels of aesthetics, the subcon-
scious, early influences in life, as well as a dimension of rational historic and
political thinking. And in trying to formulate the replies to my own questions,
I find some of the answers concerning issues such as the relations between
human societies and the sea through the course of history and, in particular,
the sea as an element of inspiration of aesthetics, myths, and religions.

Did mankind emerge from the seas?

Life began in the oceans. The world ocean was, as Marston Bates put it,
a sort of 'thin organic soup in which almost anything might happen'.[1] In
this mixture of dissolving minerals, large molecular compounds of carbon
combined and recombined, for millions of years, finally making cells. Many
more millions of years went by before the cells, passing through various
phases of evolution, had 'learned' photosynthesis, that is, the direct use of
sunlight to produce high-energy phosphates. By that time the monocellular
being had evolved the pigment chlorophyll, which made possible the trapping
of light. Thus the cell began to use sunlight to synthesize glucose, releasing
molecular oxygen as a by-product which entered the atmosphere. About
three thousand million years ago, the amount of oxygen thus released
was sufficient to make cell respiration possible. Living matter in the
oceans made its own environment as it made itself. Pre-Palaeozoic
plankton—especially the blue-green algae that are still with us today—filled
the atmosphere with enough oxygen to make possible the evolution of the
metazoan some six hundred million years ago.

From the ocean, life expanded into estuaries, rivers and lakes, where
the true finfish evolved, many of whom eventually returned to the sea:
the first wave of re-migrants, to be followed by others. Reptiles too returned,
as did birds, their proud wings shrivelling into flippers, their legs (clumsy
on land) turned into agile rudders. Even the mammals returned, whale,
dolphin, seal and otter, and readapted to fishlike forms and habits. And
man may be next.

According to some, *Homo sapiens'* land-based existence may be of brief duration, episodic. According to at least one respected scientist, Sir Alister Hardy, man's upright position, his body-hairlessness, the subcutaneous layer of fat he is prone to accumulate—all point to the possibility that his early evolution was bound, not to tree-dwelling, but to swimming in the balmy waters of tropical seas.[2]

More recent experiments, made by immersing tiny babies into deep water where they swim spontaneously and fearlessly, their eyes open without pain, emerging to breathe like dolphins or seals, would tend to confirm Sir Alister's thesis. Quite possibly, then, men could swim before they could walk upright.

The blissful gargling, smiling and kicking of a new-born baby, enveloped for the first time by the body-warm waters of the bathtub, manifests voluptuous creature comfort, as though it were a return to the maternal womb, a micro-ocean, the salinity of its fluid resembling that of primaeval waters. Every micro-ocean restages the drama of the origin of life in the gestation of each embryo, from one-cell protozoa through all the phases of gill-breather and amphibian, to mammalian evolution, to man's birth—the expulsion from the micro-ocean to dry land.

And every human, in turn, is a good bit of planet ocean: 71 per cent of his substance consists of salty water, just as 71 per cent of the earth is covered by the oceans.

A force of creation and destruction

We do not know how humankind, before the dawn of history, knew; but man knew. Mankind knew that all life began in the oceans and enshrined this knowledge in uncounted myths and legends, the world over, unconnected—establishing, as it were, a first intellectual bond uniting the entire species.

Most of us are familiar with Genesis. In the beginning, the Spirit of God moved upon the face of the waters. *Fiat lux* (Let there be light!) was His first command of creation. Next He divided the waters which were under the firmament from those which were above it; 'And God said, Let the waters under the heaven be gathered together unto one place, and let the dry land appear: and it was so. And God called the dry land Earth, and the gathering together of the waters called he Seas.' Genesis also reports that God first made the fishes in the sea and the birds in the air and the beasts on land, and, last of all, he made man. How did the ancients *know*?

A Swiss scholar, F. Morven, recently collected the legends from the sea in a beautiful volume of the same title.[3] Morven reminds us that, in virtually all the mythologies of the world, water is given primacy over the other elements. It was the first thing, after which came all others. According to the Greeks and the Aztecs, even the gods were 'born of water'. Morven then surveys a wealth of myths of creation, from Peru and Mexico, North America, India, and the land of the Kalmuck, Hindu lore and Muslim wisdom, Scandinavian sagas and Slavic traditions—one more beautiful than the next, a wreath of poetic, creative dreams.

Water, first among the elements

The Manova Sastra tells the story of the creation in the same order as the Bible: the world was obscure and confused, as if in a deep sleep. God, existing by virtue of His own powers, manifested Himself in five elements and scattered the darkness. By wielding His power, He produced first of all the waters, imparted movement to them by means of fire and created an egg, as brilliant as the sun, from which Brahma, father of all reasoning beings, emerged.   F. Morven

---

Water, Morven summarizes, contained the seeds of everything. In most traditions and cosmogonies, the earth was covered by waves, awaiting the creative agent which was to make the earth rise above them.

The oceans are not only life-giving; they also take life. They are as destructive as they are creative, and that divine wrath at man's sinfulness brought on torrential rains and tidal floods that engulfed the earth was known to the people of Israel as to those of Peru, of Babylon, of Japan, and India. The authors of the Bible even knew that, before the Flood, the continents were all in one piece; there were no islands, and it was the Flood that separated the continents—the earliest version of continental drift!

The sea, legend and power

Sea monsters and sirens, serpents and evil spirits have been thought to lurk in the oceans and on the sea bed; they lured sailors to perdition. Hindu mythology has woven the creative and destructive aspects of the oceans into one pattern of endless, cyclic recurrence.

'Oh King of Gods', Brahma said to Vishnu, 'I have seen it all perish, again and again, at the end of every cycle. At that time, every single atom dissolves into the primaeval water of eternity, whence originally all life arose. Everything then goes back to the fathomless, wild infinity of the ocean which is covered with utter darkness and is empty of every sign of animate being.'[4]

Could it be this ambivalence between good and evil that makes the ocean so 'human', the mirror of our souls

> La mer est ton miroir!
> tu contemples ton âme
> dans le déroulement infini de sa lame
> et ton esprit n'est pas un gouffre moins amer.
>
> (The sea is thy mirror!
> Thou contemplateth thy soul
> In the infinite extent of its swell,
> And thy spirit is no less bitter an abyss.)

Baudelaire says, in one of the most inspired poems of *Les fleurs du Mal*.

The sea, like a mother, provided nourishment to infant humanity which possibly fished before it hunted. The inventiveness of primitive man in fish-catching has amazed many an anthropologist. We are familiar today with the same basic equipment, as Robert C. Miller points out in his beautiful book, *The Sea*:

hand lines with baited hooks, fish traps of various kinds, and nets adopted to different types of fishing such as shore seining or fishing from boats, seem to be at least as old as recorded history.[5]

The earliest cave dwellers in the Mediterranean region in mesolithic times had become fishermen by the seventh millennium B.C. Large numbers of fishbones were found in their caves. Three thousand years before our own time, as I noted in my book *Drama of the Oceans*,

fishing had developed into a highly organized craft. Miniatures discovered in Minoan houses destroyed in an earthquake about 1500 B.C. show boats full of tackle, rods, and hooks, and divers plunging into the sea with their bags, one of them carrying what looks like a large sponge.

Since the stone age, fisheries have constituted the basis of the economies of coastal communities. Fishing encouraged shipbuilding and enhanced the spirit of science and exploration, international trade and naval power. The power and influence of the Hanseatic League, the medieval federation of north German cities that made a still-perceptible cultural impact on Hamburg, Lübeck, Bremen, Riga and Tallinn, was based on its herring fishery. When the herring fishery collapsed in the fifteenth century, so did the power of the league. It was as though sea-power followed the migrations of fish. For three centuries (until the eighteenth century), Holland was the strongest maritime power, its economy based largely on its herring fishery. England, Scotland, and Norway followed.

'He who rules the sea, rules the land' has been a conventional wisdom. The Venetians, at the peak of their power, formulated it, '*Chi xe paron del mar xe paron de la tera*' (He who is master of the sea is master of the land), and the same idea was known to the Danish kings, '*Herre over Vandet, er og Herre over Landet*'.⁶

Many a proud nation's fate was indeed sealed by the outcome of naval battle. Suffice it to mention the Battle of Salamis (480 B.C.), most skilfully planned by Themistocles, which stemmed the tide of Persian expansion in Europe. The battle is splendidly described in *The Persians* by Aeschylus, who himself had been a mariner in the Athenian navy at Salamis. Actium and Lepanto are other familiar names in the long series of naval conflicts that tilted the balance of power in favour of Rome, in favour of Christendom, respectively. Queen Elizabeth I of England and King Philip II of Spain fought at sea, and Spain went down with its Armada in 1588; Napoleon was undone at Trafalgar (1805); the Tsars' empire succumbed to ascendant Japan in the naval battle of Tsushima Strait (1905).

A brief departure from the universally accepted notion that 'Neptune's trident is the sceptre of the world' came with the theory of *geopolitics*, between the two World Wars. Borrowed by the German Karl Haushofer from the British Harold Mackinder, this held that he who rules eastern Europe, commands the heart of the earth—the Heartland—and who rules the Heartland rules the earth. 'Landpower wields the seas, not conversely.'⁷

Short-lived, geopolitics went down with the Third Reich, and after the Second World War, the strategic importance of the oceans and of naval power—now in the form of missile-carrying submarines—has been stressed more emphatically than at any time before. Technological developments (high-flying spy planes and satellites) have made the dry earth unsafe as a repository for a second-, or for that matter, a first-strike force. The balance of terror today rests hidden in the opaque waters of the deep sea. Escalating submarine power in an insane race for supremacy, the superpowers may end by destroying themselves, the rest of humanity and life in the oceans as well—closing, perhaps, one of those cycles at the end of which every single atom dissolves in the primeval waters of eternity, when everything returns to the fathomless, wild infinity of the oceans, in utter darkness, bereft of life.

'The sea made man's soul'

The ocean, that mighty body of water that both separated and united men, has been a 'Great Educator' which made peoples great. Sea-faring populations have provided history's most remarkable merchants and explorers. Sea-faring people have cherished freedom, too. They felt as free as the oceans. Republics are the creation of maritime peoples; tyranny was born inland.

Hegel has a prophetic page in his *Philosophie des Rechts* on the role of the seas in industrial societies, which he compares with the role of the earth in agricultural societies. He was aware of the culture-forming, educative influence of the ocean and invited us to compare the maritime nations, in their industriousness and enlightenment, with those nations whom destiny had denied navigation and who, 'like the Egyptians and the Indians, sank into stupor and the most horrid and shameful superstitions'.

Hegel's knowledge of the geography and culture of Egypt and India obviously was not what it should have been, but his emphasis on the teaching and culture-creating aspects of the oceans is undoubtedly correct. 'The sea made man's soul, and the waves give him intelligence', a Finnish proverb has it. Throughout recorded history, the oceans and man's relationship to the oceans have inspired architects, painters, musicians and writers.

Down to the waves in ships

Ships are crowning achievements of human civilization, yet what is a ship? *In Drama of the Oceans* I noted,

Consider what goes into the building of a ship: the whole arsenal of a people's crafts and sciences, of art and life style, of world view and collective purpose. A ship epitomizes man's attitude toward other men, and toward nature.

Ships have been a favourite subject of paintings and miniatures almost as long as there has been painting, from the Egypt of the Pharaohs to the present time. Seascapes, though not too frequent in classical painting, show wave crests stylized into almost tapestry-like patterns (*The Birth of Venus*, by Botticelli) or dwelled on the chiaroscuro of billowing mountains and vales merging with, and reflecting, an equally photodynamic cloudscape, as in Tintoretto's magnificent *Christ at the Sea of Galilee*.

But it is with the Romantics and post-Romantics, Impressionists, Fauves and Expressionists, and their new and intense relationship with nature, that the ocean becomes an inexhaustible model for the painter, a looking-glass for his soul as it mirrors firmament and stars—with Monet, Cézanne, Gauguin, Van Gogh, Turner, Watteau, the Germans, the Dutch, to mention only a few, and Hokusai in Japan. Turner painted *The Snowstorm* after living through a tempest at sea, off the coast of England. 'I got sailors to lash me to the mast to observe it', Turner recorded, bringing to mind Odysseus tied to the mast in order to listen to the song of the Sirens. 'I was lashed for hours', Turner continued, 'and did not expect to escape, but I felt bound to record it if I did.' The result, Edward Lockspeiser writes, 'was the greatest of his seascapes . . . which, said Turner, "no one has any business to like . . . I did not paint it to be understood. I wished to show what such a scene was like"'.[8]

Hokusai's famous print, the *Hollow of the Wave at Kanagawa*—which, incidentally, was the picture chosen by Debussy for the cover of the orchestra score of *La Mer*—was eloquently described by Edmond de Goncourt, in a study published in 1896.

> The design for the Wave is a sort of deified version of the sea, made by a painter who lived in a religious terror of the overwhelming sea surrounding his country on all sides: it is a design which is impressive by the sudden anger of its leap into the sky, by the deep blue of the transparent inner side of its curve, by the splitting of the crest which is thus scattered into a shower of tiny drops having the shape of animals' claws.[9]

The sea and music

Hokusai's *Wave*, nevertheless, has something rigid and static about it, revindicating Lucien Favre's statement 'La peinture est, si l'on veut, une musique des couleurs sans mouvement.' (Painting is, if you will, a music of colours without movement.)

Music, instead, can be seen as the art closest to nature; music has a time dimension so that it can capture not only the sounds of the sea, its colours and textures, but their teasing changes and variations, their rhythms in time. 'Music has this over painting', Debussy wrote, 'it can bring together all manner of variations of colour and light—a point not often observed though it is quite obvious.'[10] And Vallas, in his *Theories of Claude Debussy*, wrote:

> Although they claim to be nature's sworn interpreters, painters and sculptors can give us but a loose and fragmentary rendering of the beauty of the universe. Only one aspect, one instant is seized and placed on record. To musicians only is it given to capture all the poetry of night and day, of earth and heaven, to reconstruct their atmosphere and record the rhythm of their great heartbeats.[11]

The playing of silvery ripples, the crashing of surf are easily located in orchestration; the swelling of storm and its exhaustion find expression in established modes of crescendo and decrescendo. The rolling of waves, their eternal cadences, are readily translated into the measures of musical time. The multiple layers of ocean space, from the mysterious sea floor through submerged waves, submarine rivers, to the bobbing, scintillating surface, can be captured in counterpoint, vertically; its flux in time reflected in the duration of horizontal, melodic, development.

The sea is onomatopoeic to the highest degree, as generations of composers have known. Space does not permit us to dwell long on Richard Wagner's treatment of the sea, on the musical characterizations of *The Flying Dutchman*—'And roaring and whistling and surging round them all is the Sea', as Ernest Newman observes, 'not so much as the mere background of the drama as the element that has given it birth';[12] on the longing strains of

Tristan and Isolde reflecting a leaden Irish Sea over which the vessel must come—'*das Schiff, sahst Du's noch nicht*' (the ship, see you it not yet)—and the sensuousness of drowning in love, in the Liebestod, as in the sea:

Heller schallend,	The wafting sound
mich umwallend,	does me surround:
sind es Wellen	Is it a wave
sanfter Lüfte ?	that me doth lave,
Sind es Wolken	or rather a cloud
wonniger Düfte ?	of fragrance proud?
Wie sie schwellen,	See how it rises,
mich umrauschen,	how it mesmerizes!
soll ich atmen,	It breathes on me,
soll ich lauschen ?	it is peace to me;
Soll ich schlürfen	I sip of it,
untertauchen ?	am submerged by it.
Süss in Düften	Sweet is the balm,
mich verhauchen ?	touching me with calm.
In dem wogenden Schwall,	In the troubled current,
in dem tönenden Schall,	in the pealing torrent—
in des Welt Atmens	the breath of life
Wehen, dem All,	in all the strife—
ertrinken,	I shall break life's link,
versinken,	indeed I shall sink
unbewusst,	without light,
höchste Lust !	joyous delight!

In the long-drawn, single E-flat opening of *Rheingold*, water symbolizes the beginning of all things: water from which, through sequences of empty fifths, the flow of the Rhine takes its course, and which is the beginning of the gods and the creation of men in their complex interaction with eternity.

Nor can we do justice here to the onomatopoeic possibilities created by the new technologies of electronic synthesizers and concrete music. Suffice it to mention Alan Hovhaness' well-known *And God Created Great Whales*, a composition counterpointing the lugubrious, lonesome, unearthly song of the humpback whale, in its manifold shades and modulations, with ominous, pentatonic strains (of obvious Japanese inspiration); swelling, climaxing in confrontation and ultimate tragedy. Surprisingly, this effect seems not to have been intended. Throughout his work, prone to the seduction of Japanese styles of music, Hovhaness was apparently unaware of the drama he set up between Japanese whalers and their tragic victims. He wrote:

Pentatonic melody sounds openness of wide ocean sky. Undersea mountains rise and fall in horns, trombones and tuba. Music of whales also rises and falls like mountain

ranges. Song of whale emerges like giant mythical sea bird. Man does not exist, has
not yet been born in the solemn oneness of nature.[13]

Impressed by developments in the International Whaling Commission,
I hear the piece differently!

The composer who epitomizes the influence of the sea on music, in the
broader context of the other arts (literature and painting), is Claude
Debussy, whom Robert Godet defined as 'an island, surrounded by water
on all sides'.[14] Debussy wrote of himself: 'You may not know that I was
destined for a sailor's life and that it was only quite by chance that fate led
me in another direction. But I always retained a passionate love for [the
sea].'[15] Of his many major and minor works permeated by the sea, mimicking,
reflecting the sea, certainly the most important, the culmination and
synthesis, is *La Mer*. Hearkening back to early impressions of the sombre
North Atlantic and the more suave Mediterranean, this symphonic composition is articulated in three movements entitled 'De l'aube à midi sur la
mer', 'Jeux de vagues' and 'Dialogue du vent et de la mer' (From dawn
to noon at sea, Play of waves, and Dialogue of the wind and the sea). The
last movement reflected, as one critic put it, 'those ever delightful frolics
in which [the sea] exhausts her divine energy, and the spell of foam and
waves and spray, swirling mists and splashes of sunlight'.[16] Music, as
Baudelaire put it on another occasion, ravishes you like the sea: 'La musique
souvent me prend comme une mer.'

Music, the sea, and Thomas Mann

We find music and the sea intimately linked in the writings of my father,
Thomas Mann. Reminiscing on his childhood summers in the small Baltic
town of Travemünde, he writes in an occasional essay,

In that place the sea and music entered a sentimental union in my heart, forever, and
something was born of this union of feeling and ideas, and that is narrative, epic prose.
Epic. For me, that always has been a concept closely linked to that of sea and music,
in a way, composed of them, and as C. F. Meyer could say of his poetry that everywhere in it there was the great, calm light of the glaciers; so, I should think, that the
sea, its rhythm, its musical transcendance, somehow is present everywhere throughout
my books—even then, often enough, when it is not mentioned explicitly. Yes, I
should hope I have indicated my thankfulness to the sea of my childhood, the Bay
of Lübeck. Maybe it was its palette that I used, and if my colours have been found
opaque, without glow, abstemious—well, one may ascribe it to certain perspectives,
through silvery beeches, to the pale pastels of sea and sky on which my eyes rested
when I was a child and happy.[17]

Epic narrative, composed of sea and music! No wonder the annals of literature are as full of the oceans as those of painting and of music—from Homer and the Greek tragedians to Joseph Conrad, Baudelaire, Verlaine, Melville, Thomas Mann, and after.

Although I was not conscious of it until much later, my father's love affair with the ocean must have influenced me powerfully. Rereading his works in my mature years, when I have myself become so deeply involved with the oceans, I find his analysis of the human relationship to nature, and especially the sea, the most profound I have come across. He recognized man's awe in face of the sea's infinity and wildness, in contrast to the constraints of civilization, both equally necessary and complementary; the sea as all and nothing, damnation and redemption, longing and fear; the sea as the dark and wild element within the artist, within his characters, within himself. 'The sea', Mann wrote in another brief essay, was

Infinity! My love for the sea, whose enormous simplicity I have always preferred to the pretentious multifariousness of the mountains—my love for the sea is as old as my love for sleep, and I am fully aware of the common root of these two sympathies. I have within myself much that is Indian: a great deal of inert and heavy longing for that form or lack of form of perfection, called 'Nirvana', or nought, and even though I am an artist, I have a rather unartistic inclination towards eternity, manifesting itself in an aversion against articulation and measure. What counteracts this inclination, believe me, is correction and discipline: or, to use a more serious term: it is morale ... What is morale? What is the morale of the artist?[18]

And on another occasion he wrote:

The sea is not landscape. It is the experience of eternity, of nothingness and death: a metaphysical dream; and the same wellnigh applies to the thin-aired regions of eternal snows. Sea and high mountains are not rural, not terrestrial; they are elementary in the sense of ultimate and savage, extra-human magnificence, and it would almost seem as though the civil, civic, urbane, bourgeois artist were inclined to skip over rural landscapishness and go directly for the elementary: for it is in the face of the elementary that he feels fully justified to confess and reveal his relationship to nature for what it really is. Fear, alienation; illicit and wild adventure.

His feeling of awe, though tempered on many occasions, especially in *Travelling with Don Quixote*—by irony and self-irony, has no room for a utilitarian ocean: an ocean tamed to satisfy human needs, harnessed for progress.

It is nothing new to me that the sea, experienced from a ship, makes far less impression than experienced from the beach. The enthusiasm excited in me by the sacred crashing of its waves, on the *terra firma* on which I stand, is lacking. There is a demystification. A spell is broken by the sobering of the element into the role of road, of highway for travellers. The ocean loses its character as spectacle, dream, idea, spiritual perspective on eternity, and its becomes 'environment'.[19]

What would Mann say to the raping of the ocean, its pollution by the penetration of the industrial revolution into its depth? How would he judge our attempts to cope with this new situation by imposing a new order on the oceans? Are not 'order' and 'oceans' antonyms in his grandiose perspective?

The sea and cultural evolution

There would be much room for irony over our 'humanized ocean' but, in the last analysis, there could not be disapproval. For our humanizing efforts, our utilitarianism does not detract one whit from the concept of the enormity and wildness of the sea. It is not the oceans we want to dominate and regulate, it is human activities and human encroachment. The new ocean sciences do not demystify the oceans, just as the penetration of the universe by science does not diminish its mystery and majesty, and the greatest scientists often are the greatest mystics.

If the ocean has played an enormous role in the evolution of human life and culture, in the dreams of men, this influence is bound to grow in the coming period of history. Man is returning to the sea. Mankind's dependence on the sea is increasing dramatically: for food, metals and minerals, energy.

We are likely to witness the transformation of a marine economy based on hunting and gathering (traditional fisheries) into one based on the cultivation of aquatic plants and the husbanding of aquatic animals. The emergence of aquaculture, including mariculture, may be a development no less important in anthropological, even evolutionary, terms than was the emergence of agriculture perhaps ten thousand years ago.

If it is true—as many technicians assert—that, in the long term, ocean mining of metals and minerals may be less expensive than land-mining, while at the same time it offers other obvious advantages (such as the absence of conflict with competing land uses, less environmental impact, direct access to cheaper transportation), then we are likely to see a gradual shift from land-based mining to ocean mining. This is bound to cause displacements, internally within nations and between nations, but also offer new opportunities and challenges.

Ocean energy resources—not only petroleum and natural gas, but the untapped, infinite, renewable and non-polluting energy resources of the sea itself: the energy of tides and waves and currents, of thermal or salinity gradients, or of the huge and incredibly fast-growing biomass (kelp) of the sea—will make a major contribution to the satisfaction of the world's energy needs.

The new Law of the Sea

Be it on the basis of these economic–technological developments, be it on the foundation of immemorial dreams, or both, mankind has embarked on

the task of building a new international economic order. This—not unlike
the primeval earth in the dreams of men—appears to be emerging from the
oceans. The adoption, in 1982, of the United Nations Convention on the
Law of the Sea by an overwhelming majority of the international community
marks a breakthrough in this direction. Defective as it may be—and what is
not defective in this world of ours?—and reflecting the dream of man only
imperfectly, this convention offers a new platform to launch economic
development and new efforts for peace.

This is not the place to examine in detail the merits and demerits of the
convention and its many implications and ramifications. In a final glance at
the sea, at the sea of our dreams, we perceive the hoary concepts of ownership
and sovereignty (on which the history of the last centuries was largely based)
distorted and transformed in the reflection of teasing waves, broken up as in
the scintillating colours of the Impressionists, the dissolved harmonies of
Debussy. Now these concepts take the new form of the common heritage of
mankind, a notion transcending ownership and adding a new dimension to
sovereignty: participation in common decision-making.

Based on the dream of common heritage, we see, emerging from the sea, a
new 'ecological consciousness', a different vision—new to us though, in
some parts of the world, ancient—of man's relationship to nature in general
and to the sea in particular. We see a vision of human evolution and history,
not as confrontation with nature but as part of nature; not called by any god
to subdue her, but led, by nature herself, to co-operate.

This co-operation calls for interaction with nature, for co-operation among
human beings. For the environment in general (not only the sea), both
natural and social, is an extended mirror of man's soul. For better or worse,
just as we perceive ourselves, so we see the world around, oceans and all. ■

Notes

1. M. Bates, *The Forest and the Sea*, New York, Random House, 1960.
2. A. Hardy, *The Living Stream*, New York, Harper & Row, 1965.
3. F. Morven, *Legends from the Sea*, New York, Crescent Books, 1980.
4. E. Mann Borgese, *Drama of the Oceans*, New York, Harry Abrams, 1975.
5. R. Miller, *The Sea*, New York, Random House, 1966.
6. Morven, op. cit.
7. G. Borgese, *Common Cause*, New York, Duell, Sloan and Pierce, 1943.
8. E. Lockspeiser, *Debussy, His Life and Mind*, London, Cassell, 1965.
9. Ibid.
10. Ibid.
11. L. Vallas, *The Theories of Claude Debussy*, New York, Dover Publications, 1967.
12. E. Newman, *Wagner as Man and Artist*, New York, Vintage Books, 1960.

13. A. Hovhaness, programme notes to *And God Created Great Whales*, World Premiere Recording, Columbia Stereo Masterworks, M:30390.
14. Lockspeiser, op. cit.
15. L. Vallas, *Claude Debussy: His Life and Work*, Oxford, Oxford University Press, 1933.
16. Lockspeiser, op. cit.
17. T. Mann, 'Lübeck als geistige Lebensform', *Collected Works*, Vol. XI, Frankfurt am Main, S. Fischer Verlag, 1960.
18. Mann, 'Süsser Schlaf', op. cit.
19. Mann 'Meerfahrt mit Don Quijote', op. cit., Vol. IX, 1934.

A closing word is offered in regard to continuing research, resource applications, and a cohesive approach on a global basis to the rational exploitation of one of the world's greatest 'commons'—the ocean.

Chapter 35

Postface

Francisco F. Papa Blanco and Jacques G. Richardson

Dr. Papa Blanco is an electrical and mechanical engineer from Uruguay who left Unesco, as director of its Division of Technological Research and Higher Education, in 1984. The authors prepared the original version of this short paper as part of a working document used by Unesco's Advisory Panel on Science, Technology and Society that met at various times in Paris between 1981 and 1983 mainly to advise Unesco's Director-General on the formulation of that institution's medium-term plan for 1984-1989. This is a revision, updated, of the initial formulation.

Marine research

Oceanological research will relate increasingly to the driving mechanisms of plate tectonics (including ocean rifts and volcanism), geological correlation and the sea's influence on climate. Much additional research is required on the evolution and extinction of biota and on ocean ecosystems, including the preparation of comprehensive inventories of flora and fauna—especially in the Southern Hemisphere.

Study of marine chemistry and of the composition and properties of sea water, as well as of the formation of manganese nodules, is expected to provide better understanding of the formation of mineral deposits in general. Discovery of ore bodies associated with deep-ocean hot springs poses new questions concerning the origin of certain important ore deposits on land and about the potential of finding such ore on the sea floor.

All these areas of investigation call for an increasingly interdisciplinary method within the sciences, expanded to include terrestrial environmentalists and sociologists when subjects such as the evolution of the use and pollution of coastal zones are studied.

Instrumentation of all kinds, including that able to provide better life-support systems to man operating in the aquatic environment, needs conception, elaboration, testing and refinement. There is a growing use of robots which should greatly extend the scope and consequence of research, particularly in the depths.

Applications

The rational use of living resources calls for effective management schemes based on the most advanced methods. Novel, on-board fishery technology can be expected to raise the efficiency of harvesting the sea's edible products. Large fishery ventures from the Northern Hemisphere will turn increasingly to suppliers from the South. But the Antarctic's krill, the collection of which theoretically could reach 5 million tonnes annually, is likely to remain a largely untapped source of protein.

Exploitation of the continental shelf's and the sea bed's chief minerals should expand, and there will be growth in the rate of this exploitation as the cost of extraction is lowered in comparison with terrestrial mining. Non-conventional methods to mine dispersed resources and extract dissolved minerals will be developed on the basis of knowledge—yet to be acquired—on the properties and dynamics of marine systems and new technology.

Applied research should continue to expand in the exploration for petroleum and natural gas, as should investigations to prevent and remedy pollution originating in the transport, processing and burning of hydrocarbons and other fossil fuels, the production and use of synthetic organic compounds and pesticides, and the accumulation of heavy metals carried by industrial effluents.

Other applications potentially rich in their ultimate rewards include new systems for ocean observation and ocean transport, the ocean as a source of energy, offshore waste disposal, marine parks and tourism, and animal and plant mariculture.

In military technology, anti-submarine warfare based on acoustico-electrical detection of undersea vessels can be expected to develop, reaching the limits of efficiency imposed by the physical properties of the aquatic medium. A potential consequence would be neutralization of the nuclear forces now roaming the world's ocean fairly freely, with the political and other social implications this could have. Improved detection capabilities should find, in due time, significant application in civil undersea exploration.

Societal interactions

Continuing research and the new technologies thus developed will make it possible for more than the currently seven leading nations* to share extensively exploration and exploitation of the marine hydrosphere and its resources at increasing depths. For this, the world community needs to make it possible to conduct research and development of a scope far exceeding that limited by national jurisdiction. Implementation of the new Law of the Sea, and the new ocean regime that this implies, will require extensive international co-operation and strengthening of the appropriate international mechanisms.

Man can be expected to draw more food from the sea, but he will indeed have growing recourse to mariculture in order to raise significantly the 71 million tonnes of fish and other comestibles taken annually from the ocean. The conflict represented by the dual approaches of exploiting and preserving ocean biota may find reasonable resolution as marine scientists and engineers are joined by public opinion and other participative mechanisms to help influence governments in rational long-term policy formulation and decision-making. In coastal communities, the search for appropriate socio-economic equilibrium is likely to become a major element in societal development because, by the early 1990s, half the world's population will live on or near the littoral. ■

* The seven countries are the Federal Republic of Germany, France, Japan, Norway, the United Kingdom, the Union of Soviet Socialist Republics and the United States of America.

BIBLIOGRAPHY: To delve more deeply

(The literature on the sea, beginning with its folk tales, is enormous. There follow a few, highly selected titles of some of the most current scientific, engineering and economic literature dealing with oceanic studies.)

Aquatic Sciences and Fisheries Abstracts, a bibliographic database covering the world's literature on the science, technology and management of marine and fresh-water environments; produced monthly by Cambridge Scientific for the Food and Agriculture Organization of the United Nations. Enquiries to FAO Fisheries Division, Rome.

BOMBARD, A., PAOLINI, C., *Protégeons la mer,* Paris, Nathan, 1977.

BRAMWELL, M., *Atlas of the Oceans,* London, Mitchell Beazley, 1977.

BREWER, P. (ed.), *Oceanography, The Present and the Future,* New York-Heidelberg-Berlin, Springer Verlag, 1983.

BYKOV, V.P. (ed.), *Marine Fishes,* Balkema, Rotterdam, 1984.

CANADIAN NATIONAL COMMITTEE, SCIENTIFIC COMMITTEE ON OCEANIC RESEARCH, *Proc. Joint Oceanographic Assembly 1982,* Ottawa, Department of Fisheries and Oceans, 1983.

COUSTEAU, J.-Y., *Saumons, castors et loutres,* Paris, Flammarion, 1978.

Deep-Sea Hot Springs and Cold Seeps (theme), *Oceanus,* Vol. 27, Number 3, Fall 1984.

Deep Seabed Stable Reference Areas, Washington, National Research Council (Board on Ocean Science and Policy), 1984.

'El Niño Event: 1982,' a 12-minute silent film of GOES-infrared imagery of the equitorial Pacific Ocean. Enquiries to: IOC/TOGA Project, c/o Unesco, 7 place de Fontenoy, 75700 Paris (France).

FADIKA, Lamine Mohammed, Vers un nouvel ordre de la mer (interview), *La Sirène* (French edition of UNEP's *The Siren*), No. 26, December 1984.

GIBRAT, R., *L'énergie des marées,* Paris, Presses Universitaires de France, 1966. This work is directly related to the design, construction and operation of the world's first tidal power plant, situated near the mouth of the Rance River in northwestern France.

GORSHKOV, S., *World Ocean Atlas,* Vols. 1-3, Oxford, Pergamon, 1976-1983. The author was an admiral in the Soviet Navy.

HOLLAND, H.D., The Geology and Geochemistry of Hydrothermal Vents at the Ocean Floor (lecture), London, The Royal Society, 21 March 1985.

JANNASCH, H.W., The Chemosynthetic Support and Diversity of Life at Deep-Sea Hydrothermal Vents (lecture), London, The Royal Society, 21 March 1985.

Industry and the Ocean (theme), *Oceanus,* Vol. 27, No. 1, Spring 1984.

MARCHUK, G., KAGAN, B., *Ocean Tides,* Oxford, Pergamon, 1984.

NELSON, J.S., *Fishes of the World* (2d ed.), Chichester, John Wiley, 1984.

PARRY, J.H., *Romance of the Sea,* Washington, DC, The National Geographic Society, 1981.

RIEFENSTAHL, L., *Jardins de corail* (trans. G. Dispot), Paris, Chêne, 1978.

RIFFAUD, U., LePICHON, X., *Expédition FAMOUS, à trois mille mètres sous la mer*, Paris, Albin Michel, 1976.

RONA, Peter, A., et al. (eds.), *Hydrothermal Processes at Seafloor Spreading Centers*, New York, Plenum, 1984.

ROUBAULT, M., COPPENS, R., *La dérive des continents* (2nd ed.), Paris, Presses Universitaires de France, 1979.

ROUGERIE, J., VIGNES, E., *Habiter la mer*, Paris, Editions Maritimes et d'Outre-mer, 1978.

SCOTT, Peter, The End of Whaling? *The Siren* (UNEP), No. 26, December 1984.

Sea Riches: What Future? (in German), *Wirtschaftswoche*, 24 August 1984 .

SEARS, M., MERRIMAN, D. (eds.), *Oceanography: The Past*, New York-Heidelberg-Berlin, Springer Verlag, 1980.

SIMON, A.W., *Neptune's Revenge: The Ocean of Tomorrow*, New York, Franklin Watts, 1984.

WHITEHEAD, P.J.P., et al. (eds.), *Fishes of the North-eastern Atlantic and the Mediterranean* (Vol. I), Paris, Unesco, and Bungay, Richard Clay (The Chaucer Press), 1984.

WHITTOW, G.L., RAHN, H. (eds.), *Seabird Energetics*, New York, Plenum, 1984. A study in physiological ecology, based on a symposium held in Honolulu, August 1983.

ZWEIG, R.D., Freshwater Aquaculture in China: Ecosystem Management for Survival, *Ambio*, Vol. XIV, No. 2, 1985.

INDEX

DATE DUE

~~APR 10 '90~~			
~~NOV 15 '90~~			
DEC 11 '90			
NOV 25 '92			
~~DEC 23 '92~~			
NOV 10 '05			